ILSI Human Nutrition Reviews

Series Editor: Ian Macdonald

Already published:

Sweetness
Edited by John Dobbing

Calcium in Human Biology
Edited by B. E. C. Nordin

Sucrose: Nutritional and Safety Aspects
Edited by Gaston Vettorazzi and Ian Macdonald

Zinc in Human Biology
Edited by C. F. Mills

Forthcoming in the series:

Diet and Behaviour
Edited by G. H. Anderson, N. A. Krasnegor,
A. P. Simopoulos and G. D. Miller

Dietary Starches and Sugars in Man: A Comparison

Edited by John Dobbing

With 31 Figures

Springer-Verlag
London Berlin Heidelberg New York
Paris Tokyo

John Dobbing, DSc, FRCP, FRCPath
Emeritus Professor of Child Growth and Development, Department
of Child Health, University of Manchester, Oxford Road,
Manchester M13 9PT, UK

Series Editor
Ian Macdonald, MD, DSc, FIBiol
Department of Physiology, Guy's Hospital Medical and Dental
School, St Thomas' Street, London, SE1 9RT, UK

ISBN 3–540–19560–2 Springer-Verlag Berlin Heidelberg New York
ISBN 0–387–19560–2 Springer-Verlag New York Berlin Heidelberg

British Library Cataloguing in Publication Data
Dietary starches and sugars in man.
1. Man. Carbohydrates. Metabolism   I. Dobbing, John, *1922–*   II. Series
612'.396
ISBN 3–540–19560–2

Library of Congress Cataloging-in-Publication Data
Dietary starches and sugars in man: a comparison/ edited by John Dobbing.
p.   cm. — (ILSI human nutrition reviews)
Includes bibliographies and index.
ISBN 0–387–19560–2 (U.S.: alk. paper)
1. Starch—Metabolism.   2. Sugars—Metabolism.   3. Food—Carbohydrate
content.   I. Dobbing, John.   II. Series.
[DNLM: 1. Dietary Carbohydrates.   2. Nutritive Value.   QU 75 D565]   QP701.D53
1989   612.3'96—dc20   DNLM/DLC
                                    for Library of Congress   89-11348
                                                              CIP

QP
701
.D53
1989

Filmset by Wilmaset, Birkenhead, Wirral
Printed in Great Britain at The Bath Press, Avon

2128/3916–543210 Printed on acid-free paper

Foreword

Far more starch than sugar as such is consumed by man, and the starches and sugars that form part of the diet have metabolically much in common. However, there are differences and it is important to know if these are significant in terms of human nutrition. This is particularly important in view of the current advice to reduce dietary fat intake, and to replace this energy source by increasing the intake of dietary carbohydrate. Additionally, there is a considerable interest in both groups of compounds: starches, because of their clinical role in the regulation of blood glucose, and sugars because of the poor image they have in the media, although this has been righted in recent official reports. This volume assesses the scientific evidence on various nutritional aspects of these two groups of carbohydrates in early 1989. As is the nature of experimental biology, most of the findings referred to deal with the acute effects of ingesting starches and sugars in persons or experimental animals in good health.

The Scientific Planning Group for this monograph had hoped to include chapters on the effects of these carbohydrates on gastric and intestinal motility, as well as any effects they might have on behaviour and cognitive performance. However, it seems that there is insufficient data in these two important areas. Dental aspects of ingesting various carbohydrates are extremely important, so much so that it was decided not to include them, as they are worthy of a book equal to, or greater in length than this one, and their inclusion would detract from equally important, but less well known aspects of carbohydrate nutrition.

This volume was made possible by the support of the European Branch of the International Life Sciences Institute (ILSI). ILSI is a non-profit scientific foundation established to encourage and support research and educational programmes in nutrition, toxicology and food safety and to encourage cooperation in these programmes among scientists from universities, industry and government in order to facilitate the resolution of health and safety issues world-wide.

London
February 1989

Ian Macdonald
Series Editor

Preface

The contents of this book have passed through a rigorous peer review system similar to that of previous "Dobbing" monographs. This distinguishes the book importantly from most other multi-author works, and from the more usual "books of meetings".

All the chapters were circulated to each of the other authors, whose task was then to write considered Commentaries on them, much in the same way as referees of scientific papers are asked to do, except that in this case they were signed and referenced. The group then met together for three days and discussed intensively all the points raised in the Commentaries; and when it was felt useful new Commentaries were added.

The result of all this is that most of the original papers were considerably modified in the light of the Commentaries and of the prolonged discussions to which they gave rise, and wherever this was not appropriate the Commentaries have been printed after each chapter. Because of this the reader can have at least as much confidence in the product as he would have in a peer-reviewed journal; indeed rather more so, since few scientific papers have been scrutinised by eleven other experts in the field.

We wish to thank all the authors for their ready acceptance of this design, which made unusually great demands on their time, and sometimes on their patience. We also thank our publishers for their part in publishing the book so promptly and so well.

Editorial thanks are finally due to ILSI Europe who sponsored the endeavour, and whose Scientific Planning Group under the Chairmanship of Professor Ian Macdonald selected the participants; as well as to Dr Jean Sands who helped a great deal with the detailed scientific management.

Hayfield
February 1989

John Dobbing
Editor

Contents

Contributors

Dr. D. A. Booth
School of Psychology, University of Birmingham, PO Box 363,
Birmingham B15 2TT, UK

Dr. P. R. Flatt
Department of Biochemistry, University of Surrey, Guildford,
Surrey GU2 5XH, UK

Dr. B. Flourié
Service de Gastro-Entérologie, Hôpital Saint-Lazare, 107 rue du
faubourg Saint-Denis, 75475 Paris Cedex 10, France

Dr. M. W. Kearsley
Research and Development, Roquette Frères, 62136 Lestrum,
France

Dr. M. Laville
Hôpital E. Herriot, Endocrinologie-Diabète-Nutrition, place
d'Arsonval, 69372 Lyon Cedex 03, France

Dr. R. J. Levin
Department of Biomedical Science, The University, Western Bank,
Sheffield S10 2TN, UK

Professor I. Macdonald
Physiology Department, UDMS, Guy's Hospital, London SE1
9RT, UK

Professor V. Marks
Department of Biochemistry, University of Surrey, Guildford,
Surrey GU2 5XH, UK

Dr. S. Normand
Hôpital E. Herriot, Endocrinologie-Diabète-Nutrition, place
d'Arsonval, 69372 Lyon Cedex 03, France

Dr. T. L. Peeters
Gut Hormone Laboratory, Gasthuisberg ON, Catholic University
of Leuven, B-3000 Leuven, Belgium

Dr. S. Picard
Hôpital E. Herriot, Endocrinologie-Diabète-Nutrition, place
d'Arsonval, 69372 Lyon Cedex 03, France

Professor J. P. Riou
Hôpital E. Herriot, Endocrinologie-Diabète-Nutrition, place
d'Arsonval, 69372 Lyon Cedex 03, France

Dr. P. J. Sicard
Research and Development, Roquette Frères, 62136 Lestrum,
France

Professor D. A. T. Southgate
Nutrition and Food Quality Department, AFRC Institute of Food
Research, Colney Lane, Norwich NR4 7UA, UK

Professor C. Williams
Department of Physical Education and Sports Science,
Loughborough University, Loughborough, Leicestershire, UK

Dr. P. Würsch
Nestlé Research Centre, Nestec Ltd, 1000 Lausanne, Switzerland

Chapter 1

The Chemistry of Starches and Sugars Present in Food

M. W. Kearsley and P. J. Sicard

Introduction

In human nutrition carbohydrates play a major part in supplying the metabolic energy that enables the body to perform its different functions. Due to their availability, starch, sucrose and lactose occupy a very important position in this vast group of natural substances. The way they are consumed can, however, vary greatly as a consequence of various factors: geographical area, state of development of the country, cultural traditions, importance of health considerations, etc.

Lactose is generally used in crude form, as milk or derived products. Sucrose, the universal sweetener, must undergo extensive purification before being usable for industrial or domestic purposes. The case of starch is more complex. Despite the fact that in most cases it is consumed in the form of cereals, tubers or vegetables, in developed countries its extraction and refining has given rise to very great industrial activity, the purpose of which is to produce a wide range of thickeners and sweeteners which have important applications in the food industry [1].

Because starch processing is a capital-intensive activity, with heavy investments and rather low added value, world human consumption of starch and its derivatives has reached a plateau for traditional applications. However, new product development from starch is, and will continue to be, a very important area for expansion [2,3]. Significant developments have already been made in this area to meet the demand for health foods that has resulted from trends towards healthy eating. Novel uses for starch are being examined, examples of which include fat replacers based on maltodextrins to replace traditional fats such as butter; non-cariogenic sweets based on polyols to replace traditional cariogenic confectionery; products such as cyclodextrins which will reduce cholesterol levels in foods; and non-digestible starch derivatives to supplement dietary fibre.

There will also be developments in the production of foods with particular features to meet the demands of minority groups. Examples are products for diabetics based on fructose and polyols, and gluten-free products for coeliacs.

There is also a continuing demand for alternative sweeteners. The fine tuning of existing products to meet the exact requirements of a customer's process will be a major aspect of future development work too. These tailor-made products are already an important feature of normal production as customers seek to impose their specifications on the producers' own.

The main limitations with regard to new product development are probably legal restrictions. This is reflected in the fact that most companies now have either food legal departments, or at least someone experienced in such matters. When new products are developed, consideration must be given to existing food regulations not only in the country of manufacture but also in countries where there are potential customers for the product.

An important factor in the development of any new product is the size of its potential market, and this must be known. Products with a very specialised application may have only limited economic interest but be easier to introduce into the market than products with wider applications which may require a long time for approval. The longer this period the more costly the overall development will be. There may be a move towards products with high added value.

Where complex modifications are carried out on the starting material to produce the new product, approval will generally be more difficult to obtain. Typical considerations are:

Is the reaction simple or common or is it unusual? If simple and common, approval is more likely.

What are the reaction conditions. Is a catalyst involved?

If enzymes are involved, is the source (organism) approved or safe?

Is the reaction specific? Are toxic side products produced?

Can contamination of the final product occur from the reactants? For example, is there catalyst or enzyme carry-over into the finished product?

Does the product purification involve non-food materials such as resins or solvents?

Are there any metabolic abnormalities with the new product?

Is good manufacturing practice followed and are the analytical characteristics of the product monitored?

These are important questions for the development of any new product within the starch and sugars industry, and as more specialised products are manufactured, so the problems become more complex.

Natural Carbohydrates

There exists a vast family of naturally occurring compounds known as carbohydrates and an equally vast number of derivatives of these. Only a small proportion of them, however, are of commercial significance and used in the food industry. These compounds can be divided into three broad groups:

1. Complex macromolecules; the polysaccharides. The main member of this group of commercial interest is starch, although it includes other important compounds such as pectins, cellulose and gums [4].

2. Simple carbohydrates: the mono- and disaccharides, including glucose (dextrose), fructose, sucrose and lactose.
3. Derivatives of these simple carbohydrates, including products produced by fermentation, oxidation, reduction and isomerisation.

This section will be concerned with starch and particularly the structure and properties of native starch, whilst the second and third groups above are covered in later sections.

Starch

Starch is the main food reserve in plants and is laid down in specialised locations within the plant during periods of photosynthesis. These locations include tubers, kernels and grains; the starches of greatest commercial interest are those found in the potato tuber, the maize kernel and the wheat grain.

World starch production is between 20 and 25 million tonnes per year, of which 10–12 million tonnes of starch and its derivatives are used for food purposes each year [5–11]. In Europe the figures are respectively 5.5 million tonnes for production and 2 million tonnes for food use. Compared with sucrose production, these figures are relatively small. World production of sucrose is about 104 million tonnes per year, and that in Europe 13 million tonnes. Of this over 99% is used in foods [12].

Starch is composed only of glucose units and is thus a homopolysaccharide. A related compound is cellulose, which is a structural material as opposed to a food reserve; this is also a homopolysaccharide made up of glucose units. The difference between the two lies in the way in which the glucose units are joined together in the two compounds. These structural differences allow starch to be relatively easily hydrolysed by acids or enzymes, whilst cellulose is resistant to these agents. In the native state starch occurs in the form of granules.

The Starch Granule Structure

Although starch is composed only of glucose units, the glucose residues are joined together in two different ways and starch consists of two different polymers. One is linear with the glucose units joined by $\alpha(1-4)$ linkages; this is amylose. The other has a branched structure with glucose units joined by $\alpha(1-6)$ linkages at the branch points and $\alpha(1-4)$ linkages in the linear parts: this is amylopectin [13–15]. In cellulose the glucose units are linked by $\beta(1-4)$ linkages. The structures of the two polymers are shown in Figure 1.1.

The starch granules contain different ratios of amylose and amylopectin depending on the source, as shown in Table 1.1 [11,16]. Additionally the size and shape of the granules depend on the source of the starch [17]. Whilst starch is insoluble in cold water and forms a suspension on mixing, heating in excess water causes the familiar increase in viscosity associated with starch pastes. This is illustrated in Fig. 1.2.

Owing to their structural differences, amylose and amylopectin have different properties:

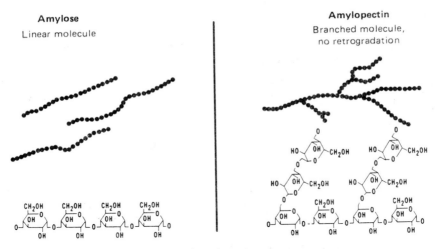

Fig. 1.1.   The structure of amylose and amylopectin.

Table 1.1.   Amylose and amylopectin content of starches

	Amylose (%)	Amylopectin (%)
Standard maize	24	76
Waxy maize	0.8	99.2
High amylose maize	70	30
Potato	20	80
Rice	18.5	81.5
Tapioca	16.7	83.3
Wheat	25	75

Amylose. Although described as a linear molecule, the nature of the tetrahedral carbon atom and the $\alpha(1-4)$ linkage results in practice in a helical structure for amylose, with one turn of the helix occurring every 6 glucose units. Amylose contains between 200 and 2000 glucose units depending on the source. The helical structure readily forms complexes with inorganic ions, the ion entering the helix [18]. Probably the best known of these reactions is that with iodine, where the brown iodine reacts with the white starch to give a blue complex [19].

Amylose is insoluble in cold water. When starch is solubilised by heating and allowed to cool, the amylose chains have a tendency towards self-association [20]. This process is referred to as retrogradation and is illustrated in Fig. 1.3. In dilute solutions, with less than 1% solids, a precipitate forms, whilst at higher concentrations a gel is produced. This gel formation can be utilised by the food manufacturer. The firmness of the gel increases with time, is accelerated by freeze/thaw conditions, and is accompanied by exclusion of water from the starch–water matrix, a process known as syneresis.

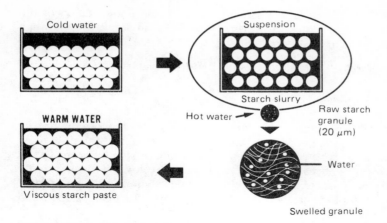

Fig. 1.2. Physical modifications undergone by a starch suspension during cooking. In cold water starch is insoluble and forms a suspension. When the water is warmed, it begins to penetrate the starch granules at a certain temperature (gelatinisation temperature). The granules swell and the viscosity of the slurry increases. It is no longer a suspension but a paste or gel.

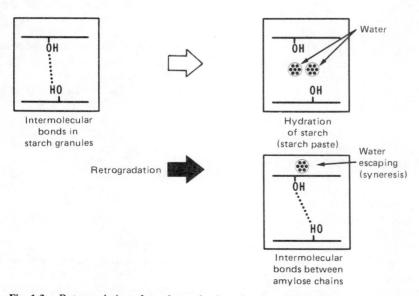

Fig. 1.3. Retrogradation of starch: mechanism of re-association of linear chains after pasting.

Amylopectin. This is the branched polymer. It contains between 10 000 and 100 000 glucose units, and is thus a much larger molecule than amylose [18,21]. Owing to the branches very little helical structure is found in amylopectin and no blue colour is given with iodine; rather, a red-brown colour is produced.

Like amylose, amylopectin is insoluble in cold water but may be solubilised by heating. In normal processing conditions the tendency of amylopectin to retrograde is weak, as the branched structure inhibits close association of the linear fragments in the molecules. However, on cooling, amylopectin can form a loose gel, so that the clear paste gradually becomes opaque with time owing to slight association of the branch ends of the molecules [22].

Physicochemical Properties of Native Starch

The size, shape and composition of the starch granule and the amylose:amylopectin ratio give different starches different properties, and these may be utilised by the food manufacturer. The phosphate content of native starch, expressed as chemically linked phosphorus (Table 1.2), though very low, can significantly affect its properties, particularly the gelatinisation temperature and profile [15,23]. The large phosphate groups effectively force the starch chains apart and allow easier access of water.

If an aqueous suspension of native starch is examined under a polarising microscope, a characteristic polarisation cross is seen owing to its well-defined microcrystalline structure. The structure is maintained by hydrogen bonding between the starch chains [17,24–27]. Dry heat will disrupt this structure although the polarisation cross still remains.

Typically, however, starch is not heated in the dry state in food systems but rather is heated in the presence of water, and under these conditions more dramatic changes occur in the structure and the polarisation cross is lost. As the water temperature increases the granule starts to absorb water as hydrogen bonds are disrupted, and a gradual increase in viscosity occurs until a viscous solution or starch paste is formed. This process is termed gelatinisation and gelatinisation temperature is characteristic of the starch source [28–30]. Pure amylose has a lower hot viscosity than pure amylopectin, and the ratio of the two polymers in the starch will therefore govern the overall viscosity of the starch.

Starch paste has characteristic properties such as clarity, viscosity, texture, stability and taste and these may be conferred on the foodstuff in which the starch is included. These properties depend on the degree of gelatinisation and it is important, therefore, that industrially this process is carefully controlled [31,32]. Maximum viscosity is attained when the granule is in its most swollen state, yet still intact. Continued heating past this point causes the granule to rupture and this is accompanied by a fall in viscosity.

Industrially the gelatinisation profile of a starch is of major importance and the process is followed routinely using the Brabender amylograph. Typical curves are shown in Fig. 1.4. which illustrates the temperature profile in the Brabender instrument and its effect on the starch viscosity. The starch suspension (fixed ratio of starch to water) is heated at a controlled rate of 1.5 °C per minute to 95 °C and held at this temperature for 15 minutes, then cooled at the same controlled rate to 20 °C (see Chap. 2). This method gives a means of comparing the behaviour of different starches during cooking [33].

At temperatures below about 65 °C the granules have not started to swell and there is no development of viscosity. Above this temperature swelling and viscosity development begin. The peak viscosity is a measure of the ability of the granule to swell before rupture, whilst the following decrease in viscosity is associated with continued stirring at high temperature, resulting in the breaking open of the swollen granules. The secondary increase in viscosity during the

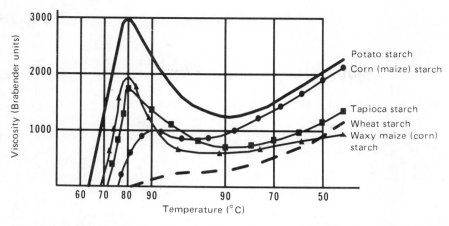

Fig. 1.4. Brabender viscosity curves of various native starches. The viscosimetric determination was carried out on 6.8% (w/v) starch suspensions in water at pH 7.0.

cooling phase is associated with the phenomenon of retrogradation, the extent of the increase being related to the amylose content of the starch.

Different starches give different Brabender profiles, as shown in Fig. 1.4. For example, at the same ratio of starch to water, potato starch develops a higher viscosity than maize starch. The properties of the different common starches are shown in Table 1.2.

Table 1.2. Characteristics of native starches (Brabender viscosity curves, 6.8% dry substance)

	Corn (maize) starch	Waxy corn (maize) starch	Potato starch	Wheat starch	Tapioca starch
Gelatinisation temperature (°C)	75	72	65	85	72
Viscosity (Brabender peak viscosity: BU)	1100	2000	3200	400 (uncooked)	1900
Retrogradation	Considerable	Little	Considerable	Considerable	Medium
Clarity (cooked and cooled)	Opaque	Clear–transp.	Clear	Opaque	Clear
Texture	Short	Very long, stringy	Long	Short	Long
Taste	Cereal	Cereal	Bland	Cereal	Bland
Phosphorus content (%)	0.015	0.003	0.06	0.05	–

The Need for Modified Starches

Starch paste functionality may be reduced by prolonged heating (boiling for example) or exposure to high temperatures (UHT processing), by too much stirring during or after cooking (high shear systems) or by exposure to low pH. This reduction occurs as a result of the destruction of the intragranule hydrogen bonds which maintain the integrity of the granule during gelatinisation. Also, exposure to freeze/thaw conditions has a detrimental effect on the starch paste, accelerating retrogradation.

In modern processes starches may be expected to perform satisfactorily under one or all of these hostile conditions, and additionally in some products they may be expected to form a viscous solution without application of heat. Whilst native starches satisfy the demands of processes where such conditions are not encountered, they break down with consequent loss of viscosity under more severe processing and of course are not dispersible in cold water. Native starches are and will continue to be widely used in the food and related industries; however, as processes become more sophisticated so native starches are not able to provide the required viscosity profiles needed in the finished product. This already prevents their use in a number of food applications and the problem is likely to increase rather than decrease in the future.

To meet the changing requirements of the food industry, and to overcome the deficiencies of native starches, it is possible to modify starches chemically and physically to provide stability under severe processing conditions, and also under freeze/thaw conditions, and to provide viscosity without heating [34,35]. Table 1.3 summarises the need for these so-called modified starches.

Table 1.3. Technological justification of the need for modified starches in the food industry

For production of:	Manufacturers look for:	It is therefore necessary to use:
Instant products	Cold water solubility	Pre-cooked starches
Gums	Decreased viscosity (concentration effects)	Converted starches (thin boiling, oxidised, dextrins)
Canned foods, salad dressing	Improved heat, shear and acid resistance	Cross-linked starches
Frozen foods, canned foods	Improved low temperature stability (retrogradation resistance)	Stabilised starches

Excluding pre-gelatinised starches, the modifications are characterised by the starch granule remaining intact throughout the process. The modifications are therefore carried out at temperatures below starch gelatinisation temperature, and since many of the reactants are toxic, a thorough washing takes place at the end of each process to remove all traces of reactant. The chemical groups substituted into the starch chains during cross-linking and stabilisation are not toxic and the modified starches are safe for consumption [36–39]. Modified starches have the same nutritive value as native starches. The types of modification permitted by existing EEC legislation [40] are discussed in the following section.

Granular Starch Modifications

Cross-Linked Starches

Maximum viscosity in a starch paste is found when the swelling of the granule is at a maximum, corresponding to the peak of the Brabender curve. Cross-linking

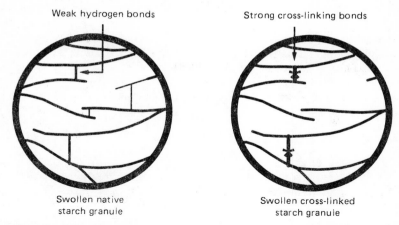

Fig. 1.5. Influence of cross-linking on the structure and stability of starch granules.

will maintain the starch granules in the swollen state without rupture, even under severe processing conditions where the hydrogen bonds in the granule are destroyed. The more the starch is cross-linked, the more severe these conditions can be without loss of granule integrity [41]. Figure 1.5 illustrates the basics of cross-linking in the starch granule.

Cross-linking may be accomplished by treating the starch with very small amounts of chemical agents, typically phosphate [23] or adipate, as shown in Fig. 1.6(a). Figures 1.7 and 1.8 show the effects of cross-linking a native starch on its viscosity in acid and high shear systems.

Starch selection for a particular process therefore requires an exact knowledge of the processing conditions in order to match the starch to the process.

Stabilisation

Whilst cross-linking gives the starch stability under severe processing conditions, it has no effect on stability at low temperatures, that is under freeze/thaw conditions. Instability at low temperatures is primarily a function of the amylose content of the starch. Stability may be incorporated into the starch granule by introduction of substituent groups [42] primarily onto the amylose chains to prevent aggregation. Amylopectin chains also react with the stabilising agent, as the reaction is not specific and the effect is to reduce the tendency of a normal maize starch to gel, and of a waxy maize starch to lose its clarity. Schematically it is possible to compare the "before" and "after" reactions with a zipper, as shown in Fig. 1.9. With no substituent groups (native starch) the amylose chains align easily (zip closes), whilst with these large groups present the amylose chains find it more difficult to align. This has the same effect as a tooth missing or broken on the zipper. Stabilising agents include acetic anhydride [43] and propylene oxide [44], which introduce acetate and hydroxypropyl groups into the starch granules respectively, as shown in Fig. 1.6(b).

The introduction of large substituent groups into the starch granule weakens the inter- and intramolecular hydrogen bonds within the granule, and this causes

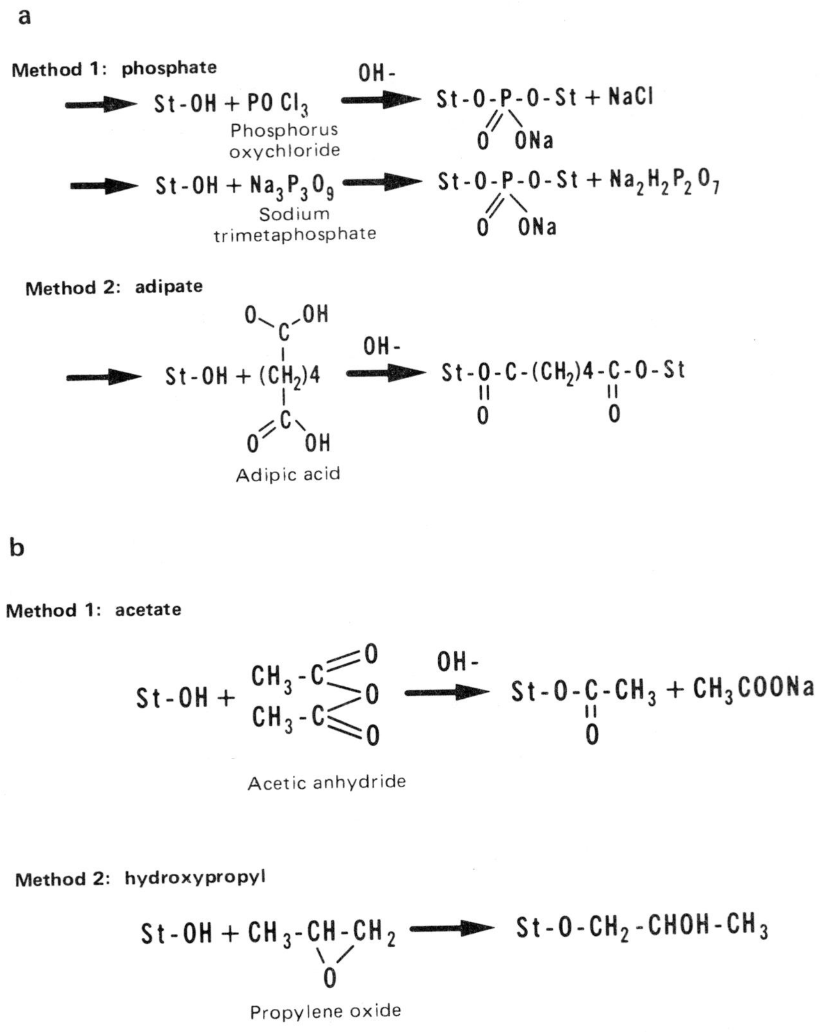

a

Method 1: phosphate

$$\longrightarrow St\text{-}OH + PO\,Cl_3 \longrightarrow^{OH-} St\text{-}O\text{-}P\text{-}O\text{-}St + NaCl$$

Phosphorus
oxychloride

$$\underset{O\quad ONa}{\overset{/\!/\,\backslash}{}}$$

$$\longrightarrow St\text{-}OH + Na_3P_3O_9 \longrightarrow St\text{-}O\text{-}P\text{-}O\text{-}St + Na_2H_2P_2O_7$$

Sodium
trimetaphosphate

$$\underset{O\quad ONa}{\overset{/\!/\,\backslash}{}}$$

Method 2: adipate

$$\longrightarrow St\text{-}OH + (CH_2)4 \longrightarrow^{OH-} St\text{-}O\text{-}C\text{-}(CH_2)4\text{-}C\text{-}O\text{-}St$$

with adipic acid structure
$$O{\sim}_{C}{\nearrow}^{OH}$$
and
$$\underset{O}{\overset{\nearrow C}{}}{\searrow}_{OH}$$

Adipic acid

b

Method 1: acetate

$$St\text{-}OH + \begin{array}{c} CH_3\text{-}C{\overset{0}{\underset{0}{\lessgtr}}} \\ CH_3\text{-}C{\overset{0}{\underset{0}{\lessgtr}}} \end{array} \longrightarrow^{OH-} St\text{-}O\text{-}\underset{O}{\overset{}{C}}\text{-}CH_3 + CH_3COONa$$

Acetic anhydride

Method 2: hydroxypropyl

$$St\text{-}OH + CH_3\text{-}CH\text{-}CH_2 \longrightarrow St\text{-}O\text{-}CH_2\text{-}CHOH\text{-}CH_3$$
$$\underset{O}{\overset{\backslash\,/}{}}$$

Propylene oxide

Fig. 1.6. Chemical modification of starch (St-OH) compatible with existing food regulations. **a** Cross-linked starches. **b** Stabilised starches.

a reduction in the gelatinisation temperature of the starch. The reduction is proportional to the amount of stabilising agent introduced.

When deep-frozen products containing native starches are thawed, they show a pronounced gelation with loss of water owing to combined retrogradation and syneresis effects. Repeated freezing and thawing accentuate these effects. Fig. 1.10 shows the results of water lost from two starch products during repeated freezing and thawing cycles: one product is made from native and one from stabilised starch. One cycle means deep freezing at $-18\,°C$ for 24 hours followed by thawing at $20\,°C$ for 8 hours.

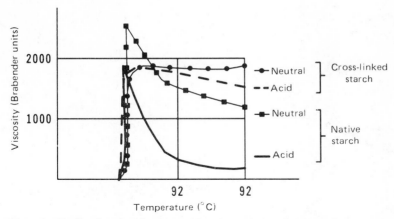

Fig. 1.7. Brabender viscosity curves of native and cross-linked waxy maize starch (6.8% w/v) in neutral (pH 6.0) and acid (pH 3.0) conditions.

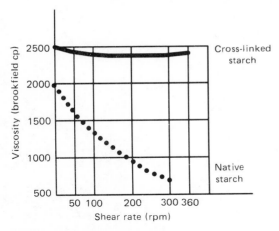

Fig. 1.8. Comparative effects of high shear (colloid mill) on the viscosity of a paste of native or cross-linked waxy maize starch.

Pre-cooked Starches

Many food processors wish to manufacture products which develop viscosity in water or other aqueous systems without the application of heat. This can be achieved by using a pre-cooked starch. The starch is dispersed in water, cooked by steam injection, dried on a drum drier, and finally subjected to a comminution process [37]. The last process can be controlled to give pre-cooked starches of different particle size and hence slightly different properties. The process is shown diagrammatically in Fig. 1.11.

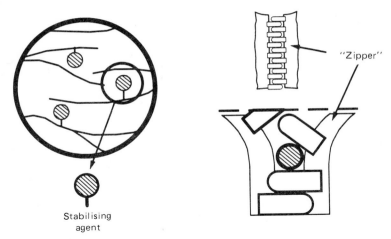

Fig. 1.9. Chemical stabilisation of starch. The role of starch substitutes in preventing the re-association of linear chains.

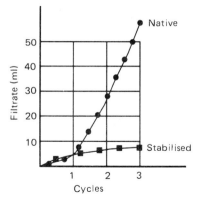

Fig. 1.10. Influence of the chemical stabilisation of waxy maize starch on its behaviour during repeated freezing and thawing cycles.

Pre-cooking of starch can also be achieved by extrusion of native granules at low water content (20%) [45–47]. In some cases, compacting at very high pressures has been claimed to produce the same physicochemical modifications of native starch.

Pre-cooked starches develop maximum viscosity at lower temperatures compared with the native starch and the same properties and advantages of the native starch can therefore be obtained without cooking. Modified starches can also be pre-cooked, and the particular characteristics of the modified starch are then similarly made available without the need to apply heat during the processing of foods.

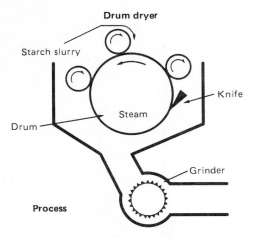

Fig. 1.11.   Industrial production of pre-cooked starches.

Converted Starches

Starch gels and pastes containing about 5% dry matter are widely used in the food industry to provide a range of textures and viscosities in finished products. At concentrations greater than 5% the starch becomes difficult to handle at high temperatures owing to the very viscous amylopectin fraction, yet there are food applications where higher concentrations would be desirable to provide chewiness, elasticity, body and palatability, for example. Typical examples are in confections where gelatine and gum arabic traditionally provided such properties. Compared with starch, however, gelatine and gum arabic are expensive commodities and if they could be replaced by starches there would be economic advantages to the food processor.

It is possible, by carrying out an internal hydrolysis reaction in the starch granule, to produce starches which at 15%–20% concentration give the same viscosity as a native starch at 5% concentration. The higher concentration gives improved textural properties however. On cooling, such starches set to a firm gel, the strength of which may be controlled by the reaction time and the starting material. Three processes may be envisaged: (1) an aqueous acid hydrolysis at subgelatinisation temperatures (30–40 °C) to produce so-called thin boiling starches; (2) a dry acid process in which the starch is heated to above 150 °C to produce the so-called dextrinised starches or dextrins; (3) an oxidation with sodium hypochlorite at subgelatinisation temperature to produce so-called oxidised starches.

Thin Boiling Starches

Reaction of starch with acid at low temperature results in a random intragranular hydrolysis of the $\alpha(1-4)$ and $\alpha(1-6)$ linkages in the amylose and amylopectin molecules. Owing to the use of subgelatinisation temperatures the starch granule remains intact and thin boiling starches require cooking to develop their viscosity

[48,49]. During the process the amylopectin chains are hydrolysed, together with the amylose, and the high hot viscosity problems associated with amylopectin are reduced. The longer the reaction period the more hydrolysis takes place. This results in a lower hot viscosity for the cooked starch and enables a greater concentration to be used to develop a fixed viscosity, but adversely a weaker gel structure is formed on cooling. When the process is carried out on maize starch, for example, this will set to give a firm gel if the reaction time is short, and correspondingly less firm gels with longer reaction times. With waxy maize starch, less firm gels are formed over the whole range of reaction times owing to the high amylopectin content. Thin boiling starches may be used to replace gelatine in the confectionery industry.

Dextrins

Similar types of reaction take place during dextrinisation as during production of thin boiling starch: hydrolysis of $\alpha(1-4)$ and $\alpha(1-6)$ linkages. Reaction temperatures are, however, significantly higher, and this leads to greater hydrolysis and some colour development, particularly with prolonged reaction time and temperature [39–50]. Compared with thin boiling starches, dextrins may be used at higher concentrations to give comparable viscosity, and are typically used to replace gum arabic in confectionery.

Oxidised Starches

Sodium hypochlorite is used as a source of the oxidising agent chlorine and reaction with starch results in random oxidation of hydroxyl (–OH) groups into carbonyl (C=O) groups and also scission of the intramolecular glycosidic bonds to form aldehydic (–CHO) and carboxyl (–COOH) groups [39]. Hydrolysis of the polymeric starch structure also occurs. The carboxy groups function in the same way as the acetate groups in stabilised starches, in that their relatively bulky structure reduces the tendency of the amylose fraction in the starch to associate to give the characteristic gel. They are therefore used where stability is required, and where transparency in the finished product is important. Such a situation is very common in confectionery, where oxidised starches generally replace gum arabic or gelatine.

Hydrophobisation

Whereas in most cases the chemical modification of starch results in an increased affinity for water, there are a few cases where it is desirable to impart a more hydrophobic character to native starch in order to give it more affinity for lipophilic substances. This can be achieved by reacting native starch with, for example, substituted cyclic dicarboxylic acid anhydrides, such as:

$$CH_3-(CH_2)5-CH=CH-CH-CO$$
$$| \qquad\qquad\qquad\qquad\qquad O$$
$$CH_2-CO$$

in the presence of sodium hydroxide [51]. The resulting starch octenylsuccinate, even at a low degree of substitution (DS) (0.02–0.03), can be used in various food applications. These include beverage emulsions, clouding agents, flavour encapsulation, vitamin protection, salad dressings and creams.

The ability of starch to encapsulate lipophilic products can equally result from an enzymatic modification which consists in a simultaneous hydrolysis and cyclisation of linear chains yielding cyclic oligomers generally known as cyclodextrins.

Enzymic Cyclisation of Starch

Enzymic cyclisation is performed in the presence of the enzyme cyclodextringlucanotransferase (EC 2.4.1.19) acting on a low dextrose equivalent (DE) liquefied starch, and results in the formation of cyclic maltodextrins [52–54].

Cyclodextrins are cyclic oligosaccharides composed of several glucose monomers bound together into a doughnut-shaped truncated cone. This configuration creates a hydrophilic rim at the top and bottom of the cone and a hydrophobic cavity in the middle, resulting in a structure capable of binding and encapsulating molecules of the appropriate size within the cavity [55]. This is known as molecular inclusion, and complexes can only form if the guest molecule has the correct steric shape completely or partially to fill the inner ring of the cyclodextrin. The cavity size depends on whether the cyclodextrin is composed of six, seven or eight glucose units: α-, β- and γ-cyclodextrins respectively. Fig. 1.12 illustrates the structure of β-cyclodextrin.

Inclusion of the guest molecule within the structure of cyclodextrin changes the physical and chemical properties of the compound and this can lead to certain advantages for these complexed products. These include reduced oxidation of flavour oils, the ability to mask odours and flavours, increased stability for labile compounds, reduced loss of volatile compounds, and solubilisation in aqueous media of lipophilic compounds.

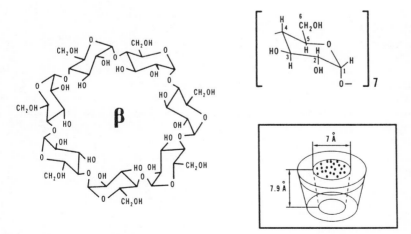

Fig. 1.12.   Different aspects of the structure of β-cyclodextrin.

Cyclodextrin/guest complexes are made using one of three methods [56]: (1) by mixing the two compounds together in aqueous solution or by adding a non-aqueous phase (oil) to a hot aqueous solution of cyclodextrin and cooling; (2) by dry mixing cyclodextrin and guest, adding water to form a paste and drying; or, (3) for gases only, by bubbling the gas through the cyclodextrin solution.

Extensive Chemical and Enzymatic Hydrolysis of Starch

Starch in the native state is insoluble in water but may be solubilised by heating or by modification. In addition to modifications in which the starch granule remains intact there are more severe types of modification in which the integrity of the granule is completely lost and the end products are freely soluble in water. These modifications initially involve hydrolysis of the starch polymers into lower molecular weight components; in later reactions structural rearrangements of these primary components are carried out to produce a further series of related products.

Starch Hydrolysis

Industrially the hydrolysis is carried out in two steps: liquefaction and saccharification.

Liquefaction

Liquefaction involves simultaneous gelatinisation and viscosity reduction. Gelatinisation is necessary because, owing to its compact granular structure, starch is resistant to rapid acid or enzyme attack. The liquefaction is essential to enable high concentrations of starch to be used without the blocking of pipes or the burning on heated surfaces which would occur with highly viscous starch pastes.

Starch is therefore heated to about 100 °C in aqueous suspension (gelatinisation) to disrupt the granular structure in the presence of either hydrochloric acid or the enzyme α-amylase (EC 3.2.1.1). The α-amylase enzyme used is a type stable at high temperatures which is produced by *B. licheniformis*.

Only a limited hydrolysis takes place during liquefaction and only small amounts of low molecular weight components are produced. The DE value generally reached at the end of the liquefaction is between 10 and 15 [57–59].

Saccharification

Saccharification involves the formation of higher concentrations of sugars and, if the process is allowed to go to completion, dextrose (glucose) will be the ultimate product. Typically, however, the reaction is not allowed to proceed to completion but is stopped at some predetermined point. The reaction mixture at this stage

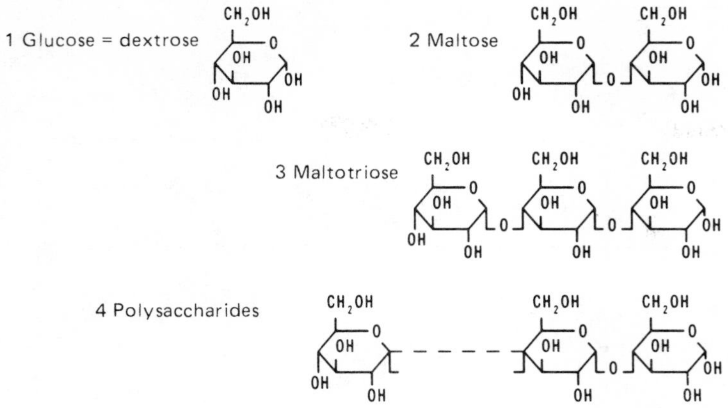

1 Glucose = dextrose

2 Maltose

3 Maltotriose

4 Polysaccharides

Fig. 1.13. Starch hydrolysis products from α-D-glucose (dextrose) to higher polysaccharides.

contains a mixture of starch hydrolysis products such as dextrose, maltose, maltotriose and fragments containing up to 20 glucose units, as shown in Fig. 1.13. These complex mixtures of glucose polymers are called glucose syrups [60,61].

Acid or different enzymes are used in the process, the latter including β-amylase (EC 3.2.1.2) and amyloglucosidase (EC 3.2.1.3). The methods of hydrolysis of the acid and different enzymes are shown in Fig. 1.14.

After liquefaction standard glucose syrups are produced using acid or α-amylase or combinations of the two, high maltose syrups using β-amylase, and syrups containing high concentrations of dextrose using amyloglucosidase. The range of products available as a result of the hydrolyses is shown in Fig. 1.15.

The term dextrose equivalent, or DE, is used to describe the extent of hydrolysis of the starch. DE is calculated as dextrose but in practice is not related

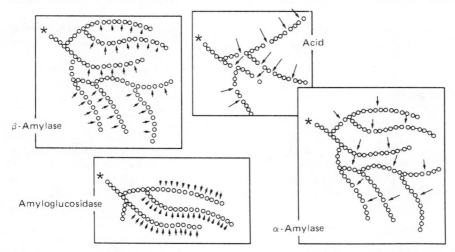

Fig. 1.14. Schematic description of various hydrolytic actions carried out on starch, with either acid (HCl) or enzymatic catalysis.

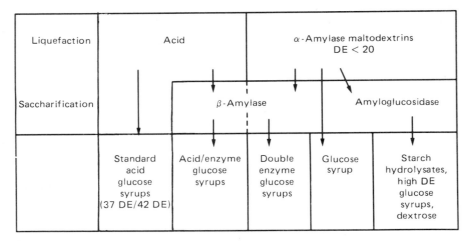

Fig. 1.15. Hydrolysis products obtained from starch by a combination of various degradation agents. Industrially these products exist as either solutions (syrups) or dried powders. The dextrose equivalent (DE) of each product is a precise indication of the extent of starch hydrolysis.

to the actual dextrose content of the product. It measures the reducing capacity of all the reducing sugars present towards Fehling's solution and expresses it as a percentage of the total solids present. The Lane and Eynon method is traditionally used to measure DE. At the extremes of hydrolysis, starch and pure dextrose, DE is a measure of the dextrose content. Starch, the starting material is designated 0 DE, whilst dextrose is designated 100 DE; intermediate values represent various stages in the breakdown of the starch.

Whilst DE is a measure of the reducing sugars present or the extent of starch hydrolysis it does not necessarily provide any information as to the composition of the product. It is possible to have two glucose syrups of the same DE but with entirely different composition and properties [62]. For example, a 25 DE syrup will contain more dextrose when it is produced by an acid hydrolysis than when it is produced using α-amylase, as Table 1.4 shows. More dramatic compositional differences are also shown for three 37 DE syrups produced using the three available techniques.

Maltodextrins

Starch hydrolysates having a DE of 20 and below are not referred to as glucose syrups but rather as maltodextrins, and are labelled as such on product packaging material. They are normally only available in spray-dried (powder) form owing to the inherent retrogradation and poor keeping qualities of the syrup forms, and they are typically used to provide bulk without excessive sweetness in food products. Examples of maltodextrin composition are given in Table 1.5.

Dried Glucose Syrups

Starch hydrolysates having a DE greater than 20 are referred to as glucose syrups. In the spray-dried form they are called dried glucose syrups and are commonly

Table 1.4. Carbohydrate composition of glucose syrups

D.E.	HYDROLYSIS	DP1 Dextrose	DP2 Maltose	DP3 to DP7	Higher than DP7
25	ACID	8% 8%	46%		38%
25	α AMYLASE	3% 9%	44%		44%
37	ACID	16% 13%	48%		23%
37	ACID (20 D.E.) + β AMYLASE	8%	30%	28%	34%
37	α AMYLASE (20 D.E.) + β AMYLASE	5%	35%	32%	28%

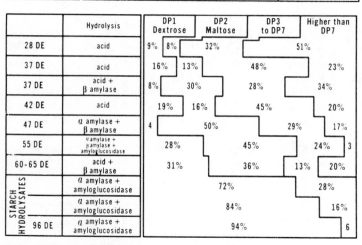

	Hydrolysis	DP1 Dextrose	DP2 Maltose	DP3 to DP7	Higher than DP7
28 DE	acid	9% 8%	32%		51%
37 DE	acid	16% 13%	48%		23%
37 DE	acid + β amylase	8%	30%	28%	34%
42 DE	acid	19% 16%	45%		20%
47 DE	α amylase + β amylase	4	50%	29%	17%
55 DE	α amylase + β amylase + amyloglucosidase	28%	45%	24%	3
60-65 DE	acid + β amylase	31%	36%	13%	20%
	α amylase + amyloglucosidase	72%			28%
	α amylase + amyloglucosidase	84%			16%
96 DE	α amylase + amyloglucosidase	94%			6

(STARCH HYDROLYSATES)

D.E., dextrose equivalent.
DP, degree of polymerisation, i.e. number of glucose units linked to one another per fragment of starch.

Table 1.5. Carbohydrate composition of maltodextrins

D.E.	DP1 DEXTROSE	DP2 MALTOSE	HIGHER THAN DP2
5-8	0.5 1		98.5%
11-14	1 2.5		96.5%
15-18	1.5 4.5		94%
18-20	2.5 7%		90.5%

DP1 and DP2, number of glucose units linked to one another per fragment of starch.

available in the range of 30–60 DE. Above about 60 DE the dried material is so hygroscopic that drying is not a practical proposition, and the syrups are only available in concentrated liquid form (normally 75%–85% solids). Glucose syrups above 30 DE are also available in liquid form. Thus a wide range of end products is available depending on the needs of the food manufacturer, as shown in Table 1.4.

Maltose Syrups

When β-amylase is used in the hydrolysis process, maltose is the main end product. Maltose syrups are characterised by resulting in less sticky products when used in foods (confectionery for example) and are slightly less sweet than their traditional acid-hydrolysed counterpart of the same DE. Typical compositions are given in Table 1.4. A specialised high maltose syrup is used to produce Lycasin (registered trademark of Roquette), a hydrogenated glucose syrup. This is discussed on p. 26.

Hydrolysates

Very high DE glucose syrups, 80–98 DE, are usually referred to as hydrolysates. Typical compositions are given in Table 1.4. Hydrolysates are mainly used as precursors for further processes, but may also find uses in the food industry. Downstream products from hydrolysates include fructose syrups and pure fructose, sorbitol, mannitol and glucono delta lactone. These will be discussed in detail in the following section. Table 1.6 shows the relationship between DE value and the properties of glucose syrups and maltodextrins (not maltose syrups).

Food manufacturers develop processes in their own way and often require very specialised glucose syrups with specific dextrose or maltose or maltotriose

Table 1.6. Properties and functional uses of glucose syrups in the food industry

Property or functional use (alphabetically)	Type of corn syrup		
	Low DE	Intermediate DE	High DE
Bodying agent	←		
Browning reaction			→
Cohesiveness	←		
Fermentability			→
Flavour enhancement			→
Flavour transfer medium			→
Foam stabiliser	←		
Freezing point depression			→
Humectancy	←		
Hygroscopicity			→
Nutritive solids	←		
Osmotic pressure			→
Prevention of sugar crystallisation	←		
Prevention of coarse ice crystals during freezing	←		
Sheen producer	←		
Sweetness			→
Viscosity	←		

concentrations, for example. Technically these can usually be produced by the syrup manufacturer although the economics of such processes may often be limiting. A solution to producing possibly hundreds of different syrups by direct hydrolysis is to produce specific syrups and then manufacture the rest by blending those. This is carried out to a certain extent already in the starch industry, and in the future the process is likely to increase.

Simple Natural Carbohydrates and Their Derivatives

Despite the fact that it ranks far behind sucrose in terms of quantities consumed by individuals and the food industry, starch has the advantage of greater versatility in that it can be used either as a macromolecule, whether chemically modified or not, or in depolymerised forms ranging from large oligomers to simple carbohydrates. The main interest in simple carbohydrates – monosaccharides, disaccharides and their derivatives – generally lies in their sweetening power and affinity for water. Moreover, they usually exhibit distinct metabolic behaviour, which can result in various uses according to the physiological requirements to be satisfied.

Monosaccharides

α-D-Glucose (Dextrose)

This hexose can be considered as the physiological sugar. It is found in the free state in fruits and, with free fructose, is the main component of honey. Present in cellulose and starch in polymerised form, it is generally obtained from starch by a process combining dual enzyme hydrolysis followed by crystallisation of α-D-glucose monohydrate from the 96–98 DE hydrolysate.

The pseudo-aldehyde group of α-D-glucose is responsible for its reducing behaviour. Moreover this group is involved in the transformation of α-D-glucose into D-sorbitol or D-gluconic acid.

Fructose

D-Fructose (laevulose) is widely distributed in nature either as a free sugar, as in honey and fruits, or in polymeric form, as in insulin which is abundant in the roots of plants such as chicory and Jerusalem artichoke. Industrially, D-fructose can be obtained from inverted sucrose or from isomerised starch hydrolysates (high fructose corn syrups: HFCS). In both cases, the first step of fructose purification involves a chromatographic enrichment so as to obtain a fraction containing more than 90% fructose (w/w) that can then be submitted to crystallisation.

Because the non-fructose chromatographic fraction can be enzymatically re-isomerised, a high yield of fructose can be obtained, whatever the starting feedstock. However, in Europe this technology is of a limited use, due to a quota system which limits the industrial use of glucose isomerase (EC 5.3.1.5) [63,64]. Depending on its environment, pure D-fructose can be significantly sweeter than sucrose, so it can contribute to a reduced calorie intake [65–67].

D-Fructose

The catalytic reduction of fructose yields equimolar amounts of D-sorbitol and D-mannitol, from which D-mannitol is easily separated in crystalline form due to its low solubility in water.

Mannose

D-Mannose is naturally present in polysaccharides called mannans. With a view to utilising it as a starting material for D-mannitol, it is possible to obtain D-mannose by chemical epimerisation of D-glucose [68].

D-Mannose

Xylose

The main interest of D-xylose is for the production of xylitol. Industrially, it is obtained through the extraction and hydrolysis of the pentosan fraction of various plant materials: corn cobs, hazelnut shells, rice bran, birch wood [69–71].

D-Xylose

Disaccharides

Sucrose

Sucrose is the most important of the simple carbohydrates and its production has been highly industrialised using two main agricultural sources: beet and cane [12,72]. From a chemical point of view, it results from the association of a glucose residue and a fructose residue through an $\alpha(1-2)$ linkage:

Sucrose

As such, sucrose is a non-reducing sugar, with a high water solubility. It can be easily hydrolysed by acids, or by the enzyme invertase (EC 3.2.1.26) [58]. By tradition, sucrose is used as a reference for sweeteners.

Besides its classical uses in all parts of the food industry, sucrose has found some new applications in the production of novel ingredients or food additives. As will also be seen, the epimerisation of sucrose yields another disaccharide, palatinose, which in turn can be converted to the polyol sweetener Palatinit®. Sucrose can equally undergo various enzymatic modifications resulting in the synthesis of a new class of sweeteners: Neo-sugar® and coupling sugar. Esterification of sucrose with fatty acids can produce a family of low-calorie fat replacers with promising applications, which are presently being reviewed by health authorities. Chlorination of sucrose can increase its sweetening power by a factor of 650. The resulting intense sweetener, Sucralose, is in fact 4,1′,6′-trichloro-4,1′,6′-trideoxygalactosucrose [73].

Lactose

Lactose is normally found in milk at a concentration of about 50 g/l. Industrially, it can be isolated from whey or UF permeates [74,75]. It consists of a galactose residue joined to a glucose residue by a $\beta(1-4)$ linkage:

Lactose

Lactose is a reducing sugar and is unusual amongst simple carbohydrates in that it has a limited solubility in water. This drawback, as well as a digestive intolerance phenomena, can be eliminated by hydrolysis with the enzyme lactase (EC 3.2.1.23), which results in syrups with increased sweetness [76,77].

Maltose

Maltose is exclusively obtained from starch, by the action of β-amylase on a low DE liquefied starch. When β-amylase is coupled with pullulanase (EC 3.2.1.41), it is possible to obtain syrups with a β-maltose content as high as 85% (w/w) [78,79].

CH₂OH   CH₂OH
   O   O OH

   OH   O   OH
HO

   OH   OH

β-Maltose

Though regarded for a long time as uncrystallisable, maltose can now be obtained industrially in a crystalline form by combining the action of pullulanase with the use of liquid chromatography.

Maltose is a diholoside and a reducing sugar. Maltose syrups are increasingly used in the food industry. Crystalline maltose, though, is generally considered uneconomical for traditional food applications and, apart from pharmaceutical applications, its main use is as a starting material for the production of crystalline maltitol.

Derivatives of Mono- and Disaccharides

Hydrogenated Derivatives (or Sugar Alcohols)

Hydrogenated derivatives are produced in high-temperature, catalytic hydrogenation reactions in which aldehydic functions are converted to primary alcohol (H–C=O ⟶ H2–C–OH) and ketonic to secondary alcohol functions (C=O ⟶ H–C–OH and C=O ⟶ HO–C–H). During the process, reducing sugars are converted to non-reducing polyols, with dramatic changes in reactions involving reducing functions. Whilst nutritional value is unchanged by the process, hydrogenation can change the way in which the products are metabolised by the body [80,81].

Sorbitol

Sorbitol is widely distributed in fruits [82]. It is produed from dextrose and is available in both syrup and crystalline form. It does not require insulin in its metabolism and thus finds uses in foods for diabetics, although its laxative properties limit excessive consumption. It does not take part in Maillard browning reactions, is not easily caramelised, and is not fermented by yeasts. Confectionery or pharmaceutical products manufactured from sorbitol are non-cariogenic [80,81].

Glucose → Sorbitol

Mannitol

Mannitol may be produced from mannose, but is commercially produced from fructose. When fructose is hydrogenated a mixture of sorbitol and mannitol is produced. Mannitol is much less soluble than sorbitol and is removed by crystallisation. It has similar properties to sorbitol and is mainly used as a dusting agent in confectionery, or in the pharmaceutical industry in tabletting appli-cations [80–82].

Fructose   Sorbitol   Mannitol

Maltitol

Maltitol is produced from maltose and is available in both syrup and crystalline forms [81,83]. It is the main component of Lycasin® [80,82]. Due to its stability and sweetening power, maltitol has rapidly developing applications in the manufacture of confectionery and gums.

Maltitol

Lycasin®

Lycasin® is produced from a high maltose/glucose syrup and contains about 8% sorbitol and 55% maltitol in addition to hydrogenated tri-, tetra-, penta- etc., saccharides. It is mainly used at present in chewing-gums and pharmaceutical syrups for its non-cariogenic properties and sweet taste [80–82]. It is available only in liquid form.

Lactitol

Lactitol is produced from lactose by conventional hydrogenation [84,85].

Lactitol

Oxidised Derivatives

Gluconic Acid

Gluconic acid is produced by oxidative fermentation of dextrose. It is easily crystallised into the lactone with loss of water. Glucono delta lactone is used as an acidulant in cottage cheese manufacture and in baking, where it replaces acid phosphate or citric acid. When initially tasted, glucono delta lactone is sweet, but as it is hydrolysed by the water in the saliva to gluconic acid, the taste becomes acidic [66,87].

Glucose   Gluconic acid   Glucono delta lactone

Epimerised Derivatives

Palatinose® (Isomaltulose)

This is produced from sucrose by an intramolecular rearrangement catalysed by an enzyme system present in bacteria such as *Protaminobacter rubrum*

[88,89]. The (1–2) bond between glucose and fructose is broken and the glucose and fructose rejoined through C1 on glucose and C6 on fructose.

CH₂OH
C=O
HO
OH
OH
O—CH₂
CH₂OH
O
HO
OH
OH
OH
Palatinose

Palatinose forms the starting material for a second reaction in which it is hydrogenated to form a mixture of glucose–sorbitol α(1–6) and glucose–mannitol α(1–6). This material is Palatinit® or Isomalt.

CH₂OH
OH
HO
OH
OH
O—CH₂
CH₂OH
O
HO
OH
OH
+
CH₂OH
O
HO
OH
OH
HO
HO
CH₂OH
OH
OH
O—CH₂

Palatinit®

Palatinit has the characteristic properties of hydrogenated sugars. In addition it presents the advantage of a low calorie content, being incompletely metabolised [90,91].

Lactulose

Lactulose is produced from lactose. It consists of a galactose residue joined to a fructose residue via a β(1–2) linkage. It has found particular use in laxative products [92].

CH₂OH
O
HO
OH
OH
O
CH₂OH
O
OH
CH₂OH
OH
Lactulose

Condensation Derivatives

Polydextrose

This complex polymer results from the thermal condensation of dextrose in the presence of citric acid and a small amount of sorbitol [93]. It consists of randomly linked dextrose units with the (1–6) bond predominating. Polydextrose is soluble in water; it is a non-crystalline, non-cariogenic material with little sweetness and, perhaps most important of all, it has a reduced calorific value (of about 1 Cal/g) compared with other sugars and derivatives. Its main use is as a bulking agent [94].

Enzymatic Transfer Derivatives

Neo-sugar®

Neo-sugar® results from the action of a fructosyl transferase from *Aureobasidium pullulans* on sucrose, which plays a dual role of fructose donor and fructose acceptor [95,96]. Neo-sugar® is a complex mixture of oligosaccharides containing mainly:

G–F	Sucrose
G–F–F	Kestose
G–F–F–F	Nystose
G–F–F–F–F	–

It is generally considered not to be digested in the small intestine and to be cariogenic, with the ability to promote the growth of *Lactobacillus bifidus* [97].

Coupling sugar

This product results from a transglucosidation reaction between sucrose (donor) and maltodextrin (acceptor), catalysed by cyclodextringlycosyltransferase, the same enzyme as used for the production of cyclodextrins [96].

Conclusions

The chemistry of carbohydrates in relation to their utilisation in food industry is complex and diverse.

Starch, due to its macromolecular organisation and its chemical and enzymatic reactivity, is probably the most sophisticated model of adding value to an agro resource.

Simple sugars and their derivatives are of obvious importance to the food manufacturer, who can make an appropriate choice according to technological, chemical (e.g. browning reactions), regulatory and economic considerations.

Polyols such as sorbitol, mannitol, xylitol and Lycasin®, as well as polydextrose, have, in addition to their physicochemical and organoleptic characteristics, the advantage of being non-cariogenic and may be used in foods intended to contribute to dental health.

The relative sweetness of major carbohydrates is shown in Table 1.7. This is one of their most important properties and they are widely used to contribute this property to foods [99].

Table 1.7. Relative sweetness of carbohydrates and related compounds

Sucrose	1[a]
Dextrose	0.7
Fructose	1.3
Mannose	0.5
Xylose	0.75
Maltose	0.4
Lactose	0.3
Sorbitol	0.7
Mannitol	0.5
Maltitol	0.75
Lactitol	0.4
Xylitol	0.9
Palatinit®	0.6
Neo-sugar®	0.6
Coupling sugar	0.6
Invert sugar	1.1
High fructose glucose syrup	1
15 DE maltodextrin	0.1
30 DE glucose syrup	0.25
43 DE glucose syrup	0.3
65 DE glucose syrup	0.6

[a]The standard is a 5% (w/v) sucrose solution.

The sugar alcohols may in some cases be suitable for use in foods for diabetics, and fructose also may be used for this application. There are obviously many other properties which could be discussed, although possibly sweetness is the most important feature of carbohydrates. It is unlikely in the foreseeable future that diets will exclude this sensation, and developments in this area will continue [100].

References

1. Beaux Y, Mazerolle P (1988) Les Céréales dans L'Agro-industrie. ITCF, Paris, pp 3–11
2. Farris, PL (1984) Economics and future of the starch industry. In: Whistler RL et al. (eds) Starch: chemistry and technology. Academic Press, New York and London, pp 11–24
3. Rexen F, Munck L (1984) Utilization of starch. In: Cereal crops for industrial use in Europe. EEC Report, EUR 9617 EN, pp 64–119
4. Dossier "Les Polysaccharides". Biofutur, Paris, September 1986: 17–31

5. Jones SF (1983) The world market for starch and starch products with particular reference to cassava starch. Report of the Tropical Development and Research Institute, G173, London
6. Debatisse ML (1987) Industries due Sucre el des amidons (CEE, USA, Japan). Skippers, Paris
7. Meyer PA (1984) Corn wet milling. Milling and Baking News, 28 February:19–20
8. Long JE (1985) United States markets for starch-based products. In: Van Beynum GMA, Roels JA (eds) Starch conversion technology. Marcel Dekker, New York, pp 335–347
9. Corn Refiners Association (1988) Annual report. Washington DC
10. Centre for European Agricultural Studies (1986) The Production and use of cereal and potato starch in the EEC. Office for Official Publications of the European Communities, Luxembourg
11. Munck L, Rexen F, Hasstrup-Pedersen L (1988) Cereal starches within the European Community: agricultural production, dry and wet milling and potential use in industry. Die Stärke 40:81–87
12. Light FO (1985) World sugar statistics. In: International sugar economic yearbook and directory. Ratzeburg, pp 1–72
13. Banks W, Greenwood CT, Muir DD (1973) The structure of starch. In: Birch GG, Green LF (eds) Molecular structure and function of food carbohydrate. Applied Science Publishers, London, pp 177–194
14. Whistler RL, Daniel JR (1984) Molecular structure of starch. In: Whistler RL et al. (eds) Starch: chemistry and technology. Academic Press, New York and London, pp 153–182
15. Tegge G (1984) Chemische Zusammenselzung und Konstitution. In: Tegge G (ed) Stärke und Stärkederivate. Behr's Verlag, pp 29–48
16. Swinkels JJM (1985) Sources of starch, its chemistry and physics. In: Van Beynum GMA, Roels JA (eds) Starch conversion technology. Marcel Dekker, New York, pp 15–46
17. Moss GE (1976) The microscopy of starch. In: Radley JA (ed) Examination and analysis of starch and starch products. Applied Science Publishers, London, pp 1–32
18. Young AH (1984) Fractionation of starch. In: Whistler RL et al. (eds) Starch chemistry and technology. Academic Press, New York and London, pp 249–283
19. Lyne FA (1976) Chemical analysis of raw and modified starches. In: Radley JA (ed) Examination and analysis of starch and starch products. Applied Science Publishers, London, pp 133–165
20. Miles MJ, Morris VJ, Orford PD, Ring SG (1985) The roles of amylose and amylopectin in the gelation and retrogradation of starch. Carbohydr Res 135:271–281
21. Inouchi N, Glover DV, Fuwa H (1987) Chain length distribution of amylopectins of several single mutants and the normal counterpart, and sugary-1-phytoglycogen in maize. Die Stärke 39(8):259–266
22. Ring SG, Colonna P, I'anson KJ et al. (1987) The gelation and crystallization of amylopectin. Carbohydr Res 162:277–293
23. Solarek DB (1986) Phosphorylated starches and miscellaneous inorganic esters. In: Wurzburg OB (ed) Modified starches: properties and uses. CRC Press, Boca Raton, pp 97–112
24. Gallant DJ, Sterling C (1976) Electron microscopy of starch and starch products. In: Radley JA (ed) Examination and analysis of starch and starch products. Applied Science Publishers, London, pp 33–59
25. Greenwood CT (1978) Observations on the structure of the starch granule. In: Blanshard JMV, Mitchell JR (eds) Polysaccharides in food. Butterworth, London, pp 129–138
26. French D (1984) Organization of starch granules. In: Whistler RC et al. (eds) Starch: chemistry and technology. Academic Press, New York and London, pp 183–247
27. Oostergetel GT, Van Bruggen EFJ (1987) The structure of starch: electron microscopy and electron diffraction. Food Hydrocolloids 1:527–528
28. Miller BS, Derby RI, Trimbo HB (1973) A pictorial explanation for the increase in viscosity of a heated wheat starch – water suspension. Cereal Chem 50:271–280
29. Blanshard JMV (1978) Physico-chemical aspects of starch gelatinization. In: Blanshard JMV, Mitchell JR (eds) Polysaccharides in food. Butterworth, London, pp 139–152
30. Ring SG (1987) Molecular interactions in aqueous solutions of the starch polysaccharides: a review. Food Hydrocolloids 1:449–454
31. De Willigen AHA (1976) The rheology of starch. In: Radley JA (ed) Examination and analysis of starch and starch products. Applied Science Publishers, London, pp 61–90
32. Zobel HF (1984) Gelatinization of starch and mechanical properties of starch pastes. In: Whistler RL et al. (eds) Starch: chemistry and technology. Academic Press, New York and London, pp 285–309
33. Tipples KH (1982) Uses and applications. In: Shuey WC, Tipples KH (eds) The amylograph handbook. AACC, St Paul, Minnesota, pp 12–24

34. O'Dell J (1978) The use of modified starches in the food industry. In: Blanshard JMV, Mitchell JR (eds) Polysaccharides in food. Butterworth, London, pp 171–182.
35. Langan RE (1986) Uses of modified starches in the food industry. In: Wurzburg OB (ed) Modified starches: properties and uses. CRC Press, Boca Raton, pp 199–212
36. Radley JA (1976) The manufacture of modified starches. In: Radley JA (ed) Starch production technology. Applied Science Publishers, London, pp 449–479
37. Tegge G (1984) Modifizierte Stärken. In: Tegge G (ed) Stärke und Stärkederivate. Behr's Verlag, pp 165–193
38. Rutenberg MW, Solarek D (1984) Starch derivatives: production and uses. In: Whistler RL et al. (eds) Starch:chemistry and technology. Academic Press, New York and London, pp 312–388
39. Fleche G (1985) Chemical modification and degradation of starch. In: Van Beynum GMA, Roels JA (eds) Starch conversion technology. Marcel Dekker, New York, pp 73–99
40. CEE (1985) Proposition de directive du conseil relative au rapprochement des législations des etats membres concernant les amidons modifiés destinés à l'alimentation humaine. JOCE 1 February 1985, No. C31/6–C31/10
41. Wurzburg OB (1986) Cross-linked starches. In: Wurzburg OB (ed) Modified starches: properties and uses. CRC Press, Boca Raton, pp 41–54
42. Radley JA (1976) The manufacture of esters and ethers of starch. In: Radley JA (ed) Starch production technology. Applied Science publishers, London, pp 481–542
43. Jarowenko W (1986) Acetylated starch and miscellaneous organic esters. In: Wurzburg OB (ed) Modified starches: properties and uses. CRC Press, Boca Raton, pp 55–78
44. Tuschhoff JV (1986) Hydroxypropylated starches. In: Wurzburg OB (ed) Modified starches: properties and uses. CRC Press, Boca Raton, pp 89–96
45. Mercier C, Feillet P (1975) Modification of carbohydrate components by extrusion-cooking of cereal products. Cereal Chem 63:240–246
46. Mercier C, Charbonniere R, Gallant D, Guilbot A (1978) Structural modification of various starches by extrusion-cooking with a twin-screw French extruder. In: Blanshard JMV, Mitchell JR (eds) Polysaccharides in food. Butterworth, London, pp 153–170
47. Doublier JL, Colonna P, Mercier C (1986) Extrusion cooking and drum drying of wheat starch. II. Rheological characterization of starch pastes. Cereal Chem 63:240–246
48. Rohwer RG, Klem RE (1984) Acid-modified starch: production and uses. In: Whistler RL et al. (eds) Starch chemistry and technology. Academic Press, New York and London, pp 529–541
49. Wurzburg OB (1986) Converted starches. In: Wurzburg OB (ed) Modified starches: properties and uses. CRC Press, Boca Raton, pp 17–40
50. Acton W (1976) The manufacture of dextrins and British gums. In: Radley JA (ed) Starch production technology. Applied Science Publishers, London, pp 273–293
51. Trubiano PC (1986) Succinate and substituted succinate derivatives of starch. In: Wurzburg OB (ed) Modified starches: properties and uses. CRC Press, Boca Raton, pp 131–147
52. Horikoshi K (1979) Production and industrial applications of beta-cyclodextrin. Process Biochem 14(5):26–30
53. Flaschel E, Landert JP, Spiesser D, Renken A (1984) The production of alpha-cyclodextrin by enzymatic degradation of starch. Ann NY Acad Sci 434:70–77
54. Sicard PJ, Saniez M-H (1978) Biosynthesis of cycloglycosyltransferase and obtention of its enzymatic reaction products. In: Duchene D (ed) Cyclodextrins and their industrial uses. Editions de Santé, Paris, pp 75–103
55. Le Bas G, Rysanek N (1987) Structural aspects of cyclodextrins. In: Duchene D (ed) Cyclodextrins and their industrial uses. Editions de Santé, Paris, pp 105–130
56. Hirayama F, Uekama K (1987) Methods of investigation and preparing inclusion compounds. In: Duchene D (ed) Cyclodextrins and their industrial uses. Editions de Santé, Paris, pp 131–172
57. Manners DJ (1978) The enzymatic degradation of starches. In: Blanshard JMV, Mitchell JR (eds) Polysaccharides in food. Butterworth, London, pp 75–91
58. Sicard, PJ (1982) Application industrielle des enzymes. In: Duran G, Monsan P (eds) Les enzymes: production et utilisations industrielles. Gauthier-Villars, Paris, pp 121–164
59. Reilly PJ (1985) Enzymic degradation of starch. In: Van Beynum GMA, Roels JA (eds) Starch conversion technology. Marcel Dekker, New York, pp 101–142
60. Fullbrook PD (1984) The enzymic production of glucose syrups. In: Dziedzic SZ, Kearsley MW (eds), Glucose syrups: science and technology. Applied Science Publishers, London, pp 65–115
61. Macallister RV, Wardrip EK, Schnyder BJ (1975) Modified starches, corn syrups containing glucose and maltose, corn syrups containing glucose and fructose, and crystalline dextrose. In:

Reed G (ed) Enzymes in food processing, 2nd edn. Academic Press, New York and London, pp 332–359

62. Birch GG (1977) The general chemistry and properties of glucose syrups. In: Birch GG, Shallenberger RS (eds) Developments in food carbohydrate 1. Applied Science Publishers, London, pp 1–17

63. Van Tilburg R (1985) Enzymatic isomerization of corn starch-based glucose syrups. In: Van Beynum GMA, Roel JA (eds) Starch conversion technology Marcel Dekker, New York, pp 175–236

64. Cronin T (1988) Starch/sugar competition for sweetener market in Europe. Zuckerindustrie 113:283–288

65. Pawan GLS (1973) Fructose. In: Birch GG, Green F (eds) Molecular structure and function of food carbohydrate. Applied Science Publishers, London, pp 65–80

66. Osberger TF, Linn HR (1978) Pure fructose and its applications in reduced-calorie foods. In: Dwivedi BK (ed) Low calorie and special dietary foods. CRC Press, Boca Raton, pp 115–123

67. Osberger TF (1986) Pure crystalline fructose. In: O'Brien Nabor L, Gelardi RC (eds) Alternative sweeteners. Marcel Dekker, New York, pp 245–275

68. Sicard PJ (1983) A new sucrochemistry from Starch. In: Heslot H, Villet R (eds) Biomass as a source of industrial chemicals. Adeprina, Tech & Doc, Lavoisier, Paris, pp 229–256

69. La Forge FB, Hudson CS (1918) The preparation of several useful substances from corn cobs. J Ind Eng Chem 10(11):925–927

70. Dunning JW, Lathrop EC (1945) The saccharification of agricultural residues. Ind Eng Chem 37:24–35

71. Voirol F (1985) Xylitol: its caries-preventive and technological properties and food applications. In: Dacosta Y (ed) Polyols and polydextrose Apria, Paris, pp 187–221

72. Nicol WM (1982) Sucrose, the optimum sweetener. In: Birch GG, Parker RJ (eds) Nutritive sweeteners. Applied Science Publishers, London, pp 17–35

73. Houch L (1977) Selective substitution of hydroxyl groups in sucrose. In: Hickson JL (ed) Sucrochemistry. ACS Symposium Series, Washington DC, pp 9–21

74. Delaney RAM, Donnelly JK, O'Sullivan O (1973) Lactose and reverse osmosis. In: Birch GG, Green CF (eds) Molecular structure and function of food carbohydrate. Applied Science Publishers, London, pp 155–176

75. Nickerson TA (1977) Lactose sources and recovery. In: Birch GG, Shallenberger RS (eds) Developments in food carbohydrate 1. Applied Science Publishers, London, pp 77–90

76. Lefevre M, Morel M (1979) Le lactose. L'Alimentation et la Vie 67:175–182

77. Williams CA (1983) Lactose hydrolysate syrups. In: Grenby TH et al. (eds) Developments in sweeteners. Applied Science Publishers, London, pp 27–50

78. Fullbrook P (1982) Malt and maltose syrups. In: Birch GG, Parker KJ (eds) Nutritive sweeteners. Applied Science Publishers, London, pp 49–81

79. Saha BC, Zeikus JG (1987) Biotechnology of maltose syrup production. Process Biochem 22:78–81

80. Sicard PJ, Leroy P (1983) Mannitol, sorbitol and lycasin: properties and food applications. In: Grenby TM et al. (eds) Developments in sweeteners 2. Applied Science Publishers, London, pp 1–25

81. Dwivedi BK (1986) Polyalcohols: sorbitol, mannitol, maltitol and hydrogenated starch hydrolysates. In: O'Brien Nabors L, Gelardi RC (eds) Alternative sweeteners. Marcel Dekker, New York, pp 165–183

82. Sicard PJ, Serpelloni M, Dupas H (1985) Le sorbitol, le mannitol et les sirops de glucose hydrogénés. In: Dacosta Y (ed) Polyols et polydextrose. Apria, Paris, pp 75–134

83. Fabry I (1987) Malbit and its applications in the food industry. In: Grenby TH (ed) Developments in Sweeteners 3. Applied Science Publishers, London, pp 83–108

84. Linko P (1982) Lactose and lactitol. In: Birch GG, Parker KJ (eds) Nutritive sweeteners. Applied Science Publishers, London, pp 109–131

85. Den Uyl CH (1987) Technical and commercial aspects of the use of lactitol in foods as a reduced-calorie bulk sweetener. In: Grenby TH (ed) Developments in sweeteners. Applied Science Publishers, London, pp 65–81

86. Bar A (1986) Xylitol. In: O'Brien Nabors L, Gelardi RC (eds) Alternative sweeteners. Marcel Dekker, New York, pp 165–183

87. Kieboom APG, Van Bekkum H (1985) Chemical conversion of starch-based syrups. In: Van Beynum GMA, Roels JA (eds) Starch conversion technology. Marcel Dekker, New York, pp 263–334

88. Weidenhagen R, Lorenz S (1957) Palatinose (6-alpha-glucopyranosido-fructofuranose), ein neues bakterielles Umwandbungsprodukt der Saccharose. Z Zuckerind 7:533–534
89. Strater PJ (1987) Palatinose. Zuckerindustrie 112:900–902
90. Siebert G, Grupp U (1979) Alpha-D-glucopyranoside-1,6-sorbitol and alpha D-glucopyranoside-1,6-mannitol (palatinit). In: Guggenheim B (ed) Health and sugar substitutes. S Karger, Basel, pp 109–113
91. Strater PJ (1986) Palatinit. In: O'Brien Nabors L, Gelardi RC (eds) Alternative sweeteners. Marcel Dekker, New York, pp 217–244
92. Hicks KB, Parrish FW (1980) New method of preparation of lactulose from lactose. Carbohydr Res 82:393–397
93. Rennhard HH (1973) Polysaccharides and their preparation. US Patent 3,876,799
94. Torres A, Thomas RD (1981) Polydextrose and its applications in foods. Food Technol, July:44–49
95. Adachi K (1982) Novel fructosyl transferase and process of preparation. Jpn Pat Spec 57–166981
96. Adachi Y (1983) Immobilized enzymes, process for preparation thereof and process for production of sweeteners by means of immobilized enzymes. Jpn Pat Spec 58–162292
97. Ziesenitz SC, Siebert G (1987) In vitro assessment of nystose as a sugar substitute. J Nutr 117:846–851
98. Kitahata S, Okada S (1975) Transfer action of cyclodextringlycosyltransferase on starch. Agric Biol Chem 39:2185–2191
99. Beck KM (1978) The practical requirements for the use of the synthetic sweeteners. In: Dwivedi BK (ed) Low calorie and special dietary foods. CRC Press, Boca Raton, pp 51–58
100. Bye P, Mounier A (1984) Produits sucrants et edulcorants. In: Les futurs alimentaires et energétiques des biotechnologies. Imprimerie de l'Ouest Cahiers de l'ISMEA, hors série no. 27:107–210

Commentary

Southgate: Carbohydrates are often formally defined as substances in which the molar ratio of C:H:O is 1:2:1. This definition fails for oligosaccharides, polysaccharides and sugar alcohols. In this book "carbohydrate" includes mono-, di- and oligosaccharides, the sugar alcohols and polysaccharides.

Würsch: It might be useful to emphasise that chemically modified starches, apart from the hydrolysed starches, contribute only a very small part of the starch consumed in foods, although their rheological role in prepared foods is determinant.

Southgate: It is important to recognise that the major part of the starch in the diet is derived from food, and not from these modified starches, despite their importance as ingredients in many processed foods.

Most of the starch in foods is not in a native state, because the majority of foods have received some form of processing and heat treatment, both during production and in domestic preparation. The sources of native starch in granular form are very limited, for example the banana and uncooked cereal foods such as mueslis and unprocessed bran. Thus the diet contains starch in a range of physical states, from granules that are substantially intact to starch gels where granular structure has been completely lost.

In developed countries starch contributes about half of the total carbohydrate intake, whereas in many developing countries it provides more than 80% of the carbohydrate intake (see reference 1, Chap. 4).

Marks: The calorific value of dietary carbohydrates in human subjects depends on the completeness of their absorption from the gut as reflected by their absence from the faeces. Carbohydrates and sugar alcohols that are incompletely absorbed into the body have an *apparently* low calorific value. However, even if they are not absorbed into the body in the same form in which they were ingested, but only as their fermentation products, their calorific value will be the same as that of glucose. It is therefore illogical and improper to describe the calorific value of sugar alcohols that are not excreted in the faeces as being less than that of glucose, fructose and other more readily absorbable carbohydrates.

Booth: It might be best if it were always pointed out that "sweetness equivalents" such as those in Table 1.7 are not reliable, except as matches for sweetness without regard to time-intensity and quality between the single sweetener in pure aqueous solution and sucrose solution at a particular concentration (e.g. 10 g in 100 ml of solution).

It is indisputable, and by no means necessarily a matter for regret, that diets will include sweetness for the foreseeable future. Nevertheless, this may be an instance in which "technology push" should be guided by more differentiated considerations than a rather broad sense of "consumer pull". The sweetness of most sugars associates the sensation with the ingestion of energy and of fermentables, in timings and frequencies that prejudice the waistline and the teeth. Thus it is to customers' advantage if beverages and foods used in this way are available in a wide variety, including traditional or new products that are less sweet or use other attractive principles. A key to the popularity of sweetness is that, unlike any other known principle, it elicits an innate acceptance reaction when presented in otherwise unfamiliar, uninteresting or unpalatable items, and the resulting acceptance tends to make the presented level of sweetness the preferred one in such an item in future. Thus, familiar items can be too sweet (as Coca Cola discovered) and new introductions can be unnecessarily high in sweetness to assure adequate diffusion through the market.

Chapter 2

Structure of Starch in Food: Interaction of Starch and Sugars with Other Food Components

P. Würsch

Introduction

Starchy foods are rarely eaten raw and must undergo a heat treatment in order to become palatable and highly bioavailable. During industrial food processing starchy foods are often subjected to further drying or freezing operations to increase the speed of preparation at home, to improve further the palatability, and to extend the shelf life, etc. These operations also affect the functional and nutritional properties. This chapter will briefly describe the modifications to the starch which occur in the course of various processing operations, the role of starch in the rheological quality and stability of the food, and its interaction with other constituents which might influence both starch digestibility and the overall nutritional properties of the food [1].

Gelatinisation, Retrogradation and Resistant Starch

Starch, which is deposited in the plant in the form of microscopic granules, is insoluble in cold water but can be solubilised by heating in excess water. The structure of the granule is altered by the loss of crystallisation of amylopectin, followed by swelling, hydration and solubilisation. This process is called gelatinisation. The increase in consistency that occurs when starch suspensions are heated above gelatinisation temperature was traditionally attributed to the granules imbibing increasing amounts of water during swelling. This was thought to increase the chance that the granules would come into contact with each other, and therefore require more work to move the granules past each other as they continued to swell. However, it has recently been found that the maximum consistency of starch suspensions occurs after most of the granules have ceased to swell. The increase in consistency is mainly due to a sticky network, formed by starch exuded from the granules, which binds the swollen granules together [2,3]. At the end of swelling and under slight shearing, the granules disrupt completely and the starch forms a continuous and complex filamentous network.

On cooling gelatinised starch, retrogradation involving amylose and amylopectin may occur. If the solutions are sufficiently concentrated, gels are formed. Ring et al. [4] and Miles et al. [5] showed that retrogradation occurs in two steps: aggregation of the polysaccharide chains with phase separation into polymer-rich and polymer-deficient phases (gelation), followed by a slow crystallisation of the macromolecules in the polymer-rich phase. The first step lasted 100 minutes for a 7% amylose solution and 5 days for a 20% amylopectin solution, while total crystallisation was achieved in 48 hours and in more than 30 days, respectively. The development of opacity of the gel when it settles appears to be related to the irreversible phase separation into polymer-rich and polymer-deficient regions. Retrogradation is enhanced at low temperatures. Crystallinity reaches a maximum development in 50%–60% gels and disappears in very dilute (10%) or concentrated (>80%) gels. It should be noted that the moisture content in bread is 35%–40%, very close to the optimum conditions for starch crystallisation.

Retrogradation involving amylose can be a rapid process, occurring within hours after cooking, as described above. This distinguishes it from staling, which is a much slower retrogradation process involving the free chains in amylopectin [6]. Staling is accelerated by the migration of water in bread from the crumb to the crust. As well as the increase in firming, there is a decrease in the proportion of soluble starch and in the starch's ability to swell in cold water. The process of staling and firming does not seem to affect the availability of starch to pancreatic α-amylase [7]. However, starch crystallisation cannot account for all the crumb firming that occurs above room temperature and it has been suggested that moisture migration from protein to starch occurs, leading to increasing rigidity of the gluten network.

Some years ago, Englyst et al. [8] showed that part of the retrograded starch is measured as dietary fibre, on account of its resistance to digestion by a mixture of α-amylase and pullulanase. In bread, resistant starch is formed immediately after baking and the level remains very stable during more than a week of storage, even after staling of the bread [9,10]; in boiled potato, in contrast, the generation of resistant starch was time-dependent in a similar fashion to the process in highly diluted starch suspensions [11]. Freshly cooked potato was reported to contain less than 2% resistant starch, but this increased rapidly during cooling to about 3% [12]. Reheating produced a temporary decrease, but cooling the potato increased further the level of resistant starch. This pattern of decrease and increase of resistant starch by successive heating and cooling was also observed with wheat starch [9].

Berry [9] determined the level of resistant starch in starches containing various amounts of amylose. A rough correlation was found between the amylose content and the yield of resistant starch after cooking and drying. Waxy starch formed very little resistant starch, whereas high amylose starch (70% amylose) formed over 30% resistant starch. The resistant starch content varies significantly from food to food and depends also on the process of cooking and drying (Table 2.1).

The resistant starch has been shown to be resistant to digestion in vivo in rats [13,14] as well as in humans [15], but is readily metabolised by the intestinal microflora, indicating that it should be included in the physiological definition of dietary fibre.

Table 2.1.   Resistant starch in processed food (% total starch)

	In vitro	In vivo
Breads	0.3–1.5 [8,13]	1.5 [15]
Breakfast cereals	0.1–1.2 [8]	–
Cornflakes	2.9–3.1 [8,11,15]	2.6 [15]
Potato:		
boiled	1.6 [11]	1.6 [12]
instant	1.1 [29]	1.1
Rice	Trace [8]	–
Pasta	0.2–0.6 [8]	–
Leguminous seed	10.5 [29]	10.3 [29]

Changes in Starch During and After Technological Treatments

The level of gelatinisation and the level and type of retrogradation of starch are very much dependent on the processing conditions used for the pre-cooking of starchy foods. These changes can occur during any type of heat treatment applied at home or by industry and can involve starch alone or starch together with the other ingredients present in the food.

Quick-cooking rice has the advantage of saving preparation time and yielding a dry, non-sticky product. The grain should hydrate very rapidly during preparation and thus a quick-cooking rice should have a porous structure. Several processes have been patented which involve soaking, boiling and/or steaming and drying in one or two steps [16]. Non-sticky grains can be obtained by a controlled retrogradation of gelatinised starch during the manufacturing process. In one process, a freezing/thawing step was introduced before drying at low temperature to accelerate retrogradation. In another process, cooking was done in two stages with intermediate holding or cooling of the pre-cooked rice. The first cooking is intended to limit swelling to such an extent that starch gelatinisation takes place inside the swollen cells. The holding period between the two cooking stages is intended to equilibrate the moisture inside the grains and to harden the gelatinised starch by stimulating its retrogradation, especially at the outer layer of the grains where the starch is more gelatinised. During the final cooking, the cell structure and firmness of the grain will be retained and solubilisation of amylose will therefore be greatly limited. The cooked rice grains are then carefully dried [17]. The parboiling process produces physical, chemical and textural modifications in the rice with several advantages, namely reduction of leaching of the vitamins, protein and starch into the cooking water [18]. Rice grains are steeped, followed by steam-drying, which produces a gelatinisation of the starch granules embedded in a proteinaceous matrix, without breakage of the grain structure. During parboiling, retrogradation of gelatinised starch also takes place in the outer layer yielding a firmer, dry product [19].

Retrogradation of starch in cooked rice may adversely affect the quality of the consumed product by making it tough and rubbery. However, this can be avoided by controlling the manufacturing of pre-cooked rice in such a way that retrogradation takes place at the right moment. Parboiling of rice, by increasing the firmness of the cooked grains, reduces their susceptibility to pancreatic α-amylase, and probably also delays the gastric emptying of food. It was observed that the glycaemic response was reduced compared with that produced by white rice [20].

An extruder cooker may be considered as a continuous reactor capable of simultaneous transporting, mixing and shearing. This process, which uses elevated temperatures and pressures with short residence times and low moisture contents (10%–20%), disentangles amylose and amylopectin. Due to the very fast cooking and drying steps, amylose has no time to retrograde. However, Mestres et al. [3] observed that amylose and amylopectin co-crystallise during storage. In drum-drying, which is another widely used technological process, starch first gelatinises and is then rapidly dried at temperatures above those which promote complex formation or retrogradation. During storage, amylose crystallises independently from amylopectin, as demonstrated in gels.

According to Linko et al. [21], moisture content is of greater concern in single-screw than in double-screw extruders, although in the latter the effect is pronounced in material containing less than 20% moisture. Particle size may also be critical in single-barrel extruders: If the material is very coarse, the starch may not be well gelatinised, although this is less important at high moisture contents. Emulsifiers and surface-active agents act at the 0.1% level as lubricants and starch complexing agents. They influence stickiness, expansion, and carbohydrate solubility and digestibility, by complexing with the amylose fraction of starch. The two variables which exert the greatest effect on gelatinisation and the general physicochemical properties are the barrel temperature, moisture and pressure (Fig. 2.1). Complete disruption of the starch granule is generally

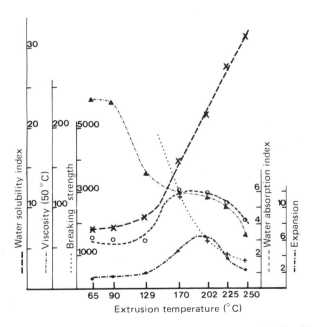

Fig. 2.1. Effects of extrusion temperature on expansion, breaking strength, viscosity at 50 °C, water absorption index and water solubility index of extruded products from corn grits. Initial moisture content before extrusion was 18.2% by weight. From Mercier and Feillet [23].

obtained by extrusion cooking, accompanied in severe conditions by partial starch breakdown and increased formation of starch resistant to intestinal digestion [22]. However, this disruption depends on the composition of the starches, that is on the ratio of amylose to amylopectin, and the amount of starch–lipid complex and the technological conditions mentioned above.

Extrusion studies on a variety of cereal starches, including corn with various amylose contents, showed that, for a given moisture, as extrusion temperature increases so the water solubility increases and water absorption and expansion achieve a maximum value before decreasing due to the chemical breakdown of starch [23]. Scanning electron microscope observations of the texture of extruded corn semolina showed that at 18% moisture and at 50 °C, particles of cell aggregates full of starch granules were formed. An aqueous protein film covered the surface. Extrusion at 107 °C produced swollen cells that expanded without rupturing. Starch granules were partially gelatinised. Loss of organised structure of the starch granules occurred above 150 °C. Cereal starch can be rendered soluble in cold water by twin-screw extrusion at more than 170 °C without formation of oligosaccharides, whereas potato starch produces oligosaccharides as small as maltose under the same conditions [23,24].

Protein–Starch Interaction

The interaction of starch with protein is essentially physical as long as the starch is not chemically or enzymatically degraded. Protein appears to form complexes with starch molecules on the granule surface, preventing the escape of exudate from the granule, and therefore interfering with the increase in consistency during heating. This interaction seems to be due to the attraction of opposite charges [25]. The association of gelatinised wheat starch and wheat protein occurs at acid and neutral pH, but is diminished at alkaline pH [26].

The effect of such an interaction is particularly significant in bread and pasta products. Bread consists of a colloidal system which is biochemically almost inactive but physically unstable. When wheat flour is hydrated to water contents above 35%, mixing leads to the formation of a coherent visco-elastic mass known as dough. Protein in the dough forms a matrix in which the starch granules are dispersed at a very high concentration. At the level of the crumb, starch granules gelatinise partially during baking, and the glutinous matrix coagulates, forming a skeleton which maintains bread structure. The degree of gelatinisation of starch in bread depends mainly on the available moisture, but also on the temperature of baking. Starch is only partially gelatinised, more so in the outer layers of the crumb than in the centre. The starch in sourdough bread was found to be less susceptible to amylolysis by α-amylase in vitro than was the starch in conventional fermented bread. It was postulated that the sourdough bread had a rigid protein matrix, which could be suppressed by pepsin pre-digestion [7].

Pasta and noodles are wheat-based products that are formed from a dough but are not leavened. The formulation consists of flour and water for pasta, and flour, water, salt and egg for Western noodles. The raw material of choice for pasta is durum semolina, which is characterised by great hardness, due to its high protein

content. During dough formation by extrusion at low temperatures, a disconti-
nuous protein matrix becomes predominant and this forms a protective coating
around the starch granules. During heat drying of the pasta, partial starch
gelatinisation occurs, and during home cooking the coagulated proteins form a
fibrillar structure which is essential for obtaining an acceptable end product [27].
Removal of the protein from spaghetti by solvent extraction produces an
extensive release of amylose during cooking, resulting in a sticky and disinte-
grated texture. The lipids in the product supplement the function of protein by
minimising other consequences of cooking, such as stickiness; the effect is
probably due to the amylose-complexing properties of the polar lipids [28]. The
firmness of the cooked pasta has an important influence on the rate of its
digestion. This is among the lowest reported [1], being in a similar range to the
rates for parboiled and leguminous seeds.

In potato, it is not protein which prevents starch leaching, but the cellular
structure of the food; it is therefore imperative when cooking potato to preserve
its cellular structure. The starch granules are stored in the cells, and, during
boiling of potato pieces, the granules gelatinise and swell until they fill the cells.
Slight shearing does not damage the cells, or the swollen starch granules, and only
a minimum of starch leaches out. If the cells are broken during shearing, though,
starch is freed and will behave as a paste, resulting in a sticky end product [30].

Starch–Lipid Interaction

Fatty acids and monoglycerides are known to form inclusion compounds with
amylose in which the linear hydrocarbon portion of the lipid is located within the
helical cavity of amylose. This complex-forming ability is reported to be very low
for lipids with two hydrocarbon chains such as lecithin, and non-existent for
triglycerides [31]. The inability of triglycerides to form complexes with amylose is
related to steric hindrance of the molecule, which prevents it from entering the
interior of the polysaccharide helix. Increasing the chain length of saturated fatty
acids has been shown to result in increased complexing, while increasing the
number of double bonds decreases complexing [32,33]. Amounts of water-
soluble carbohydrate and degree of retrogradation produced by extrusion
cooking of manioc starch decreased when oleic or linoleic acid were used in place
of stearic acid [34]. Amylopectin is also able to complex fatty acid, although the
binding is much weaker [33,35], but it does not seem to be able to complex with
monoglycerides (emulsifiers) [36]. Cereal starches contain native monoacyl
lipids, which are found in the form of inclusion complexes with 15%–25% of the
total amylose [37]. It appeared that flours submitted to technological processing,
such as drum-drying or extrusion cooking, exhibited the same percentage of
complexed amylose. Addition of long chain fatty acids or monoesters in both
processes raised the amount of complexed amylose to very high levels, and
yielded various alterations in the properties of the flour, namely a decrease in the
digestion rate of the starch and a decrease in the water solubility, in particular of
the extruded flours [22,38]. These behaviour modifications were not observed
when triglycerides were added.

Nevertheless the most important role played by the emulsifiers (monoglycerides) is their ability to prevent retrogradation of amylose. They increase the pasting temperature of a starch suspension and increase the viscosity of starch during cooking [39] by forming complexes with free amylose. This delays the swelling process of the starch granule and makes its structure more rigid, thus stabilising the swollen granules.

Interaction of Starch and Sugars with Minerals

In aqueous media, neutral polysaccharides and oligosaccharides such as starch and glucose syrups have little affinity for cations of low acidity, i.e. cations with little ability to polarise a donor atom. However, even if the affinity is low, it can have an important implication for the texture and sensory quality of foods and beverages. It can also influence the hydrolysis of starch by α-amylase [40]. However, it has not been ascertained whether the cations act on the enzyme or on the substrate.

By measuring the conductivity of cation solutions containing various carbohydrates, Cross et al. [41] found that the degree of complex formation of iron with fructose, maltose, glucose and sorbitol was higher at pH 9 than at pH 3. One of the consequences of this higher affinity was that the threshold concentration of metallic taste was considerably higher, especially in fructose solutions where the sweetness of the sugar completely disappeared. Bates [42] reported that the ferric fructose chelate exists as a low molecular weight ferric hydroxide polymer. Formation of this polymer is at a maximum in the neutral pH region, and it has a mean molecular weight of about 15 000 [42]. This ferric fructose stimulates the absorption of ferric ion in rats [43] and humans [44]. Although some complexing was measured with malto-oligosaccharides and alkaline salts, the effect was very weak and probably of no practical significance [45].

Sugars and Protein

Heat processing of protein in the presence of sugars can result in a change in the functional properties of the protein, and a loss of its nutritional quality through non-enzymatic browning reactions. Nesetril [46] showed that the presence of sugars in the dough softens the texture of the resulting bread by delaying gelatinisation of starch and denaturation of proteins. This denaturation occurs particularly in extruded products which are submitted to high temperatures and pressures over a short period of time. The chemical reactions that are taking place are not well characterised, apart from the typical Maillard reaction. Recently, Racicot et al. [47] showed that fructose as well as glucose, released during a heat treatment of sucrose, interacts with amino acid residues and forms a more hydrophilic complex, thereby increasing its solubility. Greater reactivity was observed with higher molecular weight proteins.

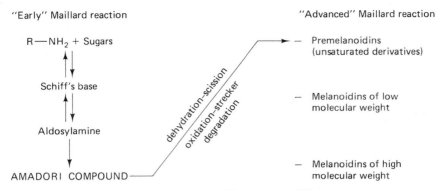

"Early" Maillard reaction   "Advanced" Maillard reaction

R—NH₂ + Sugars   — Premelanoidins
                                                               (unsaturated derivatives)

Schiff's base

                                                           — Melanoidins of low
                                                             molecular weight
Aldosylamine

                                                           — Melanoidins of high
AMADORI COMPOUND   molecular weight

dehydration-scission
oxidation-strecker
degradation

Fig. 2.2. Schematic representation of the Maillard reaction [73].

The Maillard reaction occurs when reducing sugars such as glucose, fructose or lactose, in the presence of protein, are subjected to heat (Fig. 2.2). It is affected by many factors which occur during processing and storage of food (Table 2.2) [48]. This reaction between sugars and protein is thought to start by an initial condensation between an amino acid group and a carbonyl group to form a Schiff-base and a molecule of water. Subsequent cyclisation and isomerisation under acidic conditions (Amadori rearrangement) results in a 1-amino-1-deoxy-2-ketose derivative. This component and its precursors are colourless. A series of complex reactions follows, involving the removal of the amino groups, dehydration, cyclisation and/or polymerisation, giving rise to brown complex polymers [49]. These are insoluble in 72% sulphuric acid and are therefore often estimated as "lignin" [50]. Such products are formed when wheat cereal products are toasted [51], and are probably not degraded by colonic bacteria.

The reactivity of the reducing sugars varies: aldopentoses are more reactive than aldohexoses, which are, in turn, more reactive than disaccharides (Fig. 2.3). At isomolecular levels, hexoses and pentoses destroy 42% and 66% of lysine in solution respectively [52,53]. Sugar reactivity depends also on the type of protein involved. Frangne [54] showed that the same sugar causes different lysine losses according to the protein. Lactalbumin was less sensitive than soybean globulin, but this effect was particularly evident with sucrose.

The Maillard reaction involves sugar loss as well as amino acid blockage and loss. This type of reaction has quite significant nutritional consequences, namely a loss of available amino acids, in particular the essential amino acid lysine. The Amadori compound of lysine, which is the only product formed in the early Maillard reaction, is biologically unavailable. In the advanced Maillard reaction,

Table 2.2. Factors affecting the Maillard reaction

Sugar ⎫
Protein ⎬ Type, physical state, concentration
Water content or activity
pH
Temperature
Duration
Oxygen

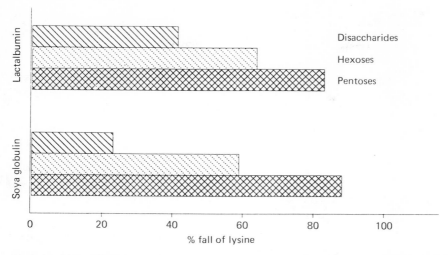

Fig. 2.3. Mean reactivity of sugars during the Maillard reaction (% fall of lysine). Sucrose is included in the disaccharides [53,54].

other amino acids are also involved, such as tryptophan, arginine and histidine [55].

Sucrose, which is a non-reducing disaccharide, does not react, but it can be thermally degraded, producing reducing molecules which can then react with free amino groups. In baking and toasting, dry heat is applied, so that the moisture content of the baked food rapidly decreases and might eventually reach the levels at which heat damage to proteins becomes important. The loss of protein value is very much dependent upon the thickness of the baked food, the heat intensity and the duration [54]. The presence of sugar is still another factor influencing heat damage in biscuits. From a study by Clark et al. [56] it can be seen that the loss in available lysine is three times as great in the presence of 30% sugar in the biscuit than in its absence. During baking of biscuits fortified with casein or lactose-containing skim milk, Clegg [57] found a 37% loss with the skim milk as opposed to 12% with casein. During extrusion cooking of a mixture of protein, starch and sucrose, the loss of lysine was 11% in cooler conditions but reached 38% when the temperature went above 200 °C [58]. The loss was lower, however, with increased moisture of the extruder feed [58,59]. The raw material initially contained only small amounts of reducing sugars, but sucrose was probably hydrolysed [59]. Others reported no loss of lysine at all at high moisture content and low temperatures [59–61], or even an increase in the in vitro protein digestibility [62] and in the net protein utilisation [63]. However, Hurrell and Carpenter [64] reported an increased loss of lysine with increased moisture content when albumin was heated with sucrose in a closed vessel.

Steinig and Montag [60] measured the degree of lysine destruction in 33 foodstuffs (Table 2.3). The highest destruction was measured in a highly toasted Zwieback (crispbread), whereas no loss was detected in toasting bread and in rye bread. The loss of lysine in wheat bread has recently been reported to be between 11% and 17% [60,65,66], with 46% in the crust and 5% in the crumb [65].

Flour contains small quantities of sugar (0.8%–2.7%: [67]), but these are not

Table 2.3. Lysine blockage in some selected foodstuffs

Foodstuff	% lysine blockage
Zwieback (crispbread)	76
Spicebread	65
Pumpernickel (wholegrain bread)	26
Flatbread	7
Ryebread	n.d.
Wholemeal bread	17
Toasting bread	n.d.
Egg noodles	28
Macaroni	17
Extrudate	n.d.
Potato chips	8
Cornflakes	8
Chocolate (milk)	13

Adapted from Steinig and Montag [60].
n.d., not detectable.

sufficient for an optimum fermentation of the dough. However, during dough preparation the natural β-amylase of the flour produces maltose needed by the yeast. Also, sucrose or fungal amylase are often added to accelerate the fermentation [65,66]. The dough, therefore, contains an appreciable quantity of reducing sugars which will then react with free amino groups during baking. The quantity of sugars found in standard bread in the UK lies between 1.8% and 3.5%, but it can reach 14.4% in currant bread and 26% in malt bread [68].

Loss of lysine and other essential amino acids in milk can also occur during processing, due to the presence of lactose. However, no loss has been found during pasteurisation and less than 2% during sterilisation and spray drying. High loss occurs during conventional sterilisation (canning) or roller-drying [69].

What is the nutritional significance of such losses in essential amino acids? A joint FAO/WHO committee estimated the amino acid requirements for infants, older children and adults [70]. For lysine, the daily requirement was 103 mg/kg for infants, 64 mg/kg for children and 12 mg/kg for adults. The adult requirement for essential amino acids falls more sharply from infancy than does the total protein requirement. The daily consumption of protein by adults in Western countries is about 1.5 g/kg, which is twice the safe level of good protein intake. For children, the safe level is about 1 g/kg, and up to 1.6 g/kg for infants. With 1%–2% lysine on average in cereal protein, 8.2% in milk and 9% in animal protein, the daily intake of adults far exceeds their requirement, and they obtain their amino acids and proteins from a wide variety of foodstuffs [71,72]. However, the situation can be critical for infants who rely on baby foods and infant formula as a sole source of nourishment.

Conclusions

The mechanisms of conversion of starch during the various processes described in this chapter have been studied for many years and knowledge has progressed with the development of the scientific tools. On the other hand it is only recently that nutritionists have become aware that starch is not simply a complex carbohydrate, but that the source, the shape and structure of the food and the interaction with other nutrients (fibre, protein, lipid) can significantly affect its nutritional properties. It is therefore essential that the source, form and treatment of the food being evaluated are thoroughly defined when studying the nutrition of starch.

Addendum

In any scientific and technical field definitions of terms tend to differ. This applies particularly to the basic starch phenomena described here. Recently a committee in the USA surveyed scientists and technologists in order to reach a consensus on the most suitable definitions for the terms "gelatinisation", "pasting" and "retrogradation" [74] (see Chap. 1). The definitions finally accepted were as follows:

Gelatinisation. Starch gelatinisation is the collapse (disruption) of molecular structures within the starch granule, manifested in irreversible changes in properties such as granular swelling, native crystalline melting, loss of birefringence and starch solubilisation. The point of initial gelatinisation and the range over which it occurs is governed by starch concentration, method of observation, granule type and heterogeneities within the granule population under observation.

Pasting. Pasting is the phenomenon following gelatinisation in the dissolution of starch. It involves granular swelling, exudation of molecular components from the granule and, eventually, total disruption of the granule.

Retrogradation. Starch retrogradation is a process which occurs when the molecules comprising gelatinised starch begin to re-associate in an ordered structure. In its initial phases, two or more starch chains may form a simple junction point which then may develop into more extensively ordered regions. Ultimately, under favourable conditions, a crystalline structure appears.

References

1. Würsch P (1989) Starch in human nutrition. World Rev Nutr Diet (in press)
2. Miller BS, Debry RI, Trimbo HB (1973) A pictorial explanation for the increase in viscosity of a heated wheat starch water suspension. Cereal Chem 50:271–280
3. Mestres C, Colonna P, Buleon A (1988) Gelation and crystallisation of maize starch after pasting, drum-drying or extrusion cooking. J Cereal Sci 7:123–134

4. Ring SG, Colonna P, I'Anson KJ, Miles MJ, Morris VJ, Orford PD (1987) The gelation and crystallisation of amylopectin. Carbohydr Res 162:277–293

5. Miles MJ, Morris VJ, Orford PD, Ring SG (1985) The roles of amylose and amylopectin in the gelation and retrogradation of starch. Carbohydr Res 735:271–281

6. Russel PL (1983) A kinetic study of bread staling by differential scanning calorimetry and compressibility measurements. The effect of added monoglyceride J Cereal Sci 1:297–303

7. Siljeström M, Björck I, Eliasson A-C, Lönner C, Nyman M, Asp N-G (1988) Effects of polysaccharides during baking and storage of bread – in vitro and in vivo studies. Cereal Chem 65:1–8

8. Englyst HN, Andersson V, Cummings JH (1983) Starch and non-starch polysaccharide in cereal foods J Sci Food Agric 34:1434–1440

9. Berry CS (1986) Resistant starch: formation and measurement of starch that survives exhaustive digestion with amylolytic enzymes during determination of dietary fibre. J Cereal Sci 4:301–314

10. Boskov H, Østergaard K, Bach Knudsen KE (1988) Effect of baking and staling on carbohydrate composition in rye bread and on digestibility of starch and dietary fibre in vivo. J Cereal Sci 7:135–144

11. Englyst H, Wiggins HS, Cummings JH (1982) Determination of the non-starch polysaccharides in plant foods by gas–liquid chromatography of constituent sugars as alditol acetates. Analyst (London) 107:307–318

12. Englyst HN, Cummings JH (1987) Digestion of polysaccharides of potato in the small intestine of man. Am J Clin Nutr 45:423–431

13. Björck I, Nyman M, Pedersen B, Siljeström M, Asp N-G, Eggum BO (1986) On the digestibility of starch in wheat bread: studies in vitro and in vivo. J Cereal Sci 4:1–11

14. Berry CS, Fisher N (1985) Effects of resistant starch on large bowel function in rats. 13th International Congress of Nutrition, Brighton, UK, 18–23 August, p 70

15. Englyst HN, Cummings JH (1985) Digestion of the polysaccharides of some cereal foods in the human intestine. Am J Clin Nutr 42:778–787

16. Roberts R (1972) Quick cooking rice. In: Houston DF (ed) Rice chemistry and technology. AACC, St Paul, Minn

17. Hellendoorn EW (1971) Aspects of retrogradation in some dehydrated starch-containing precooked food products. Stärke 23:63–67

18. Luh BS, Mickus RR (1980) Parboiled rice. In: Luh BS (ed) Rice: production and utilization. Avi Publishing Co., Westport, Conn., pp 501–542

19. Bhattacharya KR (1985) Parboiling of rice. In: Juliano BO (ed) Rice: chemistry and technology. AACC, St Paul, Minn., pp 289–348

20. Wolever TMS, Jenkins DJA, Kalmusky J et al. (1986) Comparison of regular and parboiled rice: explanation of discrepancies between reported glycemic responses to rice. Nutr Res 6:349–357

21. Linko P, Colonna P, Mercier C (1981) High-temperature, short-time extrusion cooking. In: Pomeranz Y (ed) Advances in cereal science and technology, vol IV. AACC, St Paul, Minn., pp 145–223

22. Schweizer T, Reimann S, Solms J, Eliasson AC, Asp N-G (1986) Influence of drum-drying and twin-screw extrusion cooking on wheat carbohydrates. II. Effects of lipids on physical properties, degradation and complex formation of starch in wheat flour. J Cereal Sci 4:249–260

23. Mercier Ch, Feillet P (1975) Modification of carbohydrate components by extrusion – cooking of cereal products. Cereal Chem 52:283–297

24. Doublier JL, Colonna P, Mercier C (1986) Extrusion cooking and drum drying of wheat starch. II. Rheological characterization of starch pastes. Cereal Chem 63:240–246

25. Takeuchi I (1969) Interaction between protein and starch. Cereal Chem 46:570–579

26. Dahley LK (1971) Wheat protein–starch interaction. I. Some starch binding effects of wheat-flour proteins. Cereal Chem 48:706–714

27. Feillet P, Ait Mouh O, Abecassis J (1988) Changes in pasta protein, starch and lipid composition during HT- and VHT-drying. 8th Cereal and Bread Congress, Lausanne, Switzerland

28. Dahle LK, Muenchow HL (1968) Some effects of solvent extraction on cooking characteristics of spaghetti. Cereal Chem 45:464–468

29. Schweizer TF, Andersson H, Langkilde AH, Reimann S, Torsdottir I (1989) Nutrients excreted in ileostomy effluents after consumption of mixed diets with beans or potatoes. II. Starch, dietary fiber and sugars. Am J Clin Nutr (submitted)

30. Hellendoorn EW, Van den Top M, Van der Weide JE (1970) Digestibility in vitro of dry mashed potato products. J Sci Food Agric 21:71–75

31. Krog N (1971) Amylose complexing effect of food grade emulsions. Stärke 23:206–210

32. Lagendijk J, Pennings HJ (1970) Relation between complex formation of starch with monogly-

cerides and the firmness of bread. Cereal Sci Today 15:354–356

33. Hahn DE, Hood LF (1987) Factors influencing corn starch–lipid complexing. Cereal Chem 64:81–85
34. Mercier C, Charbonnière R, Grebaut J, Guerivière JF (1980) Formation of amylose–lipid complexes by twin-screw extrusion cooking of manioc starch. Cereal Chem 57:4–9
35. Eliasson A-C, Finstad H, Liunger G (1988) A study of starch–lipid interactions for some native and modified maize starches. Stärke 40:95–100
36. Evans ID (1986) An investigation of starch/surfactant interactions using viscosimetry and differential scanning calorimetry. Stärke 38:227–235
37. Morrison WR, Laignelet B (1983) An improved calorimetric procedure for determining apparent and total amylose in cereal and other starches. J Cereal Sci 1:9–20
38. Björck J, Asp N-G, Birkhed D, Eliasson A-C, Sjöberg L-B, Lundquist I (1984) Effects of processing on starch availability in vitro and in vivo. II. Drum-drying of wheat flour. J Cereal Sci 2:165–178
39. Krog N (1973) Influence of food emulsifiers on pasting temperature and viscosity of various starches. Stärke 25:22–27
40. Sukan G, Kearsley MW, Birch GG (1979) The effect of calcium ions on the hydrolysis of starch. Stärke 31:125–128
41. Cross H, Pepper T, Kearsley MW, Birch GG (1985) Mineral complexing properties of food carbohydrates. Stärke 37:132–135
42. Bates G, Hegenauer J, Renner J, Saltman P, Spino TG (1973) Complex formation, polymerization and autoreduction in the ferric fructose system. Bioinorg Chem 2:311–327
43. Bates GW, Boyer J, Hegenauer JC, Saltman P (1972) Facilitation of iron absorption by ferric fructose. Am J Clin Nutr 25:983–986
44. Brodan V, Brodanova M, Kuhn E, Kordac V, Valek J (1967) Influence of fructose on iron absorption from the digestive system of healthy subjects. Nutr Dieta 9:263–270
45. Briggs J, Finch P, Matulewicz MG, Weisel H (1981) Complexes of copper, calcium and other metal ions with carbohydrates. Carbohydr Res 97:181–188
46. Nesetril DM (1967) Corn sweeteners: their types and uses in baking. Bakers Digest 41:28–33
47. Racicot WF, Satterlee LD, Hanna MA (1981) Interaction of lactose and sucrose with corn meal proteins during extrusion. J Food Sci 46:1500–1506
48. Saltmarch M, Labuza TP (1982) Nonenzymatic browning via the Maillard reaction in foods. Diabetes 31 [suppl]:29–36
49. Miller R, Olsson K, Pernemalm P, Theander O (1980) Studies of the Maillard reaction. In: Marshall JJ (ed) Mechanisms of saccharide polymerization and depolymerization. Academic Press, New York, pp 421–430
50. Van Soest PJ, Horvath P, McBarney M, Jeraci J, Allen M (1983) Some in vitro and in vivo properties of dietary fibres from non-cereal sources. In: Furda I (ed) Unconventional sources of dietary fiber. American Chemical Society, Washington DC, pp 135–142 (Symposium series 214)
51. Anderson NE, Clydesdale FM (1980) Effects of processing on the dietary fiber content of wheat bran, pureed green beans, and carrots. J Food Sci 45:1533–1537
52. Lewis VM, Lea CH (1950) A note on the relative rates of reaction of several reducing sugars and sugar derivatives with casein. Biochem Biophys Acta 4:532–534
53. Frangne R (1972) La réaction de Maillard. 6. Réactivité de diverses protéines purifiées. Ann Nutr 26:97–106
54. Frangne R (1972) Le comportement de la lysine et de quelques proteines au cours de la réaction de Maillard. Diplôme EPHE (Paris) (thesis)
55. Hurrell RF, Carpenter KJ (1977) Mechanisms of heat damage in proteins. 8. The role of sucrose in the susceptibility of protein foods to heat damage. Br J Nutr 38:285–297
56. Clark HE, Howe JE, Merts ET, Reitz LL (1959) Lysine in baking powder biscuits. J Am Diet Assoc 35:469–471
57. Clegg KM (1960) The availability of lysine in groundnut biscuits used in the treatment of kwashiorkor. Br J Nutr 14:325–328
58. Björck I, Noguchi A, Asp N-G, Cheftel J-C, Dahlquist A (1983) Protein nutritional value of a biscuit processed by extrusion cooking: effects on available lysine. J. Agric Food Chem 31:488–492
59. Noguchi A, Mosso K, Aymard C, Jeunink I, Cheftel J-C (1982) Maillard reaction during extrusion-cooking of protein-enriched biscuits. Lebensm-Wiss Technol 15:105–110
60. Steinig J, Montag A (1982) Studien über Veränderungen des Lysins im Nahrungsprotein. I. Ausmass der Lysinschädigung. Z Lebensm-Unters Forsch 174:453–457

61. Alid G, Yanez E, Aguilera JM, Monckeberg F, Chichester CO (1981) Nutritive value of an extrusion-texturized peanut protein. J Food Sci 46:948–949
62. Meuser F, Köhler F, Mohr G, Steyrer W (1980) Verbesserung des ernährungsphysiologischen Wertes von Maisextrudaten unter Verwendung von Kartoffelprotein. Stärke 32:238–243
63. Harmuth-Hoene AE, Seiler K (1984) Einfluss verschiedener Extrusionsbedingungen auf die Proteinqualität bei Weizenvollkorn-Extrudaten. Getreide Mehl Brot 38:245–249
64. Hurrell RF, Carpenter KJ (1975) The use of three dye-binding procedures for the assessment of heat damage to food proteins. Br J Nutr 3:101–115
65. Murata K, Takarada S, Nogawa M (1979) Loss of supplemental lysine and threonine during the baking of bread. J Food Sci 44:271–273
66. Saab RM, Rao CS, Da-Silva RS (1981) Fortification of bread with L-lysine hydrochloride: loss due to baking process. J Food Sci 46:662–663
67. Kent NL (1982) Technology of cereals, 3rd edn. Pergamon Press, Oxford, p 31
68. Holland B, Unwin ID, Buss DH (eds) (1988) Cereal and cereal products [third suppl]. The Royal Society of Chemistry and MAFF, London, pp 24–36
69. Finot PA, Deutsch R, Bujart E (1981) The extent of the Maillard reaction during the processing of milk. In: Eriksson C (ed) Maillard reactions in food. Pergamon Press, Oxford, pp 345–355 (Progress in food and nutrition science, vol 5)
70. Joint FAO/WHO/UNU Expert Consultation (1985) Energy and protein requirements. Technical Report Series 724. WHO, Geneva
71. Whitehouse RHH (1973) The potential of cereal grain crops for protein production. In: Jones JGW (ed) The biological efficiency of protein production. Cambridge University Press, Cambridge
72. Composition of foods. USDA, Washington DC
73. Hurrell RF, Finot P-A (1983) Food processing and storage as a determinant of protein and amino acid availability. In: Mauron J (ed) Nutritional adequacy, nutrient availability and needs. Birkhäuser, Basel, pp 135–156 (Experientia supplementum, vol 44)
74. Atwell WA, Hood LF, Lineback DR, Varriano-Marston E, Zobel HF (1988) The terminology and methodology associated with basic starch phenomena. Cereal Food World 33:306–311

Chapter 3

The Digestion of Starches and Sugars Present in the Diet

B. Flourié

In standard nutritional texts, the digestion and absorption of dietary carbohydrates is usually presented in simple terms. Monosaccharides are absorbed directly, whereas disaccharides are first hydrolysed in the brush border of the mucosal cells by a series of specific disaccharidase enzymes. Dietary polysaccharides have classically been considered to fall into two broad categories: those that are hydrolysed by endogenous enzymes, the available polysaccharides, and those that are not, the unavailable polysaccharides. Only starch and glycogen are considered to fall into the available category of dietary polysaccharides. This simple nutritional concept appears to run counter to many observations of the enzymatic hydrolysis of starch in vitro and in vivo.

Digestion of Dietary Carbohydrates

Digestion of Starch

Starch is composed of two polysaccharides: the linear (1–4) – linked α-D-glucose, amylose, and a highly branched form containing both (1–4) and (1–6) links, amylopectin. Both forms are localised in the plant in the form of insoluble, semi-crystalline granules. Salivary and pancreatic α-amylases act on the interior (1–4) links of starch, but cannot attack the outermost glucose–glucose linkages, so that α(1–4)-linked disaccharides (maltose) and trisaccharides (maltotriose) are produced as the final linear breakdown products.

The digestion of starch begins in the mouth with the action of salivary amylase. Salivary amylase is inactivated at the low pH of the stomach (<pH 4). However, the presence of starch and of its end products of hydrolysis protects the enzyme from inactivation at acid pH [1]. Moreover, the gastric pH is usually higher than pH 4, at least for the first hour following a meal, and possibly for a longer time

with highly buffered protein and amino acid meals. Since the stomach will have emptied a substantial portion of its contents by that time, a significant amount of salivary amylase could pass through the stomach without inactivation [2]. Thus the role of salivary amylase in the digestion of starch remains controversial, as well as the role of the non-enzymatic hydrolysis of certain carbohydrates by gastric acid [3].

In the intestinal lumen, pancreatic α-amylase cannot hydrolyse the (1–6) branching links and has relatively little specificity for (1–4) links adjacent to these branching points. Large oligosaccharides (α-limit dextrins) containing five or more glucose units (average eight), and consisting of one or more (1–6) branching links, are produced by amylase action. Active carbohydrases are present only in the mature columnar epithelial cells of the intestinal villus, being at higher concentration in the jejunum than in the ileum. Glucoamylase differs in its action from pancreatic α-amylase because it sequentially removes a single glucose from the non-reducing end of a linear α(1–4) glucosyl oligosaccharide. Alpha-limit dextrins are cleaved by the complementary action of glucoamylase (α-limit-dextrinase) and sucrase–isomaltase. Maltose and maltotriose appear to be attacked by glucoamylase or sucrase, both of which are potent maltases. Free glucose, thus released, enters the intestinal mucosal cell membrane by a carrier-mediated process (see Chap. 5). A large amount of α-amylase, secreted by the pancreas, is found in the content of the human intestine; and the efficiency of the brush border enzymatic system is very high, in the sense that the hydrolytic and absorptive capacity of the human small intestine appears to exceed dietary intakes of starch several-fold.

Digestion of Sucrose, Lactose and Oligosaccharides

Sucrose, a disaccharide composed of glucose linked to fructose, is hydrolysed to its component monosaccharides by the brush border enzyme sucrase. High dietary intake of sucrose or fructose in humans leads to an increase in the activity of this enzyme within 2–5 days; reduced consumption has an opposite effect [4]. Fructose is absorbed by a facilitated diffusion mechanism that permits an absorption rate much faster than would otherwise be expected for a water-soluble molecule the size of fructose. While humans cannot absorb fructose against a concentration gradient, the blood levels of fructose are maintained at a very low level, allowing the efficient passive absorption of this sugar (see also references 57 and 58 in Chap. 5). Nevertheless, fructose is absorbed at an appreciably slower rate than is glucose, and hydrolysis of sucrose in the gut results in luminal concentrations of fructose that are appreciably higher than that of glucose [5].

Lactose, which consists of galactose linked to glucose, is split into its component monosaccharides by the brush border enzyme lactase, which is not affected by lactose consumption (see also Chap. 5). Galactose as well as glucose is avidly absorbed by active transport. Lactose hydrolysis in normal individuals is a relatively slow process that releases insufficient glucose and galactose to ensure maximal uptake. In this case, surface hydrolysis appears to be rate-limiting for overall assimilation of the carbohydrate nutrient.

Raffinose and stachyose cannot be hydrolysed and absorbed from the small intestine. Likewise, fructo-oligosaccharides appear not to be hydrolysed in the human small intestine [6].

Digestion of Sugar Alcohols

Disaccharide sugar alcohols (maltitol, isomalt, lactitol) should only be partially hydrolysed in the small intestine by the action of intestinal carbohdrases. In vitro, mucosal homogenates of human intestine show a negligible ability to hydrolyse lactitol, with no difference between normal subjects and subjects in lactase deficiency. The activity on maltitol is about ten times higher than that on isomalt, and about 10% of the activity on maltose [7].

The absorption of monosaccharide sugar alcohols (xylitol, sorbitol, mannitol) takes place by passive diffusion, and it is generally stated that they are less well absorbed from the small intestine than is glucose. Hydrogenated glucose syrup containing maltitol may be partially hydrolysed in the small intestine by the action of the pancreatic α-amylase and the intestinal carbohydrases.

Assessment of Dietary Carbohydrate Digestibility

Our ability to assess quantitatively the efficiency of digestion and absorption of carbohydrates is surprisingly primitive. A variety of clinical tests of differing specificity, sensitivity and availability are used to assess the rate of digestion and absorption of carbohydrates in the human small intestine.

Tolerance Tests

In these tests carbohydrate loads (at least 50 g) are ingested, and blood samples obtained for sugar measurements. By comparing the increments in blood glucose brought about by equivalent amounts of carbohydrates from different foods, and then normalising these values to a baseline obtained with glucose, nutritionists are able to quantify their glycaemic potential (glycaemic index) (Table 3.1).

Although this type of test focusses attention on the factors that may alter carbohydrate digestibility, there are several drawbacks. In general there is good agreement between the rate of carbohydrate digestion in vitro and the glycaemic response for individual food, so long as potential complicating factors (such as fat) are excluded [9–11]. However, the metabolic response to a carbohydrate meal is not always as great as would be predicted from in vitro data, and there is evidence that addition of fat and protein may reduce the glycaemic impact of challenges with single carbohydrate foods [11–13]. Tolerance tests do not reflect the quantity of carbohydrate digested and they are insensitive to malabsorption of a small fraction of a carbohydrate [14]. One of the assumptions is that the rate of digestion and hence of absorption is a major determinant of the glycaemic response.

A subject of continuing debate is the relative importance of a reduced rate of gastric emptying. In an attempt to control for the total amount of carbohydrates (to equal 50 g of glucose or to increase palatability) nutritionists have either added variable amounts of other foods or modified components other than

Table 3.1. Glycaemic indices of foods in normal subjects

Food	Glycaemic index
Cereal products	
Bread, white	69
Bread, wholemeal	72
Rice, white	72
Rice, brown	66
Spaghetti, white	50
Cornflakes	80
Root vegetables	
Carrots	92
Potato, new	70
Potato, instant	80
Legumes	
Beans, navy	31
Beans, kidney	29
Beans, soya	15
Peas	33
Lentils	29
Fruit	
Apples, golden	39
Banana	62
Oranges	40
Orange juice	46
Raisins	64
Dairy products	
Milk	34
Yoghurt	36
Ice cream	36
Sugars	
Glucose	100
Maltose	105
Sucrose	59
Fructose	20

Data from [8].
The area under the 2-hour glucose curve after feeding 50-g carbohydrate portions is expressed as a percentage of the mean observed after 50 g of glucose.

carbohydrates in the meal. Co-ingested fat, protein and fibre may alter the gastric emptying of different meals. Caloric intake, viscosity and solid content of the meals are also likely to be important in modifying the rate of gastric emptying. Besides their slow rate of digestion, it is possible that the beneficial effect that leguminous seeds have in terms of their glycaemic index may relate to a reduced rate of gastric emptying due to their higher content of solids, proteins and fibres compared with other starchy foods. Fat added to a carbohydrate meal reduces the glycaemic response by slowing gastric emptying rather than by modifying starch availability [15]. The glycaemic indices are modulated by different factors (gastric motility, rate of hydrolysis, hormonal responses), which may explain why there is controversy when the glycaemic index of an individual food is applied to mixed meals [16,17].

Breath Hydrogen Analysis

Since hydrogen (H_2) is not produced by the metabolic pathways of mammalian cells, its increase in exhaled air provides an indirect marker for microbial fermentation of undigested carbohydrates in the colon. As normal subjects would completely absorb the monosaccharide moieties of carbohydrates, failure to absorb starch and lactose presumably represents incomplete digestion of these sugars to monosaccharides. Several groups have recently employed this approach to investigate a range of common foods, using H_2 production from the non-absorbable sugar lactulose as a means of calibrating the technique (Table 3.2).

Table 3.2. Digestibility of carbohydrates (CHO) determined by the breath H_2 technique in normal subjects

Food	Ingested CHO (g)	Unabsorbed CHO (%)
Cereal products		
Breads, wheat and oat white flour	50	3–4
Breads, wheat and oat whole flour	50	10–13
Cooked rice, white	75	2
Boiled noodles	75	4
Root vegetables		
Baked potatoes	50, 75, 100	8–10
Legumes		
Baked navy beans	100	23
Dairy products[a]		
Lactose in 200 ml water	12.5	40–100
Lactose in 400 ml water	20	50
Lactose in milk	18	50
Lactose in yoghurt	18	17
Sugars		
Glucose	100	0
Sucrose	100	0
Fructose	15, 20, 25, 37.5, 50	0–100
Fructo-oligosaccharides	10	100
Isomalt	20	60

Data from [6,18–25].
Excretion of H_2 after 10 g lactulose served as a standard for the amount of H_2 expected from the delivery of 10 g of carbohydrate to the colon.
[a]In lactase-deficient subjects.

There are multiple features of the breath H_2 technique that limit its accuracy in quantitating malabsorption of carbohydrates. First, the lactulose H_2 breath test appears to be a valid method for comparing carbohydrate absorption from different dietary sources in a population sample, but large individual variations preclude its use in a given subject [26]. Second, in the conversion of breath H_2 excretion to carbohydrate malabsorption, it is assumed that all malabsorbed carbohydrates are converted to H_2 with the same degree of rapidity as is lactulose. It is likely, though, that some polysaccharides from starch or fibre and sugar alcohols are converted to H_2 in vivo less completely and more slowly, which will lead to an underestimation of their malabsorption unless the total duration of excess H_2 excretion is taken into account. It is difficult to know the exact source of the unabsorbed nutrients that have served as substrate for H_2 production. The breath H_2 technique should therefore be considered to be an estimate of malabsorption of fermentable material present in a meal, i.e. starch, fibre, undigestible oligosaccharides or unabsorbed proteins. Lastly, most studies have not assessed whether the failure to digest starch persists when the starch is incorporated in a mixed meal, this being the most common manner in which it is consumed.

Analysis of Ileostomy Effluent

Ileostomy provides a valuable opportunity to study digestive physiology in man and has been used to this end mainly to assess starch digestion (Table 3.3). The subject with an ileostomy does not have a normal gut, however, and there may be a risk of physiological adaptation leading to modified digestion in such subjects. Major differences between the normal ileum and the ileostomy are the water flow rate and electrolyte absorption [31]. Moreover, it is possible that the rich flora [32,33] which inhabit the terminal ileum may break down carbohydrates by fermentation, thus invalidating ileostomy observations.

Table 3.3. Digestibility of carbohydrates (CHO) in ileostomised subjects

Food	Ingested CHO (g)	Unabsorbed CHO (%)
Starch		
White bread	62	2.5
Cornflakes	74	4
Groat oats	58	2
White bread + baked potatoes	157	2.5
Baked potatoes	25	
	50	2
	75	
	100	
Banana	1.2–19.3	70
Sugars		
Banana, sucrose	23–30	0–2
Banana, fructose	2.5–7.4	0.3
Mannitol	10	74
Raffinose	10	88

Data from [27–30].

Intestinal Intubation

Research techniques employing intestinal intubation of a gut segment, perfused at a constant rate, have provided data on the rate of carbohydrate digestion and absorption. Such techniques do not, however, measure the ability of the total gut to absorb an ingested carbohydrate load. Assuming that intestinal intubation does not of itself alter the enzymatic breakdown of food, the only means of directly quantitating the ability of the normal entire gut to digest and absorb ingested carbohydrates is by intubation of the human terminal ileum (Table 3.4). A meal containing carbohydrates and polyethylene glycol (as marker) is ingested, and the terminal ileum is either simply aspirated or constantly perfused with a second marker and aspirated by means of a double lumen tube.

Table 3.4. Digestibility of carbohydrates (CHO) determined by ileal aspirations in normal subjects

Meal	Ingested CHO (g)	Unabsorbed CHO (%)
Starch		
Mixed meal	20[a]	10
Mixed meal	61[a]	8
Mixed meal	100[b]	5
Mixed meal	300[b]	4
Lactose		
Lactose in 200 ml water:		
lactase-sufficient subjects	12.5	0–8[c]
lactase-deficient subjects	12.5	42–75[c]
Lactose in milk		
lactase-sufficient subjects	10	1[c]
lactase-deficient subjects	10	22[c]
Lactose in yoghurt		
lactase-deficient subjects	18	10[c]
Sorbitol		
Mixed meal	10	21
Maltitol		
Mixed meal	19	11[d]

Data from [20,34–38].
[a]From banana, rice, potatoes, navy beans.
[b]From white bread, baked potatoes, boiled noodles.
[c]Percentage of undigested lactose.
[d]Percentage of undigested maltitol.

The findings obtained by these different methods indicate that the digestion of sucrose and lactose is virtually complete in the small intestine of normal subjects, except in those who are lactase-deficient and who continue to drink milk. Little information is available about the digestion characteristics of sugar alcohols in vivo in the human small intestine. The fate of starch in the small intestine is less certain. Although there seems to be general agreement that a proportion of dietary starch escapes digestion in the human small intestine, quantitative estimates appear to be method-dependent. The evaluation of starch malabsorption from excess H_2 excretion in breath, using lactulose as a standard reference non-absorbable sugar, seems to overestimate the amount of starch reaching the colon. Estimates from breath H_2 analyses are indirect; they cannot distinguish between H_2 produced from fermentable cell wall polysaccharides, from undigestible oligosaccharides or from fermentable starch. They do have the advantage, though, that they can be applied to physiologically normal subjects. Conversely differences may arise from incomplete hydrolysis of starch from ileal samples. A vigorous pretreatment of the malabsorbed starch is required, and resistant starch should not be determined by incubation with amylase or fungal α-amyloglucosidase unless it is first solubilised by alkali or dimethylsulphoxide. It is likely that a small part (about 5%) of cooked starch from cereals and root vegetables is not digested and absorbed in the normal intestine. This percentage rises when diets contain ungelatinised starch (banana) or starchy foods from high-fibre cereals or legumes. Increasing the intake of starch raises the amount of this carbohydrate recovered from the terminal ileum. As the percentage of unabsorbed starch is almost the same despite the marked differences in starch intake, it may be concluded that a constant fraction of ingested starch is undigestible [28,35].

Factors Affecting the Digestion of Dietary Carbohydrates

Age and Genetic Differences

Pancreatic α-amylase in duodenal fluid is undetectable in the human infant at 1 month and shows adult activity by 2 years of age. Salivary amylase is apparently mature at birth or soon after [39–42], whereas gastric secretion is immature. A mature production rate of acid is only reached at 2 years of age. Salivary amylase, therefore, may compensate for physiological pancreatic amylase deficiency in infancy.

Fetal intestinal lactase activity develops later in gestation than that of other carbohydrases, and reaches maximal activity at, or soon after, birth. Premature infants, of gestational age 29–38 weeks, have significant breath H_2 production after ingesting lactose up to at least 50 days after birth [43]. Activities of other carbohydrases are at adult levels by the time of birth [44]. During the suckling period in mammals lactase levels remain high, but they decline (although not to zero) after weaning, between 2 years of age and puberty. While congenital alactasia is rare, primary lactase deficiency develops in most of the world's population groups (Table 3.5). In Europe, the frequency of primary adult lactose malabsorption increases in a non-linear fashion from north to south. In France the prevalence of lactose malabsorption is 20% and 40% respectively in the north and south. Family studies in numerous racial and inter-racial groups have demonstrated that persistence of lactose absorption is inherited as an autosomal dominant characteristic. Its opposite, lactase deficiency, is inherited as a single autosomal recessive gene, which, in the homozygous state, suppresses the synthesis of intestinal lactase.

Congenital sucrase–isomaltase deficiency is not as rare as originally believed. While sucrase activity is virtually absent in the mucosa of all patients with congenital sucrase–isomaltase deficiency, isomaltase activity is absent in only 80%. Studies of the affected families reveal that congenital sucrase–isomaltase deficiency is inherited as an autosomal recessive defect. Heterozygotes have enzyme deficiencies intermediate between patients and normal people, but may have sucrase intolerance. Eskimos have the highest frequency (10%) of congenital sucrase–isomaltase deficiency of any racial group.

Indirect evidence suggests that intestinal digestive and/or absorptive function may decline with advancing age. Using breath H_2 analysis after ingestion of a carbohydrate meal (55% polysaccharides, 45% mono- and disaccharides), a reduced absorption has been demonstrated in about one-third of healthy

Table 3.5. Primary lactase deficiency in adult populations

Population	Prevalence (%)
Austria	14–23
China	87
Denmark	3
England	5–30
Finland	16
France (north/south)	20–40
Greece	75
Italy (north/Sicily)	51/71
Japan	100
Sweden	3
Switzerland	17
Thailand	100
Yemen	45
Africa	
Negroids	100
Hamites/Fulani	10/20–58
Australia (White/Aboriginal)	5/70
India (north/north-west)	63/0–15
South America (Indian)	90
USA (White/Blacks)	6/73

Based on [45–54].

individuals over the age of 65 [55]. Ageing may be associated with intestinal alterations in blood flow, villus architecture, brush border membrane enzymatic activity and cellular kinetics of the mucosa [56]. Pancreatic function, however, is unaffected by advancing age.

Food Factors Affecting Digestion

Digestion of Starch

The plant of origin and the form in which they are ingested affect the digestion of starches.

Ungelatinised Starch. A wide variation in digestibility in vitro of ungelatinised starch granules has been reported. In general, cereal starches are more easily digested by pancreatic α-amylase than legume and tuber starches, and the digestibility of a starch is inversely proportional to its amylose content. This is of little significance in relation to staple foods which are eaten cooked.

Gelatinised Starch. When starch granules are heated to a critical temperature (60–70 °C) in water, they swell and lose their characteristic birefringence. The intermolecular links weaken, permitting amylose and amylopectin to form colloidal dispersions. Upon cooling, dispersed starch tends to re-associate into regions of varying polymer density. At temperatures over 60 °C the intermolecular links remain weak. Cooking increases the availability of starches for hydrolysis by amylase.

In many foods, cooking causes an almost complete gelatinisation of starch, but breakfast cereals and some baked products may contain amounts of incompletely gelatinised starch granules, because of the limited water content during processing [10,57,58]. Moreover, the extent of starch gelatinisation may also vary within a given product. It is higher in the soft part of breads than in the crust, for example. And in "French-fried" potatoes it is likely that the degree of gelatinisation on the outside is different from that on the inside. The presence of sucrose and fat in a starchy food is known to decrease starch gelatinisation [59,60].

However, even after cooking there is a variation in digestibility of gelatinised starch granules. High amylose corn (amylomaize) has poor digestibility in cooked forms, while waxy cereal starches are among the most digestible of all starches. Vegetables have a high percentage of their starch as amylose (30%–40%) and are also rich sources of dietary fibre. Compared with bread, leguminous seed starches (lentils, beans) are digested more slowly in vitro by human digestive juices [61]. The slow rate of digestion of legumes is not due to their fibre content, but may be related to the entrapment of starch in fibrous thick-walled cells, which prevents its complete swelling during cooking [62]. In vitro, the rate of starch hydrolysis in legumes is significantly increased by grinding them finely before cooking [15]. Similarly, finely milled wheat flour is digested significantly faster than its coarsely milled equivalent [63]. Thus the presence of fibre is not in itself enough to slow carbohydrate absorption, the fibre must be present in a form that restricts access of the hydrolytic enzymes to the starch [64]. As legumes are not ground before cooking, it may be supposed that dietary fibre reduces starch digestibility.

Modern methods of food processing involving high temperatures and shear stress affect the rate of starch digestion in foods. Starch in commercially canned beans (pressure-cooked at high temperature) is hydrolysed in vitro faster than that in equivalent home-cooked preparations [15]. Methods such as extrusion cooking and explosion puffing appear to make the starch from corn, rice or potato more readily digested [9,10]. In contrast, conventional cooking methods, such as boiling, involve less physical disruption and only moderate heat. They are therefore less likely to cause starch damage or complete gelatinisation.

Because of their viscous properties, some starchy foods might create a viscous micro-environment in the intestinal lumen, and so impede the diffusion of amylase to starch or the diffusion of its hydrolytic products towards the mucosa [63]. The slow rate of digestion of leguminous starch does not, however, appear to be due to viscosity [15].

Resistant Starch. During food processing (in the baking process of breads, during cooling and storage) starch may retrograde to material highly resistant to hydrolysis by α-amylase. Hydrogen bonds are re-formed and gelatinised starch may partially re-crystallise. Amylose forms strong retrogrades, stable up to 120 °C, while amylopectin retrogrades can be disrupted by gentle heating. This is what happens when staled bread is reheated: the retrogrades of amylopectin that caused the staling during storage are disrupted.

Resistant starch is undigestible in the small intestine, but easily degraded by the large bowel flora. The amount of resistant starch in food is small, being less than 1% in bread and freshly cooked potato and 3% in cornflakes. This amount can be increased during the processing of starchy foods, the increase depending on several factors such as water content, pH, temperature and duration of heating, number of heating and cooling cycles, freezing, and drying [65,66]. Englyst and

Cummings [66] studied the digestion of starch from potato cooked and treated in various ways in ileostomised subjects. Starch from freshly cooked potato was well digested, only 3% being recovered; however, 12% of starch from cooked and cooled potato (kept overnight at 5 °C) escaped digestion in the small intestine. Digestibility of starch made resistant to α-amylase by cooling improved on reheating (7%–8% recovered), suggesting that retrograded amylopectin was partly responsible for the incomplete digestion of cooled potato. In other starchy foods there is evidence that resistant starch is strongly retrograded amylose [27,65,67]. This is likely to be particularly important in high amylose foods such as legumes and long-grain rice.

Complexes and Interactions with Other Food Components. Inclusion complexes between amylose helixes and polar lipids are formed at processing and probably occur naturally in cereal starches. Extrusion cooking of foods leads to the formation of amylose–lipid complexes, which have a relative resistance to amylase in vitro. Extrusion cooking of a high-fibre cereal product, however, does not change the amount of starch recovered in ileostomised subjects [68]. Amylose content rather than amylose–lipid complex appears to be the factor altering starch digestibility [67].

Bread made from wheat flour is incompletely absorbed in man, in contrast to bread made from low-gluten wheat flour and rice flour (which contains no gluten) [23]. Since bread made from low-gluten flour with added gluten is completely absorbed, a physical or chemical interaction between these two fractions in the native flour has been postulated to explain the malabsorption. Mechanical barriers such as the protein matrix (i.e. gluten) that encapsulates gelatinised starch granules might limit access to amylase and reduce starch availability. The high gluten content of spaghetti may explain its lower glycaemic index than that of other wheat products such as bread. There is no difference in the amount of unabsorbed starch recovered from the human ileum by aspiration after ingestion of two meals containing 75 g starch in the form of either regular rusks or rusks made from high-gluten wheat flour [69].

The digestibility of starch may be decreased by non-enzymatic browning reactions (Maillard). The degree of browning depends on process times and temperature, and the presence of moisture, proteins and reducing substances such as sugars [70].

A number of other factors have been shown or suggested to impair starch digestion. These include amylase inhibitors and components such as lectins, phytic acid and tannins [71]. There is little evidence that such interactions cause starch malabsorption in vivo.

The Digestion of Sugars

The digestion of sugars may be affected by the form in which they are ingested. The low absorption of fructose given as the monosaccharide contrasts sharply with the very high absorption capacity of fructose ingested as sucrose [24]. Glucose may increase fructose absorption capacity by acting on its intestinal transport system or increasing its luminal concentration by stimulating water absorption. It cannot be ruled out, however, that this phenomenon may, at least in part, be related to the reduced gastric emptying induced by glucose in sucrose.

The concomitant gastrointestinal events appear, indeed, to play a major role on the digestion rate of disaccharides.

Gastrointestinal Factors Affecting Digestion

The degree to which a carbohydrate is digested and absorbed varies with the food source. Furthermore, in subjects receiving similar loads of a carbohydrate or consuming a standardised meal, there are marked variations between individuals in the extent to which carbohydrates are digested and absorbed [20,34,36]. This suggests that other factors must influence absolute intestinal digestive and absorptive capacity, at least for the osmotically active sugars (mono- and disaccharides). Individuals do not normally ingest a large load of a carbohydrate alone at one time, but rather eat smaller amounts of the carbohydrate within the context of a meal or incorporated into a food product containing other nutrients. Occasionally people drink large quantities of fruit juices or soft drinks which contain substantial amounts of sugars. Although it is important to know the digestive rate of any food component, it is also important to know the context in which the food is normally ingested and in which gastrointestinal function can modulate its digestion.

1. There is evidence featuring the relationship between gastrointestinal function and the digestion of osmotically active sugars. Gastric emptying and intestinal transit appear to play a critical role in the assimilation of these carbohydrates. Ingested carbohydrate load is gradually released from the stomach, and several factors in a mixed meal may reduce the rate of gastric emptying: osmolarity, duodenal pH, fat and caloric intake, viscosity and solid content of the meal. The concentration at which mono- and disaccharides enter the small bowel is important. Unabsorbed sugars cause net fluid accumulation in the intestinal lumen and decrease transit time. With the osmotically active sugars, the contact between luminal nutrients and the digestive and absorptive surface would be limited not only by transit time but also by dilution in the larger volume resulting from inhibition of fluid absorption.

Studies on lactase-deficient subjects consuming dairy products illustrate the role of gastrointestinal function in carbohydrate digestion. Lactose malabsorption in response to equivalent amounts of this sugar in lactase-deficient subjects is reduced firstly when lactose is taken as whole milk, compared with skim milk; secondly when it is taken with a meal, compared with fasting; and thirdly when osmolarity is higher, as in chocolate-flavoured milk compared with plain milk [72–75]. Addition of dietary fibre decreases the increase in breath H_2 after lactose ingestion [76]. It is noteworthy that the decrease in H_2 production is associated with a slowing in the mouth-to-colon transit time [73,74,76,77]. Thus delayed gastric emptying and slowed intestinal transit time allow a greater magnitude of hydrolysis by the residual intestinal lactase in lactase-deficient subjects. Lactose in yoghurt is better digested than that in milk [21]. Besides the bacterial lactase activity, other factors such as intestinal transit time and the high buffering capacity of yoghurt might explain this difference.

Likewise, large discrepancies exist in the literature reports relating to the intestinal capacity to digest and absorb similar loads of sugar alcohols or fructose. After analysis of the experimental conditions under which these studies have been performed, many discrepancies can be explained by changes in factors that control the gastrointestinal function, such as osmolarity of tested solutions and ingestion in a mixed meal containing fat.

2. The digestion of starch is less affected by gastrointestinal function than is that of dietary sugars. Very few studies of the gastric emptying of mixed test meals are available to assess the role of gastric emptying on starch hydrolysis. Increased particle size of ingested starch may affect the rate of gastric emptying. When normal subjects swallow rice, maize or lumps of potato without chewing, the glycaemic response is greatly reduced [78]. However, changing the physical form of a potato meal by sieving has no effect on the overall digestive process of starch [66]. Likewise, drug-induced alterations in intestinal transit time appear to modify only slightly the degree to which starch is digested [28,79]. Moreover, the variability in digestion and absorption of a potato meal between individuals does not appear to be explained by differences in the mouth-to-caecum transit time [80].

Conclusion

In the normal human intestine, various combinations of different factors result in partially digested and absorbed carbohydrates entering the colon. The rate of digestion of starchy foods depends mainly on the nature of the starch, form of the food, processing, and dietary fibre. The variety of starch sources and manufacturing processes is so great that the factors that affect starch digestion will probably vary for each starch-containing food. The rate of digestion of disaccharides seems to be affected mainly by age of the subject, ethnic differences and gastrointestinal function.

References

1. Rosenblum JL, Irwin CL, Alpers DH (1988) Starch and glucose oligosaccharides protect salivary-type amylase activity at acid pH. Am J Physiol 254:G775–G780
2. Fried M, Abramson S, Meyer JH (1987) Passage of salivary amylase through the stomach in humans. Dig Dis Sci 32:1097–1103
3. Alpers DH (1987) Digestion and absorption of carbohydrates and proteins. In: Johnson LR (ed) Physiology of the gastrointestinal tract. Raven Press, New York, pp 1469–1487
4. Rosensweig NS, Herman, RH (1968) Control of jejunal sucrase and maltase activity by dietary sucrose or fructose in man. J Clin Invest 47:2253–2262
5. Hopfer U (1987) Membrane transport mechanisms for hexoses and amino acids in the small intestine. In: Johnson LR (ed) Physiology of the gastrointestinal tract. Raven Press, New York, pp 1499–1526
6. Stone-Dorshow T, Levitt MD (1987) Gaseous response to ingestion of a poorly absorbed fructo-oligosaccharide sweetener. Am J Clin Nutr 46:61–65
7. Nilsson U, Jägerstad H (1987) Hydrolysis of lactitol, maltitol and Palatinit® by human intestinal biopsies. Br J Nutr 58:199–206
8. Jenkins DJA, Wolever TMS Taylor RH et al. (1981) Glycemic index of foods: a physiological basis for carbohydrate exchange. Am J Clin Nutr 34:362–366

9. Brand JC, Nicholson PL, Thorburn AW, Truswell AS (1985) Food processing and the glycemic index. Am J Clin Nutr 42:1192–1196
10. Ross SW, Brand JC, Thorburn AW, Truswell AS (1987) Glycemic index of processed wheat products. Am J Clin Nutr 46:631–635
11. Collier G, McLean A, O'Dea K (1984) Effect of co-ingestion of fat on the metabolic responses to slowly and rapidly absorbed carbohydrates. Diabetologia 26:50–54
12. Collier G, O'Dea K (1983) The effect of co-ingestion of fat on glucose, insulin, and gastric inhibitory-polypeptide responses to carbohydrate and protein. Am J Clin Nutr 37:941–944
13. Estrich D, Ravnick A, Schlierf G, Fukayama G, Kinsell L (1967) Effects of co-ingestion of fat and protein upon carbohydrate-induced hyperglycemia. Diabetes 16:232–237
14. Thorburn AW, Brand JC, Truswell AS (1987) Slowly digested and absorbed carbohydrate in traditional bushfoods; a protective factor against diabetes. Am J Clin Nutr 45:98–106
15. Wong S, Traianedes K, O'Dea K (1985) Factors affecting the rate of hydrolysis of starch in legumes. Am J Clin Nutr 42:38–43
16. Coulston AM, Hollenbeck CB, Liu GC et al. (1984) Effects of source of dietary carbohydrate on plasma glucose, insulin and gastric inhibitory polypeptide responses to test meals in subjects with noninsulin-dependent diabetes mellitus. Am J Clin Nutr 40:965–970
17. Wolever TMS, Nuttall FQ, Lee R et al. (1985) Prediction of the relative blood glucose response of mixed meals using the white bread glycemic index. Diabetes Care 8:418–428
18. Levitt MD, Hirsh P, Fetzer CA, Sheahan M, Levine AS (1987) H_2 excretion after ingestion of complex carbohydrates. Gastroenterology 92:383–389
19. Evard D, Trylezinski A, Cosnes J, Antoine JM, Le Quintrec Y (1987) Digestibilité in vivo des amidons du régime sans résidus. Gastroenterol Clin Biol 11:203A (abstract)
20. Bond JH, Levitt MD (1976) Quantitative measurement of lactose absorption. Gastroenterology 70:1058–1062
21. Kolars JC, Levitt MD, Aouji M, Savaiano DA (1984) Yogurt – an autodigesting source of lactose. N Engl J Med 310:1–3
22. Bond JH, Levitt MD (1972) Use of pulmonary hydrogen (H_2) measurements to quantitate carbohydrate absorption. J Clin Invest 51:1219–1225
23. Anderson IH, Levine AS, Levitt MD (1981) Incomplete absorption of the carbohydrate in all-purpose wheat flour. N Engl J Med 304:891–892
24. Rumessen JJ, Gudman-Hoyer E (1986) Absorption capacity of fructose in healthy adults. Comparison with sucrose and its constituent monosaccharides. Gut 27:1161–1168
25. Fritz M, Siebert G, Kasper H (1985) Dose dependence of breath hydrogen and methane in healthy volunteers after ingestion of a commercial disaccharide mixture, Palatinit®. Br J Nutr 54:389–400
26. Flourié B, Florent C, Etanchaud F, Evard D, Franchisseur C, Rambaud JC (1988) Starch absorption by healthy man evaluated by lactulose hydrogen breath test. Am J Clin Nutr 47:61–66
27. Englyst HN, Cummings JH (1985) Digestion of the polysaccharides of some cereal foods in the human small intestine. Am J Clin Nutr 42:778–787
28. Chapman RW, Sillery JK, Graham MM, Saunders DR (1985) Absorption of starch by healthy ileostomates: effect of transit time and of carbohydrate load. Am J Clin Nutr 41:1244–1248
29. Englyst HN, Cummings JH (1986) Digestion of the carbohydrates of banana (*Musa paradisiaca sapientum*) in the human small intestine. Am J Clin Nutr 44:42–50
30. Saunders DR, Wiggins HS (1981) Conservation of mannitol, lactulose, and raffinose by the human colon. Am J Physiol 241:G397–G402
31. Phillips SF, Giller J (1973) The contribution of the colon to electrolyte and water conservation in man. J Lab Clin Med 81:733–746
32. Gorbach SL, Nahas L, Weinstein L, Levitan R, Patterson JF (1967) Studies of intestinal microflora. IV. The microflora of ileostomy effluent: a unique microbial ecology. Gastroenterology 53:874–880
33. Finegold SM, Sutter VL, Boyle JD, Shimada K (1970) The normal flora of ileostomy and transverse colostomy effluents. J Infect Dis 122:376–381
34. Stephen AM, Haddad AC, Phillips SF (1983) Passage of carbohydrate into the colon. Direct measurements in humans. Gastroenterology 85:589–595
35. Flourié B, Leblond A, Florent C, Rautureau M, Bisalli A, Rambaud JC (1988) Starch malabsorption and breath gas excretion in healthy humans consuming low- and high-starch diets. Gastroenterology 95:356–363
36. Debongnie JC, Newcomer AD, McGill DB, Phillips SF (1979) Absorption of nutrients in lactase deficiency. Dig Dis Sci 24:225–231
37. Marteau P, Flourié B, Franchisseur C, Pochart P, Desjeux JF, Rambaud JC (1988) Role of the

microbial lactase activity from yogurt on the intestinal absorption of lactose. Gastroenterology 94:A284 (abstr)

38. Beaugerie L, Flourié B, Verwaerde F, Franchisseur C, Dupas H, Rambaud JC (1988) Tolerance and absorption along the human intestine of large chronic loads of three polyols. Gastroenterology 94:A29 (abstr)

39. Mobassaleh M, Montgomery RK, Biller JA, Grand RJ (1985) Development of carbohydrate absorption in the fetus and neonate. Pediatrics 75 [Suppl]:160–166

40. Lee PC, Nord KS, Lebenthal E (1981) Digestibility of starches in infants. In: Lebenthal E (ed) Textbook of gastroenterology and nutrition in infancy. Raven Press, New York, pp 423–433

41. Rossiter MA, Barrownan JA, Dand A, Wharton BA (1974) Amylase content of mixed saliva in children. Acta Paediatr Scand 63:389–392

42. Sevenhuysen GP, Holodinsky C, Dawes C (1980) Development of salivary amylase in infants from birth to five months. Am J Clin Nutr 39:584–588

43. Maclean WC, Fink BB (1980) Lactose malabsorption by premature infants: magnitude and clinical significance J Pediatr 97:383–388

44. Henning SJ (1987) Functional development of the gastrointestinal tract. In: Johnson LR (ed) Physiology of the gastrointestinal tract. Raven Press, New York, pp 285–300

45. Dahlqvist A (1983) Digestion of lactose. In: Delmont J (ed) Milk intolerances and rejection. S Karger, Basel, pp 11–16

46. Gray GM (1981) Carbohydrate absorption and malabsorption. In: Johnson LR (ed) Physiology of the gastrointestinal tract. Raven Press, New York, pp 1063–1072

47. Burgio GR, Flatz G, Barbara C et al. (1984) Prevalence of primary adult lactose malabsorption and awareness of milk intolerance in Italy. Am J Clin Nutr 39:100–104

48. Cook GC, Al-Torki MT (1975) High intestinal lactase concentrations in adult Arabs in Saudi Arabia. Br Med J iii:135–136

49. Cuddenec Y, Delbrück H, Flatz G (1982) Distribution of the adult lactase phenotype – lactose absorber and malabsorber – in a group of 131 army recruits. Gastroenterol Clin Biol 6:776–779

50. Kretchmer N, Ransome-Kuti O, Hurwitz R, Dungy C, Alakija W (1971) Intestinal absorption of lactose in Nigerian ethnic groups. Lancet II 392–395

51. O'Morain C, Loubière M, Rampal L, Sudaka L, Delmont J (1978) Etude comparative de l'insuffisance en lactase de deux populations adultes différentes. Acta Gastroenterol Belg 41:56–63

52. Brand JC, Darnton-Hill I, Gracey MS, Spargo RM (1985) Lactose malabsorption in Australian aboriginal children. Am J Clin Nutr 41:620–622

53. Cavalli-Sforza LT, Strata A, Barone A, Cucurachi L (1987) Primary adult lactose malabsorption in Italy: regional differences in prevalence and relationship to lactose intolerance and milk consumption. Am J Clin Nutr 45:748–754

54. Ferguson A, MacDonald DM, Brydon WG (1984) Prevalence of lactase deficiency in British adults. Gut 25:163–167

55. Feibusch JM, Holt PR (1982) Impaired absorptive capacity for carbohydrate in the aging human. Dig Dis Sci 27:1095–1100

56. Thomson ABR, Keelan M (1986) The aging gut. Can J Physiol Pharmacol 64:30–38

57. Asp NG, Björck I, Holm J, Nyman M, Siljeström M (1987) Enzyme resistant starch fractions and dietary fibre. Scand J Gastroenterol 22 [Suppl 129]:29–32

58. Holm J, Lundquist I, Björck I, Eliasson AC, Asp NG (1988) Degree of starch gelatinization, digestion rate of starch in vitro, and metabolic response in rats. Am J Clin Nutr 47:1010–1016

59. Wootton M, Bamunarachchi A (1980) Application of differential scanning calorimetry to starch gelatinization. III. Effect of sucrose and sodium chloride. Starch 32:126–129

60. Hoseney RC, Atwell WA, Lineback DR (1977) Scanning electron microscopy of starch isolated from baked products. Cereal Foods World 22:56–60

61. Jenkins DJA, Wolever TMS, Taylor RH et al. (1980) Rate of digestion of foods and postprandial glycaemia in normal and diabetic subjects. Br Med J 280:14–17

62. Würsch P, Del Vedovo S, Koellreutter B (1986) Cell structure and starch nature as key determinants of the digestion rate of starch in legume. Am J Clin Nutr 43:25–29

63. Heaton KW, Marcus SN, Emmett PM, Bolton CH (1988) Particle size of wheat, maize, and oat test meals: effects on plasma glucose and insulin responses and on the rate of starch digestion in vitro. Am J Clin Nutr 47:675–682

64. O'Dea K, Snow P, Nestel P (1981) Rate of starch hydrolysis in vitro as a predictor of metabolic responses to complex carbohydrate in vivo. Am J Clin Nutr 34:1991–1993

65. Berry CS (1986) Resistant starch: formation and measurement of starch that survives exhaustive

digestion with amylolytic enzymes during the determination of dietary fibre. J Cereal Sci 4:301–314

66. Englyst HN, Cummings JH (1987) Digestion of polysaccharides of potato in the small intestine of man. Am J Clin Nutr 45:423–431
67. Behall KM, Scholfield DJ, Canary J (1988) Effect of starch structure on glucose and insulin responses in adults. Am J Clin Nutr 47:428–432
68. Sandberg AS, Andersson H, Kivistö B, Sandström B (1986) Extrusion cooking of a high-fibre cereal product. Br J Nutr 55:245–254
69. Florent C, Flourié B, Maurel M, Pfeiffer A, Rongier M, Bernier JJ (1987) Le gluten modifie-t-il le transit et l'absorption dans l'intestin grêle des amidons panifiés? Gastroentérol Clin Biol 11:202A (abstract)
70. Hurrell RF, Carpenter KJ (1977) Maillard reactions in foods. In: Hoyem T, Kvale O (eds) Physical, chemical and biological changes in food caused by thermal processing. Applied Science Publishers, London, pp 168–184
71. Jenkins DJA, Jenkins AL, Wolever TMS, Thomson CH, Rao AV (1986) Simple and complex carbohydrates. Nutr Rev 44:44–49
72. Solomons NW, Garcia-Ibanez R, Viteri FE (1979) Reduced rate of breath hydrogen (H_2) excretion with lactose tolerance tests in young children using whole milk. Am J Clin Nutr 32:783–786
73. Solomons NW, Guerrero AM, Torun B (1985) Dietary manipulation of postprandial colonic lactose fermentation. I. Effect of solid foods in a meal. Am J Clin Nutr 41:199–208
74. Martini MC, Savaiano DA (1988) Reduced intolerance symptoms from lactose consumed during a meal. Am J Clin Nutr 47:57–60
75. Welsh JD, Hall WH (1977) Gastric emptying of lactose and milk in subjects with lactose malabsorption. Dig Dis Sci 22:1060–1063
76. Nguyen KN, Welsh JD, Manion CV, Ficken VJ (1982) Effect of fiber on breath hydrogen response and symptoms after oral lactose in lactose malabsorbers. Am J Clin Nutr 35:1347–1351
77. Ladas S, Papanikos J, Arapakis G (1982) Lactose malabsorption in Greek adults: correlation of small bowel transit time with the severity of lactose intolerance. Gut 23:968–973
78. Read NW, Welch IML, Austen CJ et al. (1986) Swallowing food without chewing: a simple way to reduce postprandial glycaemia. Br J Nutr 55:43–47
79. Holgate AM, Read NW (1983) Relationship between small bowel transit time and absorption of a solid meal. Influence of metoclopramide, magnesium sulfate, and lactulose. Dig Dis Sci 28:812–819
80. Thornton JR, Dryden A, Kelleher J, Losowsky MS (1987) Super-efficient starch absorption. A risk factor for colonic neoplasia? Dig Dis Sci 32:1088–1091

Commentary

Würsch: 1. Nilsson and Jägerstad (reference 7, Chap. 3) measured the maltitol hydrolase activity using maltitol syrup, which contains 15% maltotriitol. The rate therefore seems to be too high, since the brush border enzymes hydrolyse maltotriitol very efficiently. Lower activity with human intestinal enzymes has also been reported [1].

2. The gastric emptying of legumes such as beans or lentils is reduced compared with that of potato if they are eaten in particulate form [2], but it was shown that when both foods were eaten in puréed form with a meal the gastric emptying was the same, although the blood glucose and insulin responses elicited by the bean meal were considerably lower. This suggests that the low glycaemic response was predominantly due to the low digestion rate of its starch [3].

3. Table 3.1 presents a list of glycaemic indices established by Jenkins et al. (reference 8, Chap. 3), with glucose as a reference (=100). Since then many other

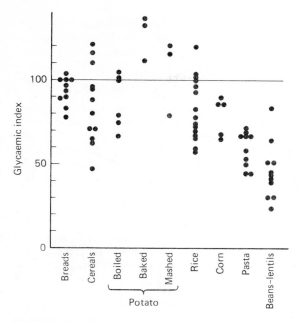

Fig. C.1. The glycaemic index of foods adjusted to that of white bread (= 100) in normal and diabetic subjects. From Würsch P (1989) Starch in human nutrition. Wld Rev Nutr Diet 60 (in press).

foods have been tested on normal and diabetic subjects, as shown in Fig. C.1. The large scatter in the glycaemic index obtained for each group of foods is due to many factors inherent in the source of the food, its form when eaten, the manner in which it is processed, and chewed, and also to the physiological and metabolic response of human subjects, especially of diabetics, presenting a wide intra- and interindividual variability.

References

1. Zeisenitz SC, Siebert G (1987) The metabolism and utilisation of polyols and other bulk sweeteners compared with sugar. In: Development in sweeteners, vol 3, Elsevier Applied Science, Amsterdam, pp 109–154
2. Torsdottir I, Alpsten M, Andersson H et al. (1984) Effect of different starchy foods in composite meals on gastric emptying rate and glucose metabolism. Hum Nutr Clin Nutr 38C:329–338
3. Torsdottir I, Alpsten M, Andersson H, Schweizer TF, Tölli TK, Würsch P (1989) Gastric emptying and glycemic response following ingestion of mashed bean or potato flakes in composite meals. Am J Clin Nutr (in press)

Riou: Interpretation of the different glycaemic indices for different starchy foods should be very cautious. Glucose appearing in the blood is of both endogenous and exogenous origin. The relative flux of each is entirely unknown in most "glycaemic index" studies. If, for example, the exogenous flux is greater with one food than with another, this could have a different effect on gastrointestinal hormone secretion, or on endocrine and exocrine pancreatic function; and this might influence the size of the glucose store in the liver, insulin sensitivity, etc.

Therefore to say that one food has a better glycaemic index than another does not mean that less carbohydrate of exogenous origin is available. Kinetic measurement of the appearance of glucose from the food eaten is necessary if we want a reliable glycaemic index.

Marks: The concept of the glycaemic index has come under severe criticism because of a number of failings, which have been summarised elsewhere in this volume (Chap. 8). Probably the greatest objection to it is the failure of the "area under the curve" (AUC), upon which the concept depends, to reflect the dose of pure glucose taken in varying amounts by healthy volunteers, let alone different doses of different foods containing different carbohydrates.

Peeters: Starch digestion can also be studied using breath tests. Recently the effect of starch structure on starch digestion has been studied using the production of $^{13}CO_2$ [1]. It was concluded that crystalline and waxy starch were digested at the same rate; that thermal processing accelerated digestion; and that the addition of bran had no effect.

Reference

1. Hiele M, Ghoos Y, Rutgeerts P, Vanttappen G (1989) Starch digestion in normal subjects and patients with pancreatic disease using a $^{13}CO_2$ breath test. Gastroenterology 96:503–509

Macdonald: One explanation of why fructose absorption is more rapid when eaten as sucrose, and in my view a very likely possibility, is that as sucrose is hydrolysed on the brush border there is a much higher concentration of fructose at the site of absorption than if it is consumed as equimolar glucose plus fructose. Amounts of fructose (with or without glucose) greater than 100 g give rise to intestinal discomfort, whereas up to 500 g of sucrose as a single meal does not.

Booth: You mention as a possibility that the greater capacity for absorption of fructose ingested as sucrose than as free fructose arises from a slowing of gastric emptying by the glucose in sucrose. Yet was not the osmotic action of fructose identified by J. N. Hunt as capable of exerting some restraint on gastric emptying, albeit at a higher threshold? Perhaps you mean that glucose is relatively more effective than fructose, not that it is the only effective component of sucrose?

Chapter 4

The Role of the Gut Microflora in the Digestion of Starches and Sugars: With Special Reference to Their Role in the Metabolism of the Host, Including Energy and Vitamin Metabolism

D. A. T. Southgate

In the context of the conventional descriptions of the digestion of the sugars and starches in foods, the role of the microflora is regarded as a minor scavenging one, and one that moreover operates only under exceptional, usually pathological conditions[1]. The small-intestinal digestion and absorption of these food carbohydrates is assumed to be virtually complete, and the interaction of the microflora with the undigested components of the plant cell wall (dietary fibre) is assumed to be quantitatively of much greater importance.

Recent studies of the physiological effects of the non-starch components of the plant cell wall show that it is no longer correct to regard the starches as a homogeneous group of polymers that are completely digested in the small intestine, because many foods contain enzymatically resistant starch that passes into the large intestine and provides a substrate for the microflora [2]. Calculations based on the yield of bacterial mass produced each day indicate that the amount of carbohydrate substrate required is substantially in excess of that provided by the non-starch polysaccharides ingested [3].

In addition, many components of foods that may be classified as sugars, for example sugar alcohols, are also poorly absorbed in the small intestine and provide substrates for the microflora.

In this chapter I will compare the metabolism of these substrates by the microflora, and discuss the implications of their microbial metabolism for the host.

In practical terms, most diets provide a mixture of sugars and starches, and it is rarely possible to make an absolute distinction between the effects of the two sources of carbohydrate. Such a distinction can be misleading when extrapolated to diets eaten by most human populations, because only when semi-synthetic diets are consumed can the effects of carbohydrate source be separated from the effects of the wide variety of other components associated with both sugars and starches in foods.

Classification of Carbohydrates in Foods

It is not appropriate to discuss the detailed chemistry of the carbohydrates in foods, as earlier chapters have defined most that concern us. It is, however, useful to make some particular distinctions for the purpose of this chapter. Firstly, all classifications of the carbohydrates are to some extent arbitrary, particularly in distinguishing between simple and complex carbohydrates. For the present purposes sugars are taken to include all simple sugars and their derivatives, and oligosaccharides containing up to five monosaccharide residues. The starches include all polysaccharides with 1–4 and 1–6 α-glucosidic linkages. Although the term polysaccharide is not defined in terms of monosaccharide residues, in practice all native starches are of high molecular weight, and the present discussion specifically excludes chemically modified starches. In discussing the metabolism of the two groups it is convenient initially to consider sugars and starches separately, and to make comparisons later.

Sugars

The food sugars that are discussed are listed in Table 4.1. The systematic names are given earlier; for the present discussion the trivial names are more convenient. The relative importance of the various components in the diet is dependent

Table 4.1.   Sugar derivatives, presence in food and conditions producing malabsorption [1,4]

Principal categories	Individual components	Conditions resulting in intestinal malabsorption	Relative importance in the diet
Monosaccharides	Glucose	Transient immaturity of	Major
	Galactose	transport systems	Minor: fermented milk products only
	Fructose		Major
Disaccharides	Maltose	Inherited disaccharidase	Minor: glucose syrups
	Sucrose	deficiency	Major
	Lactose	Loss of lactase activity	Milk products only
Trisaccharides	Raffinose	Slow passive diffusion	Minor: present in many vegetables
	Maltotriose	Disaccharides deficiency	Minor: glucose syrups
Tetrasaccharides	Maltotetraose	Disaccharidase deficiency	Minor: glucose syrups
	Stachyose	Slow passive diffusion	Minor: present in many vegetables
Pentasaccharide	Verbascose		Minor
Sugar Alcohols Mono	Sorbitol		Minor components almost
	Xylitol	Slow passive diffusion	exclusively present when used
	Mannitol		as food ingredients
Di	Lactitol		
	Isomaltitol	Low activity of disaccharidases	
	Maltitol	towards sugar alcohol	
	Platinit		

on the foods making up the diet [4]. The major monosaccharides glucose and fructose are derived, in part, from the hydrolysis of sucrose in the processing of foods. Free galactose is found only in fermented milk products such as yoghurt. The tri- and tetrasaccharides raffinose and stachyose are present in many vegetables. Maltotriose and tetraose are minor components of glucose syrups. A number of sugar alcohols are used as ingredients in processed foods, where they have demulcent properties and are said to have benefits as lower energy replacements.

Absorption

The intestinal absorption of the sugars is discussed in detail in Chap. 5 and is mentioned here only to indicate whether the sugar is a significant contributor to the microflora.

Glucose and Galactose

Glucose and galactose are actively transported and under normal conditions absorption is not limited. They are therefore only available to the microflora under pathological conditions, as will be discussed later under carbohydrate intolerances (p. 75).

Fructose

Dietary fructose in normal amounts is absorbed by a separate mechanism, and again absorption is normally complete (see Chap. 5).

Disaccharides

Disaccharides are absorbed as the monosaccharides after hydrolysis by brush border disaccharidases. Absorption is normally complete, except in disaccharidase deficiency.

Higher Oligosaccharides

These oligosaccharides are only absorbed in very small amounts, slowly by passive diffusion. The brush border does not contain α-galactosidases and these higher oligosaccharides are therefore virtually unabsorbed and provide a substrate to the microflora.

Sugar Alcohols

The monosaccharide alcohols are absorbed by passive diffusion, and when intake exceeds a threshold of about 0.1 g per kg body weight per hour the alcohols provide substrate to the microflora. Some intestinal hydrolysis of the disaccharide alcohols occurs, but usually at a slower rate in comparison with the corresponding disaccharide.

Sugars Available to the Intestinal Microflora

The sugars available to the microflora of the caecum are therefore potentially derived in two ways. Firstly, there are those sugars where absorption is by passive diffusion and which, when intake exceeds the maximal rates of absorption, will spill over into the large intestine: these include tri- and tetrasaccharides and the monosaccharide sugar alcohols. Secondly, there are those sugars where the activity of the brush border disaccharidases is insufficient to hydrolyse the amounts ingested: these include the disaccharide sugar alcohols in normal subjects, and specific disaccharides in congenital disaccharidase insufficiencies. While these conditions are strictly pathological, lactase deficiency is sufficiently widely distributed to consider it as normal and physiological in many adult populations. The disaccharidase deficiencies also provide an experimental model which illustrates the extreme effects of interactions between sugars and the intestinal microflora.

Quantitative Aspects

Detailed information on normal levels of intake of the tri- and tetrasaccharides is very limited. Figures for the consumption of sugar alcohols are analogously difficult to obtain. Both limitations arise because values for these components are only rarely included in tables of food composition [5]. Some limited values are given by Southgate et al. [6]. These show that in raw beans raffinose concentrations of around 0.4–0.8 g/100 g occur, and stachyose may be present at concentrations of over 3 g/100 g. In leafy vegetables the concentrations of these oligosaccharides rarely exceed 2 g/100 g. These values indicate that one should rarely expect total intakes of these higher oligosaccharides to exceed 5 g per head per day.

The intakes of sugar alcohols are even more difficult to assess. Probably the most widespread use is in diabetic foods and in low cariogenic confectionery, and intakes of these substances by the non-diabetic population are small. One must therefore assume that at the present time sugar alcohols make only a minor contribution to the sugar substrates in the large intestine. The proposed wider use of the disaccharide alcohols as alternative "low-calorie" sweeteners, however, justifies their consideration here.

Starches

The starches in foods contain two major types of polysaccharide: amylose, a linear 1–4α-glucan of molecular weight of the order of 10 000–60 000, and amylopectin, a branched structure consisting of long lengths of 1–4α chains with 1–6α branch points. The molecular weights of amylopectins may range up to 1 million daltons. Small amounts of intermediary forms, that is linear 1–4α molecules with a very few 1–6 branches, have been found in many isolated starches [7].

In the natural uncooked form the starch in foods is present in granules. The physical form and structure of the granules are characteristic of the plant source, and there is a very large number of distinctive starch granules [1]. They usually contain protein and lipid in addition to starch, are birefringent, and show varying degrees of crystallinity.

When starchy foods are subjected to treatment with water and heat, the starch granules absorb water, lose their crystallinity, and the granule swells. On further heating the granule structure collapses, the amylopectin forms a colloidal solution, and the starch is said to gelatinise. If the heated food is then allowed to cool, the starch retrogrades and forms an insoluble aggregated or crystalline form. In this process the linear amylose shows the highest tendency to retrograde.

The starch in ingested foods usually has a mixture of forms: native granules in uncooked foods, swollen but intact granules, and completely disrupted starch gels. In many foods the preparation of which involves a fermentation (such as bread), some enzymatically degraded materials are also present. Some processes, such as extrusion cooking, produce depolymerisation by the strong shear forces generated during the process.

Starch as consumed in foods cannot therefore be regarded as a homogeneous material, because processing has produced a range of physical states from the two major molecular structures present.

The bonds in starch, such as 1–4αD and 1–6αD and the 1–4α bonds, are hydrolysed by both salivary and pancreatic α-amylases producing maltose and isomaltose (from the 1–6α branches), α-limit dextrins, and a little glucose. In the small intestine the maltose and isomaltose are hydrolysed by the brush border disaccharidases. Starch therefore only becomes available as a substrate to the microflora when it is resistant to the endogenous enzymes, or when a maltase or isomaltase disaccharidase deficiency is present.

Resistant Starch

Early studies of the hydrolysis of starch in vitro demonstrated that complete hydrolysis was difficult to achieve, and these studies talk of limit dextrins as a residual core of the starch molecule [8]. The use of enzymatic hydrolysis for the determination of starches often resulted in poor recovery, due to the resistance of the starches to hydrolysis [4]. Many analytical protocols included pre-treatment with alkali, dimethylsulphoxide or acid in order to achieve satisfactory analytical performance. Detailed studies of the analysis of dietary fibre showed that a significant portion of starch was analysed with the plant cell wall (dietary fibre) polysaccharides [9], due to the presence of enzymatically resistant starch: and this, coupled with the estimates of carbohydrate substrate required to support the daily production of bacterial mass [3,10], has led to a growing awareness of the potential importance of starch as a substrate for the intestinal microflora.

Englyst has developed a scheme for classifying the starch in foods, and for identifying the reasons why starch may be resistant to small-intestinal digestion

[11]. Table 4.2 shows that, on the basis of this scheme, starch can become available to the intestinal microflora for a variety of reasons, that result from the physical structure of the starch in the food limiting enzymatic access to the molecule. The principal such structure is retrograded amylose, but other forms also contribute.

Table 4.2.   Effect of physical form of starch on resistance to hydrolysis and on availability as a source of substrate for the microflora

Physical form of starch in food	Examples of foods	Susceptibility to hydrolysis in vitro	Source of substrate for microflora in large intestine
Raw			
Granules: A structure[a]	Most uncooked cereals	Readily hydrolysed	Minor
Granules: B and C[b]	Potato and banana	Partially resistant to hydrolysis	In part
Cooked			
Physically inaccessible	Granules within intact cell walls: whole or broken grains, many legumes	Resistant to hydrolysis	In part
Dispersed amorphous	Freshly cooked foods	Readily hydrolysed	Very little
Retrograded amylopectin	Cooled cooked potato	Partially resistant to hydrolysis	In part
Retrograded amylose	White bread and many processed cereals	Resistant to hydrolysis	Major component of resistant starch
Gelatinised starch dried at high temperature	Some cooked cereal foods	Readily hydrolysed	Minor

[a]X-ray diffraction patterns of the A type are characteristic of cereals and manioc (cassava) starches.
[b]X-ray patterns of the B type are characteristic of tubers and legumes; amylomaize is also a B type. C is apparently a mixture of A and B patterns.

Quantitative Aspects

Englyst et al. [9] provide data on the retrograded amylose form of resistant starch in a range of cereal foods, and Crawford [12] gives values for a wide range of processed foods (see Chap. 3). These provide some approximate estimates. Bingham et al. [13], in a paper on dietary fibre consumption in Britain using the new non-starch polysaccharide (NSP) data, provide indirect evidence of resistant starch intakes. They estimate that intakes of NSP are of the order of 12 g per head per day. Earlier estimates [14] of around 20 g per head were based on values that included resistant starch and lignin; these suggest an intake of resistant starch of about 7–8 g per head per day. However, this value does not include the partially resistant or slowly hydrolysed starches postulated by Englyst and Cummings [2], and must represent the minimum contribution that starch makes to the substrates available to the large-intestinal microflora. Thus starch may be as important a substrate as the non-starch polysaccharides.

The Intestinal Microflora

There are a number of extensive reviews of the intestinal bacterial microflora in man [15], and it is neither appropriate nor necessary to consider the detailed composition of the microflora in this context. Some general observations will provide a basis for discussing the interactions of the flora with the sugars and starches that escape digestion in the small intestine. In considering the metabolism of these carbohydrates, it is necessary to draw on the large body of knowledge regarding the fermentation of carbohydrates in the rumen [16], partly because of the many analogies between this organ and the large intestine, and also because of the relative paucity of detailed studies on fermentation in the large bowel.

The significance of oral flora in the fermentation of carbohydrates in the initial stages of ingestion has been recently reviewed by Curzon [17] and will not be considered further.

Flora are sparsely distributed in the human stomach, duodenum and jejunum, and the concentrations rise towards the terminal ileum. The rate of passage of digesta through the small intestine does not encourage the growth of organisms, and it is only in the relatively slowly moving colon that concentrations approach 10^{11} to 10^{12} per gram wet weight of contents, a concentration strictly comparable with that seen in the rumen. It is the interactions of this flora with the unabsorbed carbohydrates that are of concern here.

Composition of the Flora

The range of species that have been isolated from the intestinal tract is very large. Moore [18] estimates that between 400 and 500 species are present, and the metabolically active cells constitute about half the mass of large intestinal contents. The composition of the flora appears to be relatively stable and characteristic of the individual [19], and studies on the effects of dietary variables on the composition do not produce unequivocal evidence of diet-induced changes. However, it must be recognised firstly that the intestinal microflora are difficult to study in the caecum of intact human subjects, and secondly that the quantitative methods used cannot distinguish changes that are less than an order of magnitude. It is possible, however, to be reasonably confident that changes in substrate levels, even if they do not produce qualitative changes in the range of species present, will influence the pattern of metabolic responses by the flora.

Metabolism of Carbohydrates

As a consequence of the selective effects of the colonic environment, the microflora produce a wide range of glycosidases and virtually no carbohydrates are resistant to, at least partial degradation by the gut microflora [1]. The bacterial metabolism of all carbohydrates follows the general scheme given in Fig. 4.1. Under anaerobic conditions the major products are short chain fatty acids (acetate, propionate, butyrate) and gases (carbon dioxide, hydrogen and

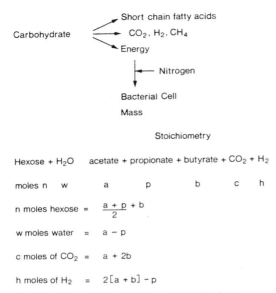

Stoichiometry

Hexose + H_2O acetate + propionate + butyrate + CO_2 + H_2

moles n   w   a   p   b   c   h

n moles hexose $= \dfrac{a + p + b}{2}$

w moles water $= a - p$

c moles of $CO_2 = a + 2b$

h moles of $H_2 = 2[a + b] - p$

Fig. 4.1.   Bacterial metabolism of carbohydrate.

methane), and the process provides energy for bacterial cell growth, for which a source of nitrogen is required. The overall stoichiometry of the fermentation of a typical monosaccharide has been established for the process in the rumen, where the fermentation is closely analogous to that occurring in the colon [20]. The equation for the reaction described by Miller and Wolin [21] does not include the substantial production of hydrogen, and also assumes a complete recovery of the carbohydrate carbon in the products of fermentation. When bacterial mass is being formed, it is clear that other stoichiometries apply. Nevertheless the general features of the reactions are well established (Fig. 4.2).

While it is probable that the composition of the microflora is an important determinant of the products, there is considerable evidence that the course of the reaction is dependent on the actual carbohydrate substrate, and in making a

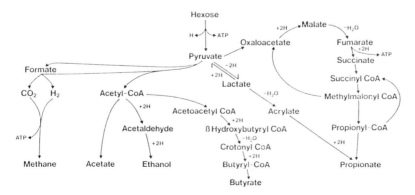

Fig. 4.2.   Anaerobic fermentation of carbohydrate.

comparison of the interactions between sugars and starches it is important to review this evidence. Two other factors also need to be considered in this comparison. The first is the physical state of the substrates as presented to the flora. The sugars are freely water-soluble and will be dispersed through the aqueous phase of the intestinal contents; the starches on the other hand will be present in a variety of states as granules within plant cellular structures, as intact or partially damaged granules, and as insoluble aggregations of retrograded starch. The second factor concerns the location of the glycosidases in the bacteria. Many of these are cell-bound and not secreted into the medium [22], and there are persuasive competitive arguments to support the idea that the products of hydrolysis are not normally released into the medium [12].

Higher Oligosaccharides

The oligosaccharides of the raffinose series (trisaccharide, raffinose; tetrasaccharide, stachyose; and the pentasaccharide verbascose) are not hydrolysed by the endogenous enzymes, and absorption is very slow; under normal conditions these sugars are the only source of sugars to the large intestinal microflora. The activity of α-galactosidases is low in the faecal homogenates, but β-fructosidase activity is higher, suggesting that the initial reaction is a hydrolysis of the fructose component [23]. In an extensive review of the intestinal metabolism of these sugars [24] major emphasis is placed on the production of gas.

Gas production in vitro with these sugars shows a lag phase which is consistent with the requirement for a preliminary hydrolysis. The composition of the gas produced by cultures from dog colon is similar for both monosaccharides and oligosaccharides, being about 32%–40% carbon dioxide and 58%–60% hydrogen.

Human studies have been limited to the measurement of volumes of gas produced, but all are broadly consistent in indicating that the major observed effect of the bacterial degradation is the rapid production of carbon dioxide and hydrogen, within 4–7 hours of ingestion. Little attention has been given to the assumed associated production of short chain fatty acids.

Monosaccharide and Disaccharide Sugars

The monosaccharide and disaccharide sugars only become available to the microflora as a consequence of deficiencies in the transport systems, or, more usually, specific disaccharidase deficiencies. Fermentation is usually extremely rapid, and produces short chain fatty acids and gas. The classical picture presented is that of an osmotic diarrhoea, accompanied by intestinal discomfort and frothy, bulky stools. The rationale for the production of osmotic diarrhoea is that the malabsorbed sugars effectively result in more water being drawn into the terminal ileum, and that the degradation of the sugars to short chain fatty acids increases the osmolality of the contents of the large intestine. The production of carbon dioxide and hydrogen results in the characteristic frothy stools. The condition can be demonstrated experimentally by producing malabsorption by treatment with acarbose [25], which results in a rapid rise in breath hydrogen, production of short chain fatty acids and retention of water in the contents of the large intestine.

There is, however, considerable evidence that short chain fatty acids can be absorbed at a significant rate from the large bowel, and it is therefore improbable that the osmotic effects of their production *per se* are responsible for osmotic diarrhoea [26,27]. Absorption of the short chain fatty acids would result in increased absorption of sodium and water and secretion of bicarbonate, thus tending to return the osmolality of the stools to the normal range. It is possible that it is the rate of short chain fatty acid production that is the key factor.

Lactase Deficiency

Disaccharidase deficiency, leading to malabsorption of disaccharide sugars, is in general an inherited metabolic disorder. However it is reasonable to consider lactase deficiency separately, as intestinal lactase activity disappears in the adults of a number of ethnic groups, especially of Asian origin, so that one could reasonably regard lactose malabsorption as the norm [28].

There is some evidence that lactose absorption in the young infant is limited, and Southgate and Barrett [29] found unabsorbed carbohydrates in the faeces of both breast-fed and bottle-fed infants. The intestinal flora of the young infant is particularly sensitive to the type of feeding. The characteristic organism in the breast-fed infant is *Lactobacillus bifidus*, and it has been postulated that its presence confers benefits on the infant [30]. The organism characteristically generates lactate. Southgate and Barrett [29] observed that the faeces from breast-fed infants, when dried, had a sugary, caramel-like odour, whereas those from infants fed cows' milk preparations had a short chain fatty acid odour.

The symptoms of lactase deficiency are colic and diarrhoea, with the frothy stools characteristic of carbohydrate malabsorption. The conventional lactose load used in evaluating lactose intolerance is 50 g per adult and 2 g/kg body weight in infants. After such a load the glucose levels in capillary blood should rise above 1.5 mmol/l. Osmotic diarrhoea can be precipitated by as little as 10 g lactose in some individuals.

Sugar Alcohols

The monosaccharide alcohols (sorbitol, xylitol and mannitol) are only slowly absorbed, and provide substrates for the microflora. The disaccharide alcohols (lactitol, isomaltitol, maltitol and platinit) are partially hydrolysed by the brush border enzymes, and the constituent sugar alcohol may similarly form a substrate for the microflora.

The sugar alcohols are metabolised by the intestinal microflora to short chain fatty acids and gases, the fermentation being analogous to that of sugars. At high levels of intake this produces osmotic diarrhoea. Xylitol is better tolerated than sorbitol, and is said to produce virtually no gastrointestinal distress at up to 90 g/day [31] while in comparative studies sorbitol produced adverse symptoms at less than 75 g/day. At the suggested limit of 60 g/day the effects of these alcohols on gastrointestinal function are acceptable; in their role as sucrose replacers, intakes above this level are likely only in very unusual circumstances.

Starch

While the utilisation of sugars by the microflora is somewhat exceptional, and in many cases restricted to pathological conditions or unusual diets, there is growing evidence that some starch utilisation is the normal condition. There are a number of strands of evidence for this, the resistance of many food starches to complete enzymatic hydrolysis in vitro being but one. A number of studies with ileostomy patients have confirmed the hypothesis that enzymatic resistance to hydrolysis is a significant factor in vivo [32–35].

Information relating to the fermentation of polysaccharides in the rumen is very extensive [16], and as these polysaccharides include starch, it is of value in discussing colonic fermentation in man. It is important to note one major difference between the rumen and the colon, though, because this will be important later when considering fermentation in relation to energy balance. The difference is that, whereas in the rumen most of the hydrogen generated in the fermentation is consumed in the reduction of carbon dioxide to methane, methane production in man is limited and highly individual [36]. Tadesse et al. [37] found significant elimination of methane in only one-third of the individuals they studied, and this seems to be a general finding. There was no correlation between breath methane and administered non-absorbable oligosaccharides, in contrast to the changes found in breath hydrogen levels.

The pattern of short chain fatty acids produced by fermentation in the rumen is given in Table 4.3, which also shows the patterns observed when various polysaccharides are fermented with faecal homogenates [38]. It is probable that the pattern is an integration of the composition of the active flora present and the substrate. The table does show that, using faecal organisms, a pattern somewhat different from that seen in the rumen is produced. The stoichiometry of the reaction is such that high acetate levels should be associated with more hydrogen production and high propionate with minimal hydrogen production. Studies of the fermentation of monosaccharides in vitro with faecal incubation systems show increased production of butyrate from sorbitol, galacturonate and glucuronic acid, and increased production of propionate from rhamnose, arabinose, xylose, ribose and the uronic acids [39].

Table 4.3. Proportions of short chain fatty acids derived from fermentation of polysaccharides

	Acetate (%)	Propionate (%)	Butyrate (%)
Mixed polysaccharides fermented in rumen	62.5	22.4	16.1
Individual polysaccharides fermented in vitro			
Starch	50	22	28
Arabinogalactan	50	42	7
Xylan	82	15	3
Pectin	84	14	2

The fermentation of starch does not appear to be associated with undesirable gastrointestinal effects, although gas production occurs and fatty acids are generated. There is a report of increased faecal bulking on high starch intakes [40], which implies that the effects of dietary fibre on faecal bulking may be confounded by the associated starch intakes.

Studies with faecal homogenates and washed bacterial cell preparations in vitro [38] have shown that the soluble amylases in the faecal material hydrolyse starch rapidly, but that the highly resistant retrograded amylose is only hydrolysed by bacterial cell preparations, indicating a cell-bound enzyme.

Implications of the Interactions with the Microflora

In discussing the interactions with the microflora, one immediate difference between the sugars and the starches is apparent. The amounts of sugar that become available to the microflora from the normal diet, such as that eaten in the UK, are very small. Those vegetables in the UK which are the major contributors of higher oligosaccharides for example [41], would provide less than 5 g per head per day. Lactose intakes are of the order of 15 g per day, so that in the unlikely event of a lactase-deficient person consuming the average quantity of milk products, only small amounts of fermentable sugars would reach the microflora. In contrast, starch appears to be a normal part of the substrates available to the microflora and may be quantitatively more important in the provision of carbohydrate substrate than the non-starch polysaccharides [2,23,35,40]. Since the provision of substrates is one of the attributes of dietary fibre [42,43], several workers argue that enzymatically resistant starch should be considered part of dietary fibre [44–46].

Thus there is an element of paradox, in that in diets where simple sugars form the major part of the carbohydrates, the non-starch polysaccharides of dietary fibre are of major importance in determining colonic function, whereas in diets where the complex carbohydrates are predominant, the non-starch polysaccharides are of less importance. This, incidentally, may account for some of the inconsistencies in the epidemiological data [47,48]. The true significance of the physiological role of enzymatically resistant starch in relation to health and disease remains, however, to be established, as indeed does that of dietary fibre [49]. The interaction of starch with the microflora has some more immediate and quantifiable effects. The most important of these relates to effects on the energy value of carbohydrates and the provision of other nutrients by this fermentation.

The Energy Value of Carbohydrates Fermented in the Large Bowel

This is currently an area of considerable debate in relation to the energy value of dietary fibre components, and especially of novel semi-synthetic carbohydrates which are not hydrolysed in the small intestine. Such substances are attractive as food ingredients, since they permit lower energy value claims in nutritional labelling. In the context of the normal diet, because starch is the only component that is fermented to any extent, the discussion will concentrate on the energy value of starch, although the arguments apply to all carbohydrates that are fermented in the large intestine.

Systems for the Calculation of Metabolisable Energy

In the UK the energy value of foods is calculated following the the modified Atwater system of energy conversion factors [50]. This is a *convention* that is based on the original experimental evidence of Atwater [51], following the experimental re-evaluation by Southgate and Durnin [52]. In this convention the available carbohydrates are assigned an energy value of 3.75 kcal/g (15.7 kJ/g), which is the heat of combustion of monosaccharide sugars. It is assumed that sugars and starches are completely absorbed and utilised as carbohydrate. The convention also implicitly assigns an energy value of zero to unavailable carbohydrates (the non-starch polysaccharides), although McCance and Lawrence [53] were aware that these unavailable carbohydrates produced short chain fatty acids in the large bowel, and that these might provide a minor source of energy [50,54].

However, the recognition that a significant amount of starch is not "available" in the sense used by McCance and Lawrence [53] requires that the effect of this fermentation be considered in detail. At this point it is important to recognise that the analytical basis for the boundary between "available" and "unavailable" carbohydrates has always been an enzymatically determined one, so that enzymatically resistant starch was included in estimates of unavailable carbohydrate by McCance et al. [55] and by Southgate and co-workers [50,56,57]. This will minimise the magnitude of the error produced by the assumption implicit in the convention. Nevertheless the system for calculating metabolisable energy using conversion factors makes a number of assumptions that are no longer scientifically tenable, and a more rational basis for the calculations is required [5,52,58–60].

Contributions to Metabolisable Energy from Fermentation

The equation for fermentation of carbohydrate in the rumen [16] shows that 6.5% of the gross energy appears as heat and 18.0% as methane, giving the energy value of the short chain fatty acids as some 75.5% of the gross energy of the carbohydrate available as metabolisable energy when absorbed. On this basis the energy value of fermented carbohydrate is of the order of 2.83 kcal/g (11.85 kJ/g). Such a calculation assumes that all the carbohydrate carbon appears as fermentation products. In the rumen approximately 22% [16] of the carbon is not found in the fermentation products, having presumably been incorporated in bacterial cell material. On this basis a metabolisable energy yield of 2.21 kcal/g (9.24 kJ/g) would be predicted. Using the fatty acid pattern observed by Englyst and Macfarlane [38] the stoichiometry is slightly different, equivalent to 2.89 kcal/g (12.1 kJ/g). However, only 60% of the fermented starch was recovered in the short chain fatty acids, equivalent to 1.71 kcal/J (7.2 kJ/g).

A more precise estimate of the potential metabolisable energy of starch fermented in the large intestine, and indeed of any other carbohydrate, requires a more complete balance of the carbon in the reaction than is currently available. The yields of bacterial cell material from fermented carbohydrate are of the order of 30%–40% in the rumen, and similar yields have been seen from the fermentation of plant cell wall material in the colon [42].

The significance of the fermentation of starch and the production of short chain fatty acids depends on the proportion of ingested starch entering the colon. This has been estimated [35,61] to be between 2% and 18%, depending on the food source of the starch. The change to the starch being a source of short chain fatty acids rather than glucose would probably result in reduced glucose absorption and reduced insulin secretion, and thus may benefit the non-insulin-dependent diabetic. In the context of energy balance, if, assuming the "worst case", a maximum of 20% of ingested starch were fermented, and fermentation were complete, the energy conversion factor for starch would fall from 3.75 kcal/g (15.75 kJ/g) as glucose to 3.58 kcal/g (15.04 kJ/g) and with 60% production of short chain fatty acids, 3.34 kcal/g (14.03 kJ/g) with starch intakes of the order of 150 g per head per day. In the normal situation, where about 5–6 g (4% of ingested starch) passes into the colon, the effect on the energy conversion factor is negligible. It is equivalent to energy losses of 27 kcal (113 kJ) and 62 kcal (260 kJ) respectively, errors that are well within the limits of errors of measurement of food intake in most human dietary studies.

The short chain fatty acids may, however, have more subtle effects. For example, butyrate is a preferred substrate for the colonic mucosa, and the acetate and propionate would be utilised by peripheral tissues and the liver respectively. The availability of carbohydrate substrates would tend to lower the pH of the colonic contents, and also, since nitrogen would be required for bacterial growth, would tend to lower colonic ammonia levels [43]. The amounts of starch entering the colon are lower than the amounts of sugar alcohols which provoke osmotic diarrhoea, and infusion of up to 50 g of wheat starch into the colon did not produce any diarrhoea [63].

Other Products of Fermentation

Incubation of faecal homogenates with mixtures that may approximate to the undigested material reaching the colon produce a range of fatty acids, but these are believed to be the products of amino acid fermentations [39]. Stimulation of the production of B-group vitamins would also be expected, but there is little evidence that the large-intestinal synthesis of vitamins, particularly thiamin, riboflavin, biotin and vitamin K, makes any real contribution to the vitamin status of the host. In the rat, unless coprophagy is prevented, a substantial part of vitamin requirements can be obtained from intestinal synthesis. In man this is clearly not significant [64], although there is some speculation about the absorption of vitamin K [65] and vitamin B_{12} [66] across the large-intestinal mucosa. The review by Suttle [67] suggests that bacterially derived forms of phylloquinone can be found in the liver of man, which is a possible indicator that vitamin K activity may be derived from intestinal synthesis. There is no real evidence that intestinal synthesis makes a significant contribution to vitamin B_{12} status, and the efficient reabsorption of the bilary secretion of B_{12} may account for the ability of vegetarians to maintain their B_{12} status on very low intakes [68]. There is therefore no evidence to link starch *per se* with effects on vitamin status, apart from the general effect of starch-rich foods on the provision of the vitamin supply in the diet.

Conclusions

The interactions of the intestinal microflora with sugars and starch present considerable contrasts. First, the interaction with a proportion of ingested starch

– possibly up to one-fifth, depending on the specific foods in the diet – is the normal situation. This starch is readily fermented and produces short chain fatty acids, which lower caecal pH and provide the mucosal cells with butyric acid, and the liver and peripheral tissues with propionate and acetate. The interaction has a minor effect on the energy supplied by starch and may alter the insulinogenic effects of the starch.

In contrast, sugars, except for small amounts of higher oligosaccharides, rarely provide substrates for the micraflora, except under pathological conditions, when an osmotic diarrhoea is the usual consequence. Sugar alcohols are also fermented, but osmotic diarrhoea is only associated with high levels of ingestion. The products of fermentation are short chain fatty acids and these make a contribution to the metabolisable energy supply of the host.

In many respects these carbohydrate fermentations are strictly analogous to the fermentation of the non-starch polysaccharides of dietary fibre. This is one of the major physiological effects of dietary fibre, and has been linked with protection against large bowel cancer.

In many diets starch may be the major carbohydrate substrate for the microflora, and some of the beneficial effects of diets rich in dietary fibre may be attributable to the starch that is also present in the majority of them [49].

References

1. FAO (1980) Carbohydrates in human nutrition. Food and Agriculture Organisation of the United Nations, Rome (Food and nutrition paper 15)
2. Englyst HN, Cummings JH (1987) Resistant starch, a "new" food component: a classification of starch for nutritional purposes. In: Morton ID (ed) Cereals in a European context. Ellis Horwood, Chichester, pp 221–223
3. Metz G, Gassull MA, Drasar BS, Jenkins DJA, Blendis LM (1976) Breath hydrogen test for small intestinal bacterial colonisation. Lancet I:668–669
4. Southgate DAT (1976) Determination of food carbohydrates. Applied Science Publishers, London
5. Greenfield H, Southgate DAT (1989) Guidelines for the production management and use of food composition data. United Nations University (in press)
6. Southgate DAT, Paul AA, Dean AC, Christie AA (1978) Free sugars in foods. J Hum Nutr 32:335–347
7. Banks W, Greenwood CT (1975) Starch and its components. Edinburgh University Press, Edinburgh
8. Pigman WW, Goepp RM (1948) Chemistry of carbohydrates. Academic Press, New York, pp 582–594
9. Englyst HN, Anderson V, Cummings JH (1983) Starch and non-starch polysaccharides in some cereal foods, J Sci Food Agric 34:1439–1440
10. Stephen AM, Cummings JH (1980) The microbial contribution to human faecal mass. J Med Microbiol 13:45–56
11. Englyst HJN (1985) Dietary polysaccharide breakdown in the gut of man. PhD thesis, University of Cambridge
12. Crawford C (1987) Survey of resistant starch in processed foods. Bulletin of the Flour Milling and Baking Research Association 2:59–64
13. Bingham SA, Williams DRR, Cummings JH (1985) Dietary fibre consumption in Britain: new estimates and their relation to large bowel cancer mortality. Br J Cancer 52:399–402
14. Southgate DAT, Bingham S, Robertson J (1978) Dietary fibre in the British diet. Nature 274:51–52

15. Drasar BS, Hill MJ (1974) Human intestinal flora. Academic Press, London
16. Hungate RE (1966) The rumen and its microbes. Academic Press, New York
17. Curzon MEJ (1988) Dietary carbohydrate and dental caries. In: Dobbing J (ed) A balanced diet? Springer-Verlag, Berlin Heidelberg New York, pp 57–74
18. Moore WEC (1978) Some current concepts in intestinal bacteriology. Am J Clin Nutr 31:S33–S42
19. Barnside GH (1978) Stability of human fecal flora. Am J Clin Nutr 31:S141–S144
20. Cummings JH (1981) Dietary fibre. Br Med Bull 37:65–70
21. Miller TL, Wolin MJ (1979) Fermentation by saccharolytic intestinal bacteria Am J Clin Nutr 32:144–172
22. Salyers AA, Palmer JK, Wilkins TD (1978) Degradation of polysaccharides by intestinal enzymes. Am J Clin Nutr 31:S128–130
23. Taeuful K, Ruttloff H, Krause W, Taeuful A, Vetter K (1965) Zum intestinen Verhalten von Galakto-oligosacchariden beim Menschen. Klin Wochenschr 43:268–272
24. Cristafaro E, Mottu F, Wuhrmann JS (1974) Involvement of raffinose family of oligosaccharides in flatulence. In: Sipple HL, McNutt CW (eds) Sugars in nutrition. Academic Press, New York, pp 313–336
25. Caspary WF, Lembcke B, Elsenhans B (1981) Bacterial fermentation of carbohydrates within the gastrointestinal tract. In: Read NW (ed) Diarrhoea: new insights. Clin Res Rev 1:[Suppl]107–117
26. McNeil NI, Cummings JH, James WPT (1978) Short chain fatty acid absorption by the human large intestine. Gut 19:819–822
27. Ruppin H, Bar-Meir S, Soergal KH, Wood CM, Schmitt MG (1980) Absorption of short-chain fatty acids by the colon. Gastroenterology 78:1500–1507
28. Dahlquist A (1974) Enzyme deficiency and malabsorption of carbohydrates. In: Sipple HL, McNutt CW (eds) Sugars in nutrition. Academic Press, New York, pp 187–214
29. Southgate DAT, Barrett IM (1966) The intake and excretion of calorific of milk by infants. Br J Nutr 20:363–372
30. György P (1974) Effects of carbohydrate on the intestinal flora. In: Sipple HL, McNutt CW (eds) Sugars in nutrition. Academic Press, New York, pp. 215–226
31. Armada F, Eisenstein A (1970) Cited by Bron and Miller, in Sipple HL, McNutt CW (eds) Sugars in nutrition. Academic Press, New York.
32. Sandberg AS, Anderson H, Hallgren B, Hasselblad K, Isaksson B, Hulten L (1981) Experimental model for in vivo determination of dietary fibre and its effects on the small bowel absorption of nutrients. Br J Nutr 45:283–294
33. Englyst HN, Cummings JH (1986) Digestion of the carbohydrate of banana (*Musa paradisiaca sapientum*) in the human small intestine. Am J Clin Nutr 44:42–50
34. Englyst HN, Cummings JH (1987) Digestion of the polysaccharides of potato in the small intestine of man. Am J Clin Nutr 45:423–431
35. Jenkins DJA, Cuff D, Wolever TMS et al. (1987) Digestibility of carbohydrate foods in ileosomate relationship to dietary fiber in vitro digestibility and glydemic response. Am J Gastroenterol 82:709–717
36. Tadesse K, Eastwood MA (1978) Metabolism of dietary fibre components in man assessed by breath hydrogen and methane. Br J Nutr 40:393–396
37. Tadesse K, Smith D, Eastwood MA (1980) Breath hydrogen (H_2) and methane (CH_4) excretion in normal man and in clinical practice. Q J Exp Pathol 65:85–97
38. Englyst HN, Macfarlane GT (1986) Breakdown of resistant and readily digestible starch by human gut bacteria. J Sci Food Agric 36:699–706
39. Mortensen PB, Holtug K, Rasmussen HS (1988) Short chain fatty acid production from monosaccharides and disaccharides in a faecal incubation system: implications for the colonic fermentation of dietary fiber in humans. J Nutr 118:321–325
40. Shetty PS, Kurpad AV (1986) Increasing starch intake in the human diet increases fecal bulking. Am J Clin Nutr 43:210–212
41. Ministry of Agriculture, Fisheries and Food (1988) Household Food Consumption and Expenditure 1986. Report of the National Food Survey Committee. Her Majesty's Stationery Office, London
42. Stephen AM, Cummings JH (1980) Mechanism of action of dietary fibre in the human colon. Nature 284:283–284
43. Cummings JH, Stephen AM, Branch WJ (1981) Implications of dietary fiber breakdown in the human colon. Banbury Report 7: Gastrointestinal cancer: endogenous factors. Cold Spring Harbor Laboratory, New York
44. Asp NG, Johanssen GG (1984) Dietary fibre analysis. Nutr Abstr Rev 54A:735–752
45. Berry CS (1986) Resistant starch: formation and measurement of starch that survives exhaustive

digestion with amylolytic enzymes during the determination of dietary fibre. J Cereal Sci 4:301–314

46. Jenkins DJA, Jenkins AL, Wolever TMS, Thompson LH, Rao AV (1986) Simple and complex carbohydrates. Naringsforskning [Suppl] 23:11–16
47. Bingham SA, Williams DRR, Cummings JH (1985) Dietary fibre consumption in Britain: new estimates and their relation to large bowel cancer rates. Br J Cancer 52:399–402
48. Bingham SA (1987) Definitions and intakes of dietary fiber. Am J Clin Nutr 45:1226–1231
49. Southgate DAT (1988) Dietary fibre and the diseases of affluence. In: Dobbing J (ed) A balanced diet. Springer, Berlin Heidelberg New York, pp 117–139
50. Paul AA, Southgate DAT (1978) In: McCance and Widdowson's The composition of foods, 4th edn. HMSO, London
51. Merrill AL, Watt BK (1955) Energy value of foods basis and derivation. US Department of Agriculture, Washington DC (Agriculture handbook 74)
52. Southgate DAT, Durnin JVGA (1970) Calorie conversion factors: an experimental reassessment of the factors used in the calculation of the energy value of human diets. Br J Nutr 24:517–535
53. McCance RA, Lawrence RD (1929) The carbohydrate content of foods. (Medical Research Council Special report 135)
54. Widdowson EM (1960) Note on the calculation of the calorific value of foods and diets. In: McCance RA, Widdowson EM The composition of foods, 3rd edn. HMSO, London, pp 153–159
55. McCance RA, Widdowson EM, Shackleton LRB (1936) The nutritive value of fruits, vegetables and nuts. HMSO, London (Medical Research Council special report 213)
56. Southgate DAT (1969) Determination of carbohydrates in foods. II. Unavailable carbohydrates. J Sci Food Agric 20:331–335
57. Southgate DAT, Bailey B, Collinson E, Walker AF (1976) A guide to calculating dietary fibre intakes. J Hum Nutr 30:303–313
58. Southgate DAT (1971) Assessing the energy value of the human diet. Nutr Rev 29:131–134
59. Southgate DAT (1975) Fibre and other unavailable carbohydrates and energy effects in the diet. Proceedings of the 4th Western Hemisphere Nutrition Congress
60. Livesey G (1988) Energy from food – old values and new perspectives. British Nutrition Foundation, London (Nutrition bulletin 52(13):9–28)
61. Stephen MA, Haddad AC, Phillips SF (1983) Passage of carbohydrate into the colon: direct measurement in humans. Gastroenterology 85:589–595
62. Pomare EW, Branch WJ, Cummings JH (1985) Carbohydrate fermentation in the human colon and its relation to acetate concentrations in venous blood. J Clin Invest 75:1448–1454
63. Flourié B, Florent C, Jouany JP, Thivend P, Etanchoud F, Rambaud JC (1985) Caloric metabolism of wheat starch in healthy humans: effects on fecal outputs and clinical symptoms. Gastroenterology 90:111–119
64. Kaspar H (1986) Effects of dietary fibre on vitamin metabolism. In: Spiller GA (ed) Handbook of dietary fibre in human nutrition. CRC Press, Boca Raton, pp 201–208
65. Anonymous (1980) Intestinal microflora, injury and vitamin K deficiency Nutr Rev 38:341
66. Albert MJ, Mathan VJ, Baker JJ (1980) Vitamin B_{12} synthesis by human small intestinal bacteria. Nature:233–781
67. Suttle JW (1984) Vitamin K. In: Machlin LJ (ed) Handbook of vitamins. Marcel Dekker, New York, pp 147–98
68. Ellenbogen L (1984) Vitamin B_{12}. In: Machlin LJ (ed) Handbook of vitamins Marcel Dekker, New York, pp 497–547

Commentary

Flourié: Fructose absorption in the human small intestine appears to be incomplete [1].

The author states that changes in substrate levels will influence the pattern of metabolic responses by the flora. There is evidence that such adaptation of the metabolic activity of the flora can occur [2,3]. Is there any advantage in their adaptation in terms of energy value?

References

1. Ravich WJ, Bayless TM, Thomas M (1983) Fructose: incomplete intestinal absorption in humans. Gastroenterology 84:26–29
2. Hill MJ (1983) Bacterial adaptation to lactose deficiency. In: Delmont J (ed) Milk intolerance and rejection. Karger, Basel, pp 22–26
3. Florent C, Flourié B, Leblond A, Rautureau M, Bernier JJ, Rambaud JC (1985) Influence of chronic lactulose ingestion on the colonic metabolism of lactulose in man. J Clin Invest 75:608–613

Author's reply: It is possible to envisage that an adaptation by the flora to the substrates entering the large intestine would be advantageous to the host, but the rate of adaptation has not been studied. The inaccessibility of the caecum in man would make such studies extremely difficult.

Levin: It is postulated that the cause of the diarrhoea induced by the production of short chain fatty acids in the colon may be their rapid rate of synthesis by the colonic bacteria overloading their normal rate of removal by absorption. In rats, however, some short chain fatty acids can elicit secretion and increased motility of the colon. It may be that these actions aid, or are the basis of, the diarrhoea [1,2].

References

1. Yajima T (1985) Contractile effect of short chain fatty acids on the isolated rat's colon. J Physiol 368:667–678
2. Yajima T (1988) Luminal propionate induced secretory response in rat distal colon in vitro. J Physiol 403:559–575

Flatt: The change to starch being a source of short chain fatty acids rather than glucose would markedly decrease meal-induced insulin secretion. Glucose is a potent stimulus for insulin release, whereas short chain fatty acids themselves are virtually ineffective at physiological concentrations [1]. In addition, enteroinsular hormones, such as GIP, secreted in response to carbohydrate and fat, are not found in the colon [2]. The involvement of gut microflora in the digestion of starches and sugars would be expected to play havoc with endocrine mechanisms normally regulating the metabolic response to food ingestion, provided that the metabolic products are absorbed in significant amounts.

References

1. Morgan LM, Flatt PR, Marks V (1988) Nutrient regulation of the enteroinsular axis and insulin secretion. Nutr Res Rev 1:79–97
2. Brown JC (1982) Gastric inhibitory polypeptide. Springer, Berlin Heidelberg New York

Macdonald: Mention should be made of the fact that sugar alcohols may give rise to a simple osmotic diarrhoea *per se*, rather than through the production by the microflora of short chain fatty acids. The onset of diarrhoea following large doses of sugar alcohols is about the same as that after the purgative Epsom salts ($MgSO_4$).

Flourié: The production of hydrogen and methane in the large intestine leads to losses of the energy content of the carbohydrates entering the colon. Some species of bacteria convert hydrogen to methane, and from 30% to 80% of people in different countries exhale methane in the breath. In this case, where methane is produced, are the energy losses supposed to be higher?

Author's reply: Reduction of carbon dioxide to methane by the hydrogen produced by fermentation would not have any further effect on energy losses from carbohydrates entering the large intestine.

Marks: Humans can tolerate much larger amounts of xylitol than the 90 g/day cited, provided they are introduced to it gradually. Makinen and Scheinen [1] cite one individual ingesting more than 400 g/day without untoward effects, and many who ate 200 g/day or more without detriment. We have observed the same phenomenon. Long-term feeding with xylitol [2] and sorbitol [3] changes the nature of the colonic microflora, and this may have something to do with the dietary adaptation that takes place.

There is evidence that small amounts of alcohol occur regularly in the blood of total abstainers, and that this originates in the gastrointestinal tract from microbial fermentation. Ordinarily it is of no great importance but, in at least some cases of fungal overgrowth, the ingestion of a carbohydrate-containing meal can lead to a very large and rapid rise in blood alcohol and alcoholic intoxication [4]. All the authentic cases of which I am aware have emanated from Japan, but the existence of this disorder has left the door open to those who claim not to drink alcohol but who have gross alcoholaemia. It has also been exploited by charlatans under the title "The Yeast Connection". It must therefore be given our attention.

References

1. Makinen KK, Scheinen A (1975) Turku sugar studies. VI. The administration of the trial and the control of the dietary regimen. Acta Odont Scand 33 [Suppl 70]:105–127
2. Salminen S, Salminen E, Koivistoinen P, Bridges J, Marks V (1985) Gut microflora interactions with xylitol in the mouse, rat and man. Fd Chem Toxic 23:985–990
3. Salminen S, Salminen E, Bridges J, Marks V (1986) The effects of sorbitol on the gastrointestinal microflora of rats. Z Ernahrungswiss 25:91–95
4. Kaji H, Asanuma Y, Yahara O, Shibue H, Hisamura M, Saito N, Kawaakakawi Y, Murao M (1984) Intragastrointestinal alcohol fermentation syndrome: report of two cases and review of the literature. J Forensic Sci Soc 24:461–471

Würsch: Consumption of polyols is, on average, very low, but since they are now used as sugar substitutes in several countries, acute large intake of as much as 20 g in sweets, chocolate or jam can occur, and this can produce intestinal disorders, especially in children [1].

Reference

1. Abraham RR, Yudkin J (1981) Controlled clinical trial of a new non-calorigenic sweetening agent. J Hum Nutr 35:165–172

Flourié: The author states that dietary starch may be the major carbohydrate substrate for the microflora. However, indirect evidence suggests that carbohydrate substrates from endogenous sources play an important role in the metabolism of the colonic flora. First, there is a large discrepancy between estimates of unabsorbed dietary carbohydrates and calculations based on the yield of bacterial mass (50–70 g/day). Secondly, a significant part of the daily hydrogen produced in the colon does not appear to result from the bacterial metabolism of dietary starch [1]. In other words, carbohydrates from endogenous substrates of colonic origin may provide a major part of the energy requirements of the colonic flora.

Reference

1. Flourié B, Leblond A, Florent C, Rautureau M, Bisalli A, Rambaud JC (1988) Starch malabsorption and breath gas excretion in healthy humans consuming low- and high-starch diets. Gastroenterology 95:356–363

Author's reply: My statement referred to dietary sources of fermentable carbohydrate. There is, as Flourié says, a substantial amount of mucus available as a substrate for the intestinal microflora.

Levin: One of the obvious questions to ask on the effects of somatostatin when given to humans is: What was the concentration achieved in the blood compared with that measured in vivo during a normal meal? Experimenters often argue that they need to use higher levels in the infusion studies than those normally found, because they are trying to mimic the high levels presumably found at the sites where somatostatin is released. While this might be the rationale for them to use high levels, it does obviously complicate interpretation of the effects they find, because they are then studying the actions of high levels of somatostatin in plasma and tissues.

Peeters: After a meal, and following clearance of the bowel of digesta, a motor pattern develops which has been called the "migrating motor complex", and which periodically clears the stomach and small intestine of meal remnants, basal secretions, cellular debris, etc. Its function has been described as that of an intestinal "housekeeper", sweeping the gut and preventing the accumulation and stasis of fluids [1]. In fact small-intestinal bacterial overgrowth has been reported to be associated with the absence of this motor pattern [2].

References

1. Vantrappen G, Jaussius J, Peeters TL (1981) The migrating motor complex. Med Clin N Am 65:1311–1329
2. Vantrappen G, Jaussius J, Hellemius J, Ghoes Y (1977) The MMC of normal subjects and patients with bacterial overgrowth of the small intestine. J Clin Invest 59:1158–1166

Chapter 5

Dietary Carbohydrate and the Kinetics of Intestinal Functions in Relation to Hexose Absorption

R. J. Levin

Introduction

The small intestine can be described as an "active interface" between the external environment and the blood and lymph that distribute the absorbed materials for the metabolic needs of the body. The activity of the interface is created by the intestine's lining of epithelial cells whose three major functions are digestion, absorption and secretion. The population is ever-changing, being constantly renewed by cells dividing in the intestinal crypts and replacing those enterocytes lost into the lumen of the bowel from the extrusion zone at the tips of the villi [1]. In most mammals the duration of the enterocyte's life is about 48 hours, and it has been colourfully described as a "short life but a merry one". The intestinal epithelium is truly an active interface for it can respond to the demands placed upon it by adapting its structure and function. Changes in the amount and type of food entering the tract can induce adaptations in its function. These occur by three basic mechanisms:

1. Changes in the number of functional enterocytes.
2. Changes in the carrier, enzymatic or metabolic processes of the enterocytes.
3. Changes in the maturation of function in the enterocytes as they migrate up from the crypts to the extrusion zone at the villus tip.

Examples of each of these categories occur when the level of dietary carbohydrate is altered significantly. The present chapter reviews the kinetics of hexose absorption and illustrates how dietary intake of carbohydrate can change the kinetics.

The Handling of Carbohydrate by the Mammalian Small Intestine

Because dietary carbohydrate in man and omnivorous mammals is a major nutrient, the alimentary tract has effective mechanisms for its digestion and for the subsequent absorption of the resultant end products. In the Western human adult diet, carbohydrate accounts for approximately half the energy intake. It is the major source in developing countries. About 60% of this ingested carbohydrate is in the form of polysaccharides, mainly starch and glycogen, but disaccharides such as sucrose and lactose represent 30% and 10% respectively. In a number of Western societies there can also be a significant dietary intake of monosaccharides such as glucose and fructose, especially in manufactured foods and drinks. Other oligosaccharides, such as raffinose and stachyose, are present in small amounts in various legumes, and, although they are unavailable for digestion by intraluminal and cellular enzymes, they are broken down by luminal bacteria, especially in the colon.

The polysaccharides are digested by enzymatic hydrolysis, initially salivary amylase and then pancreatic amylase. This occurs mainly in the lumen of the upper small bowel (Fig. 5.1). The oligosaccharides formed by bulk phase intraluminal or cavital digestion are produced rapidly in the upper intestine [2], although there can be a previous small amount of hydrolysis in the stomach. It appears that starch digestion is mostly intraluminal in man [3], and there is little of the so-called contact or membrane digestion proposed by Ugolev [4], where the adsorption of amylase onto the brush border surface of the enterocytes facilitates its enzyme activity.

Table 5.1. Major glycosidases of mammalian enterocyte brush border

Glycosidase	Complex	Enzyme activity
Maltase–sucrase Maltase–isomaltase	Sucrase–isomaltase	80% of maltase; some of α-limit dextrinase; all of sucrase; most of isomaltase
Maltase–glucoamylase (2)	Glucoamylase	All glucoamylase; most of α-limit dextrinase; 20% maltase; small percentage isomaltase
Trehalase		All trehalase
Lactase	β-Glycosidase	All neutral lactase and cellobiose
Glycosyl-ceramidase (phloridzin hydrolase)		Most of aryl-β-glycosidase

After Dahlqvist and Semenza [10].

Further breakdown of the formed glucosyl oligosaccharides resides in the brush border surface of the enterocytes. The carbohydrases are built into the surface membrane of the mature enterocyte's microvilli, closely applied to the transporting sites for the released monosaccharides [5–7]. Because of this arrangement, Crane [8] described the brush border of the enterocytes as the "digestive-absorptive surface". It is dogma that oligosaccharides can only be assimilated when hydrolysed at this active interface into their constituent monosaccharides glucose, galactose and fructose. Sugar molecules larger than monosaccharides are said to be unable to pass across the intestinal barrier. However, studies with various molecular sizes of polyethylene glycols (PEG), much larger than oligo- and disaccharides, indicate that small amounts of PEG can pass across the gut and be excreted in the urine [9]. Their passage across the gut wall probably occurs between the enterocytes, or through open channels in the extrusion zones at the villous tips. Presumably little oligosaccharide is available for these paths, because of the powerful enzymes available for its breakdown. Their range is listed in Table 5.1. This enzymatic breakdown is an extremely efficient process, apart from the hydrolysis of lactose, which in adults is relatively slow and becomes the rate-limiting step for the absorption of its constituent monosaccharides glucose and galactose [11]. In the case of all the other dietary disaccharides, the rate-limiting step appears to be the transport process across the enterocyte [11].

Transport Mechanisms for the Monosaccharides

Monosaccharides can pass across the epithelial lining of the small intestine by three processes: simple diffusion, facilitated diffusion and "active transport". The first can occur across or between enterocytes (paracellular or intercellular route), but the second and third can only occur across the cell. As illustrated in Fig. 5.1, transfer across the enterocyte from the lumen of the intestine involves:

1. Movement from the bulk phase across the "unstirred layer" to the entero-cyte's surface.
2. Movement across the brush border membrane.
3. Movement across the cytoplasm.
4. Exit via the basolateral membrane.

Movement from the Bulk Phase

Monosaccharides that are in their free form in the bulk phase of the luminal fluid will move in accordance with their concentration gradient by simple diffusion through the fluid, until they reach the surface of the enterocyte. To do this they have to pass across the so-called unstirred layer, which basically consists of the more structured layers of fluid near and at the surface of the enterocytes and the glycocalyx of the microvilli. The latter might represent a resistance to diffusive movement. This unstirred layer is an important contributing factor to the difficulty of making accurate measurements of absorption kinetics, and can influence the results both qualitatively and quantitatively (see p. 97 et seq.). It may also play a role in confining the products of hydrolysis to the cell surface. Once at the surface of the enterocyte, the hexoses can cross the membrane by diffusion (see below) and/or they can diffuse between the cells. There are no data on the quantitative importance of the two pathways.

Movement Across the Brush Border

The simple diffusive movement of hydrophilic monosaccharides across the lipid brush border membrane is relatively poor and slow and would thus be inefficient

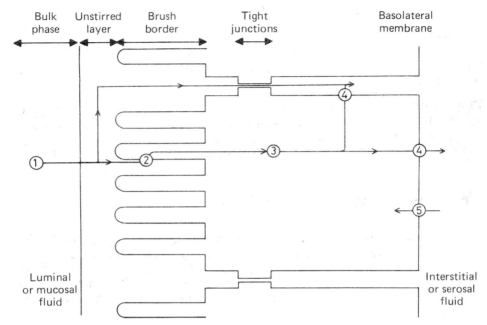

Fig. 5.1. Diagrammatic representation of the basic pathways of solute movement from the bulk phase (1) of the luminal (in vivo) or mucosal (in vitro) fluid through the unstirred layer, either for transfer across the brush border membrane (2) into the intracellular compartment (3) or to diffuse through the tight junctions into the intercellular space. The solute diffuses across the intracellular compartment (3) to the basolateral membrane (4), to be transferred by either diffusion or facilitated diffusion and enter the interstitial (in vivo) or serosal fluid (in vitro). In vivo, it is then cleared from the milieu of the enterocytes by the blood and lymph streams. At the basolateral membrane there is an entry process (5) for glucose from the blood. This may have components of facilitated diffusion and active transfer.

for the rapid assimilation of the large loads of liberated hexoses produced after a carbohydrate-rich meal, and for the transfer of very low concentrations. In order to overcome this difficulty, special transfer mechanisms of "facilitated diffusion" and of "active transfer" have evolved. In both cases the hexose is moved across the brush border membrane at a rate faster than can be accounted for by simple diffusion, presumably by the agency of a specific carrier molecule in the membrane. Only in the case of active transfer, however, can the hexose be transported "uphill" against a concentration gradient.

The Glucose/Na⁺ Co-transporter in the Brush Border Membrane

Laboratory studies with rodent intestine incubated in vitro showed that entero-cytes could transfer the dietary hexoses glucose and galactose, and a number of other glucalogues such as α-methylglucoside and 3-O-methylglucose, from the luminal fluid into the serosal fluid against their concentration gradient. This activity was described as active transport and it was shown to be dependent on cell metabolism, and especially on the presence of the Na⁺ ion in the luminal fluid

[12]. Crane [12] postulated that this Na^+-driven active transfer of hexoses is mediated by a membrane carrier, now called a co-transporter, containing two binding sites: one for the hexose and one for the Na^+ ion. The Na^+ ion, when bound to the co-transporter, enhances the affinity of the hexose site for glucose and galactose. On being exposed to the inside of the enterocyte, the Na^+ on the co-transporter diffuses away into the low Na^+ concentration of the inside of the cell. This then makes the hexose binding site less attractive to the hexose, which also diffuses off the co-transporter into the cell. The empty co-transporter, when re-exposed to the luminal fluid, can then undertake another transfer cycle. Thus the energy of the extracellular-to-intracellular Na^+ ion gradient can move the hexose uphill against its concentration gradient.

Since the formulation of this brilliantly simple concept for co-transport, numerous studies have been published confirming the mechanism, not only for hexoses but also for amino acids and other nutrients [13]. Many features of the co-transporter function, such as its specificity for particular hexose structures and particular ions, its binding kinetics, its Na^+ stoichiometry, its electrogenicity and its reversible binding of the sugar transport inhibitor phloridzin, have been examined first using everted sacs of intestine, then sheets, then isolated villi and cells and, more recently, vesicles of the brush border [14,15].

Isolation of the Glucose/Na⁺ Co-transporter

Attempts to isolate the actual co-transporter by solubilisation followed by recombination have proved difficult, but the application of photoaffinity labelling, negative purification and monoclonal antibodies led to the isolation of a polypeptide of approximately 80 kilodaltons that has behaviour characteristics suggesting that it is either the co-transporter *per se* or at least an important part of the molecule [16].

Cloning the Glucose/Na⁺ Co-transporter

Because of the obvious technical difficulties of purifying the carrier from membranes, where it constitutes less than 1% of the protein, a totally new approach to isolating the co-transporter was taken by Hediger et al. [17]. They injected the poly(A)$^+$-RNA prepared from rabbit enterocytes into oocytes from *Xenopis laevis*, which resulted in the expression of Na^+-dependent, phloridzin-sensitive uptake of α-methylglucoside, a glucalogue, in the egg. The mRNA coding for the co-transporter molecule was found, and the full length cDNA was prepared, isolated, and placed in a plasmid vector. A single clone was isolated that increased the glucalogue's Na^+-sensitive uptake by more than a 1000-fold but had no effect on the Na^+-independent uptake. The results indicated that the cDNA for the "classical" intestinal brush border glucose/Na^+ co-transporter had been cloned.

From the nucleotide sequence of the cDNA, Hediger et al. deduced the amino acid sequence of the co-transporter, allowing them to propose a model structure. This contained 11 membrane-spanning sequences. Two of these on either side of the membrane were highly charged areas, and were thought to be involved in the glucose binding with multiple lysine residues. A tyrosine residue is located at or near the Na^+ binding site, which is some 3–4 nm away from the glucose site. Clearly Na^+ binding to the co-transporter produces a long-range conformational change that increases the affinity of the glucose site for its substrate. Comparison of the structure of the glucose/Na^+ co-transporter and its DNA with those of other transport proteins showed no homology, which suggested that it represents a novel class of transport protein distinct from those for channels and pumps [17].

Electrical Activity Associated with Hexose Transfer

Because the glucose co-transporter carries charge (Na^+ ions) across the brush border membrane without a counter-ion, an electrical potential difference is created across the membrane, the enterocyte, and subsequently the intestinal wall. The active transfer of hexoses in this context can be said to be electrogenic (potential producing) or rheogenic (current producing). Initially this current and potential was thought to be supported only by the binding and de-binding of the Na^+ ion to the co-transporter, and its subsequent pumping out of the enterocyte by the transport ATPase at the basolateral membrane, but it is now known that lanthanides (rare-earth elements) will bind to the site and support hexose transport [18]. The lanthanides are trivalent cations of the element series lanthanum (atomic number 50) to ytterbium (atomic number 70). The best activators were sumarium (atomic number 62), europium (atomic number 63) and thulium (atomic number 69).

The generation of electrical activity by the movement of Na^+ ions via the co-transporter has been, and still is, of inestimable value in the assessment of the kinetics of active electrogenic hexose absorption in animals and man. It allows an on-line measure of what is happening across the brush border, enterocyte and intestinal wall, and has been used in many studies [19]. Increasing concentrations of activity transported hexoses in the luminal fluid induce increasing potentials, called transfer potentials, which follow saturation or Michaelis–Menten kinetics, and can be used to characterise the system by the operational parameters of "apparent K_m" and the maximum potential difference generated (PD_{max}) (Table 5.2). It is also possible to use the currents generated by the linked Na^+ and hexose transfer, measured as the short-circuit current, for the same purpose, although this can only be undertaken in vitro; the potentials, on the other hand, can be measured in vivo even in conscious man and neonates [20–22].

Studies have been undertaken that show that the data on hexose transfer obtained from electrical measurements are a valid assessment of hexose kinetics. Chemical measurements undertaken at the same time as electrical measurements have shown that the two sets of transfer parameters obtained are identical in vitro [23] and under in vivo conditions, as long as a correction is made for the non-electrogenic diffusive movement of the hexose [24].

Multiple Hexose Carriers

Initial studies on hexose absorption always assumed that there was only one hexose/Na^+ transporter present in the enterocytes, and that the dietary hexoses,

Table 5.2. Apparent K_m and maximum potential difference (PD_{max}) for electrogenic glucose transport and unstirred layer thickness obtained in human intestine in situ

	Intestinal segment	Apparent K_m (mM)	PD_{max} (mV)	Unstirred layer thickness (µm)
Rask-Madsen et al. [104]	(Fasted) jejunum	16 ± 2	12.4 ± 2.1	–
Rask-Madsen et al. [104]	Sucrose/glucose perfused jejunum	48 ± 2	11.2	–
Read et al. [21]	Adult jejunum	36 ± 6	7.6 ± 0.6	632 ± 24
Mcneish et al. [105]	Neonate jejunum	39.6	16.8	–
Igarishi et al. [22]	Infant jejunum	14.4 ± 2.8	11.4 ± 2.3	–
Sparso et al. [106]	Adult duodenum	53	9.9	–
Sparso et al. [107]	Adult duodenum	25.8*	1.1*	351 ± 13
Flourié et al. [108]	Upper jejunum (adult)	–	–	667 ± 32

All authors used the basic electrical recording techniques of Read et al [20,21] to obtain the "Apparent K_m" and unstirred layer thickness. Results are given as mean $\pm$ SE.
[a] Data corrected for unstirred layer effects.

glucose and galactose, and all the actively transported glucalogues, utilised this single carrier [25]. Various results in a number of studies, however, suggested that there may be more than one carrier in the membrane [26]. Strong evidence that multiple hexose carriers were present in the intestine was obtained for the rat [26] and for the hamster [27], using totally different experimental approaches. The former study showed that when rats were fasted and then allowed to drink different sugars, the feeding sugar often had different effects on the kinetics of specific transferred hexoses [26]. If only one carrier existed this should not have happened. In the hamster, the longitudinal distribution of the transport of various glucalogues was different along the intestine [27]. Subsequent studies by other workers have confirmed the finding of multiple carriers in intact rabbit intestine in vitro [28], and with vesicles from animals and from fetal and adult human intestine [29–31].

Apart from the diffusive movement, glucose transport in rabbit vesicles [29] is mediated by a major, low-affinity system ($K_m = 0.3$ mM, $J_{max} = 11$ nmol/(mg/min)) and a minor, high-affinity system ($K_m = 0.03$ mM, $J_{max} = 3$ nmol/(mg/min)). Similarly, in normal human fetal jejunum [30] there appears to be a low-affinity, high-capacity system (apparent $K_m = 4.2$ mM, $J_{max} = 31$ nmol/(mg/min)) and a high-affinity, low-capacity system (apparent $K_m = 0.4$ mM, $J_{max} = 8$ nmol/(mg/min)). In the fetal ileum, however, there was only one glucose carrier (apparent $K_m = 1$ mM, $J_{max} = 5$ nmol/(mg/min)).

Studies by Syme and Levin [32] on the effects of hypothyroidism and fasting on electrogenic hexose transport in rat jejunum in vitro, indicated that it was necessary to postulate a minimum of four different systems for hexoses to accommodate all the changes in the transport parameters of apparent K_m and PD_{max} observed for the two dietary hexoses glucose and galactose and the three glucalogues. Interestingly, none of the changes in the kinetic parameters for the glucalogues identically matched those for glucose or galactose. Thus use of glucalogues could not characterise what happened to the transfer mechanism for the dietary sugars, at least in the fasting and the hypothyroid rat.

The heterogeneity in the transfer mechanisms for the dietary hexoses and for glucalogues was discussed by Levin [33] and by Debnam and Levin [26]. It was suggested that the postulated different carriers may be examples of glucose isocarriers, the concept of isocarriers being analogous to that of isoenzymes. This speculative concept has lain studiously ignored in the backwoods of the literature.

Movement Across the Cytoplasm of the Enterocyte

Once the hexose enters the enterocyte, either by the co-transporter or diffusing into the cell, it presumably remains free (i.e. osmotically active), able to diffuse across to the basolateral membrane. Few studies have focused on the intracellular transport of the monosaccharides. They have been reviewed previously [33]. Earlier work in the late 1960s indicated that when enterocytes undertake active transfer of hexoses (or amino acids) the cells increase in volume [34], strongly suggesting that the transferred hexose is osmotically active (i.e. in free solution) during its passage across the cell. One author speculated that hexoses are bound to an intracellular carrier [35], but no confirmatory studies have ever been published.

Exit via the Basolateral Membrane

The basolateral membrane represents a barrier to the free movement of hexoses into and out of the enterocytes. Hexoses move across by diffusion and a facilitated transfer process that is Na^+-independent and is blocked by the inhibitors phloretin or cytochalasin B [15,29]. It is a mechanism similar to that present in many other cells such as erythrocytes, fibroblasts and adipocytes [15]. Studies on the carrier systems have been greatly facilitated by the preparation of sealed vesicles of the basolateral membrane [36]. Changes in the transfer of hexose across the basolateral membrane can be induced by artificially increasing the glucose level in plasma [37,38].

Characterisation of Transport Mechanisms by Kinetic Analysis

The need to characterise the mechanisms responsible for the absorption of hexoses across the small intestine led to the use of kinetic analysis. As hexose absorption is concentration-dependent, the measurement of absorption at one particular arbitrary concentration becomes idiosyncratic, and can lead to misinterpretations of the underlying processes. Early experiments indicated linear or diffusion kinetics when high concentrations of hexoses were absorbed [39,40]. Lower concentrations indicated carrier-type kinetics. The description of the transfer of glucose and galactose across isolated small intestine in vitro, by application of Michaelis–Menten saturation kinetics [41], gave a new dimension to the analysis of transport mechanisms. It allowed them to be characterised by the operational kinetic parameters of "apparent K_m" and J_{max}.

The former parameter is crudely used in transport studies as an index of the affinity of the mechanism (carrier?) for the hexoses, while J_{max} is the transport equivalent of the maximum velocity of the enzyme. These parameters allow comparisons to be made of the effects of various experimental conditions on the transport mechanisms. While enzyme kinetics purists will be appalled – rightly so in some respects – at the cavalier way that transport workers employ these operational kinetic parameters to describe the affinity and transport maximum of the absorption processes, their guarded use has been of great value in quantifying and characterising such processes. A number of authors have described the difficulties in obtaining valid kinetic data and in their interpretation [42,43]. This brief review is not the place to rehearse such complications.

Perhaps the most important factor influencing the kinetics of intestinal transport is the effect of unstirred layers. The following section will briefly deal with this.

The Effects of Unstirred Layers on the Kinetics of Absorption

The unstirred layers present in the small intestine (viz. the structured water layers close to the mucosal surface and the glycocalyx or fuzzy coat of the microvilli) have a profound effect on the estimation of the kinetic parameters for saturation kinetics and passive permeability coefficients. It is essential to correct transport data obtained experimentally for these unstirred layer effects in order to obtain the most accurate values for these parameters. The papers of Winne [44,45], Wilson and Dietschy [46] and Dugas et al. [47], and the review by Levin [19], should be consulted for theoretical and experimental details. Parsons [48] formulated a useful, simplified equation relating the "real K_m" of the carrier to the "apparent K_m", J_{max} (maximum transfer of the system), the thickness of the unstirred layer (d) and the diffusion coeficient (D) of the solute being transferred:

$$\text{"Apparent } K_m\text{"} = \text{real } K_m + J_{max}\, d/D$$

Thus, if J_{max} is large, or the unstirred layer is thick, or both, then the uncorrected apparent K_m measured for the system will be significantly larger than the real K_m of the carrier. A visual illustration of the dramatic effect of different values of J_{max} on the apparent K_m when working across unstirred layers of increasing thickness is shown in Fig. 5.2, taken from the data of Read et al. [21]. The real K_m of the carrier was set at 5 mM. If J_{max} is 10 nm/(cm^2/s), and the unstirred layer 400 μm thick, then the apparent K_m that will be measured, without correction, would be approximately 100 mM. This example clearly shows the danger of using uncorrected absorption data to obtain the apparent K_m of a transport mechanism. Values for the unstirred layer thickness in man are given in Table 5.2. Methods for correcting transport data for unstirred layers vary in complexity [19,45,47, 48]. A recent new approach using a two-dimensional laminar flow model cannot give statistical errors for the true K_m and V_{max}, because of the enormous time required for only a single iterative calculation [49].

Many comparisons of kinetic parameters that have been made in the older literature on absorption mechanisms did not correct for unstirred layer effects, so their usefulness is questionable. Typical examples are comparisons of the K_m and J_{max} for jejunum and ileum, that do not take into account the fact that the

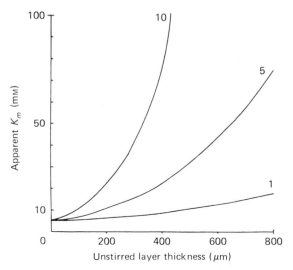

Fig. 5.2. The effects of unstirred layer thickness on the "apparent K_m" of an active transfer mechanism at different values of J_{max} (in nm/(cm²/s)). A computer analysis from Read et al. [21]. The "true" K_m of the carrier was set at 5mM.

unstirred layers and the values of J_{max} of the two areas may be different. Thus the quoted differences in the values of apparent K_m for hexose absorption for the jejunum and ileum in the rat (53 mM and 20 mM [50]) and in man (73 mM and 31 mM [51]) may be artefactual. A technique for correcting absorption data for unstirred layers was developed by Levin et al. [52] for use in experimental animals. A flow chart for the technique is shown in Fig. 5.3. When applied to data from the jejunum and ileum of the domestic fowl, the jejunal K_m for glucose (11 ± 2 mM, mean ± SE) was significantly greater than that of the ileum (4 ± 1 mM). The possible greater affinity of the ileal carrier suggested that the enterocytes in this area were acting as scavengers for luminal glucose.

The importance of the unstirred layer as a rate-limiting step for the diffusion of solutes from the luminal bulk phase to the surface of the enterocytes under physiological conditions of digestion and absorption has been questioned by a study published as yet only in abstract form [53]. The authors implanted chronic balloon cannulae in the rat jejunum and allowed the rats to recover. Then the 30-cm area between balloons was infused with probes the uptake of which was limited primarily by diffusion through the luminal contents (5.5. mM glucose, labelled warfarin and carbon monoxide). Their absorption was measured in this and in the standard "laparotomised" (anaesthetised?) rat preparation and was said to be 5 to 9 times faster in the conscious compared with the anaesthetised rat, while the unstirred layer thickness was only 80–110 μm, much thinner than that observed previously in vivo. The authors claim that normal gut motility induced excellent luminal stirring, disrupting the laminar flow seen in the anaesthetised, laparotomised preparations, and that the unstirred layer in vivo has been drastically overestimated. While it is clear that normal motility of the gut will increase luminal stirring and reduce unstirred layer thickness, insertion of balloons into rat jejuna can itself hardly be called "normal", especially as it is well

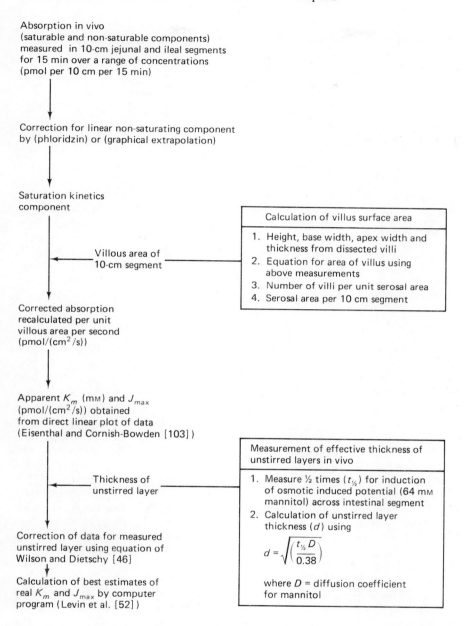

Absorption in vivo
(saturable and non-saturable components)
measured in 10-cm jejunal and ileal segments
for 15 min over a range of concentrations
(pmol per 10 cm per 15 min)

Correction for linear non-saturating component
by (phloridzin) or (graphical extrapolation)

Saturation kinetics
component

Calculation of villus surface area
1. Height, base width, apex width and thickness from dissected villi
2. Equation for area of villus using above measurements
3. Number of villi per unit serosal area
4. Serosal area per 10 cm segment

Villous area of
10-cm segment

Corrected absorption
recalculated per unit
villous area per second
$(pmol/(cm^2/s))$

Apparent K_m (mM) and J_{max}
$(pmol/(cm^2/s))$ obtained
from direct linear plot of data
(Eisenthal and Cornish-Bowden [103])

Measurement of effective thickness of unstirred layers in vivo
1. Measure ½ times $(t_{½})$ for induction of osmotic induced potential (64 mM mannitol) across intestinal segment
2. Calculation of unstirred layer thickness (d) using

Thickness of
unstirred layer

Correction of data for measured
unstirred layer using equation of
Wilson and Dietschy [46]

$$d = \sqrt{\left(\frac{t_{½} D}{0.38}\right)}$$

Calculation of best estimates of
real K_m and J_{max} by computer
program (Levin et al. [52])

where D = diffusion coefficient
for mannitol

Fig. 5.3. A flow chart of the technique developed by Levin et al. [52] for correcting absorption data for unstirred layers.

known that inflated balloons stimulate intestinal motility. Until the study is published in full it may be prudent to refrain from calling the balloon preparation motility "normal".

Fructose Absorption

Although the mechanisms for the transfer of fructose have not been examined in as much detail as those for other hexoses, a number of studies indicate that its absorption across the enterocytes in both man and animals is by a mechanism separate from the glucose/Na$^+$ co-transporter. Glucose and galactose generate transfer potentials during active transfer via the co-transporter, but fructose does not. Experimentally, the co-transporter can be inhibited by the specific blocker phloridzin. This prevents the active transport of glucose and galactose, but that of fructose is unaffected. In some rodents, such as guinea pig and hamster, a large proportion of the absorbed fructose is converted into glucose during its passage across the enterocyte [54–56], but in the rat [55] and in man [57,58], it is absorbed largely unchanged.

It was originally thought that fructose was not actively transported by enterocytes, but was absorbed by a carrier-mediated, facilitated diffusion [25]. Studies on fructose transfer across everted jejunum from starved rats [59] revealed that the fructose was accumulated against its concentration gradient by an energy- and Na$^+$-dependent process which followed saturation kinetics (apparent $K_m = 0.9$ mM). Similar findings were found independently [60].

In the rare human disorder of glucose–galactose malabsorption, which is inherited as an autosomal recessive trait, the brush borders of the sufferers have only approximately 10% of the normal concentration of the glucose/Na$^+$ co-transporter, and they thus transport glucose and galactose extremely poorly; the absorption of fructose, however, is unaffected [61]. This suggests that in the human enterocyte, the carrier for fructose is different from the glucose/Na$^+$ co-transporter, and is not under control of the same gene. The four individuals who have been found to have a specific disorder of delayed fructose absorption [62] may be illustrating a deficiency or absence of this fructose carrier. However, the transfer of fructose by biopsies of intestine from these subjects was not studied, so that no details are available as to possible biochemical or cellular causes of the delayed transport.

The absorption of fructose from ingested sucrose loads is less prone to symptoms of gastrointestinal distress and malabsorption than when pure fructose loads are ingested [63]. The absorption of fructose fed as a single hypertonic sugar solution appears to be enhanced by the addition of glucose [63,64]. It was suggested, because of this observation, that there were possibly two absorption pathways for fructose, one glucose-dependent and one glucose-independent [63], but no direct experimental evidence for these exists. In the light of recent experiments showing that glucose-stimulated fluid absorption has a solvent-drag effect on small molecules of the size of fructose (see later for details), it is most likely that the glucose-induced enhancement of fructose absorption is mediated through such a mechanism. It would also explain why fructose from sucrose is more readily absorbed: the glucose from the hydrolysed sucrose stimulates fluid absorption and thus enhances solvent-drag of the fructose.

Are There Any Remaining Controversies in Hexose Absorption?

At a time when the glucose/Na^+ co-transporter can apparently be cloned and inserted into the egg cells of choice, it would appear that there are few remaining problems about the basic mechanisms of hexose absorption. Every laboratory that has studied active hexose transfer in vitro has confirmed its electrogenicity, sodium and phloridzin sensitivity, and hexose specificity, so there is little or no controversy over the basic facts, although there may still be controversy about their interpretation – for example the actual Na^+ stoichiometry with the transferred hexose [15,29] or the actual mechanism of binding [15,16, 29].

The importance of the Na^+-dependent active transport mechanism, with an apparent K_m of about 1–5 mM for glucose, has never been questioned when assessing hexose absorption in vitro, but its importance in relation to the absorption of glucose from a carbohydrate-rich meal under physiological conditions has been questioned more than once [65–67]. It is assumed that, under physiological conditions, the hydrolysis of dietary carbohydrate would give rise to high levels of free hexoses in the luminal fluid. Luminal concentrations of glucose well above the plasma level of approximately 5 mM have been measured in situ in human (75 mM) and rat (112 mM and 35 mM) intestine after conventional meals [68–70]. Such levels indicate that the electrogenic co-transporter for hexose, with a K_m of 1–5 mM, would be saturated, and that until the luminal concentration approached that in the plasma diffusion or facilitated diffusion would be the major process by which the hexose left the lumen. Moreover, as the free luminal concentration of hexose is likely to be lower than at the carbohydrase sites at the brush border surface, the above account is even more likely. An assessment of the ratio of the movement of glucose from the lumen of rat jejunum via the active transport and diffusive systems, at increasing luminal concentrations, is portrayed graphically in Fig. 5.4 [24].

One possible way to test the importance of the electrogenic co-transporter would be to block it off with the specific inhibitor phloridzin and then see what happens to the absorption of hexoses. Unfortunately, the ethics of using the high doses of phloridzin needed to block off the carriers in the intestine of human volunteers precludes such experiments, but experimental studies in the rat give some clues about the possible importance of the electrogenic transfer mechanism.

In the jejunum of the anaesthetised rat in vivo it is possible to choose a concentration of phloridzin that will just prevent the generation of glucose transfer potentials but will still allow the absorption of glucose from the lumen [24]. Because increases in the luminal concentration of glucose caused linear increases in the absorption of glucose from the lumen without any transfer potentials being generated, it was inferred that this absorption was taking place either by diffusion or by a carrier-mediated facilitated diffusion process that did not generate transfer potentials. In other words it was not a Na^+-linked co-transporter and was less sensitive to phloridzin blockade.

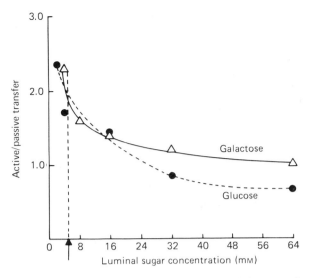

Fig. 5.4. The active/passive transfer ratio for glucose and galactose across rat jejunum in vivo calculated at different luminal concentrations. The vertical dashed line marked by the arrow is drawn at the approximate concentration for plasma glucose (5 mM). The graph is modified from Debnam and Levin [24]. The luminal concentration where the active transfer becomes greater than passive transfer is approximately 30 mM glucose, which agrees well with the 35 mM value reported by Murakami et al. [70].

Further evidence for a diffusive or facilitated diffusive mechanism for glucose and galactose absorption in the phloridzinised intestine came from studies with sorbose. This sugar, of the same molecular weight as the dietary monosaccharides, has always been regarded as a hexose that moves across the small intestine by simple diffusion only, as it is unaffected by Na^+ ion concentration or phloridzin, and has simple diffusion kinetics [25]. In the phloridzinised rat intestine, the rates of absorption of glucose, galactose and sorbose were identical, and followed diffusion kinetics [24]. Interestingly, it is claimed that there is even a Na^+-independent diffusive movement of glucose across brush border vesicles [71] and across human jejunal biopsies [72]. Thus, from these studies, simple diffusion should not be discarded as a process by which glucose and galactose can leave the intestinal lumen. Indeed in numerous measures of the kinetics of glucose absorption there is a clear-cut linear component as well as a saturative one (see [24] for references). It is tacitly, but not always openly assumed that this linear component represents a diffusive movement, although it is not easy to distinguish kinetically between diffusion and a carrier with a very high K_m (say in the order of 100 mM).

As described previously in relation to the fructose carrier, those humans suffering from glucose–galactose malabsorption have a brush border deficient in the glucose/Na^+ co-transporter. They can still split disaccharides but can only absorb approximately 10% of the resulting monosaccharides. The first sign of the malabsorption occurs usually when the subjects are fed milk. But if there can be a substantial diffusive movement of hexoses across the small intestine in experimental animals such as the rat, why do the subjects with glucose–galactose malabsorption suffer from malabsorption? Despite the recent clear-cut finding of glucose diffusion across human jejunal biopsies [72], it would appear that in the human intestine in situ, simple diffusion is unlikely to be a candidate mechanism for the transfer of nutritionally significant amounts of dietary hexose.

If simple diffusion has to be discounted as a residual safety mechanism for hexose absorption, facilitated transfer using the Na^+-independent carrier system is still left as a possibility. Is this second carrier system present in the intestines of subjects with glucose–galactose malabsorption? It may be that the 10% residual absorption of glucose observed in those with the condition is being transferred by such a carrier [61]. Interestingly, it has been reported that the absorption of galactose is even more seriously disturbed than that of glucose [61], so that the residual system present appears to show some specificity – though this suggestion has been challenged [73]. Thus subjects with glucose–galactose malabsorption present an unsolved problem. Research using vesicles, biopsies and intact animal preparations appears to indicate that diffusion-type movement makes a highly significant contribution to glucose absorption, yet in subjects without a glucose/Na^+ co-transporter it does not appear to cope with normal levels of monosaccharides resulting from the hydrolysis of dietary carbohydrates. It may be that there are more changes in the enterocyte brush border membrane of individuals with glucose–galactose malabsorption than simply the absence of the glucose/Na^+ co-transporter? For example its hexose permeability and the facilitated glucose carrier may also be abnormal.

A possible explanation for the lack of "diffusive" hexose absorption in glucose–galactose malabsorption may be identified in the recent studies of Pappenheimer and his colleagues [74–76] on the mechanism of glucose absorption. They proposed from their experiments on rat small intestine that solvent drag of fluid passing through the paracellular channels was the principal route for the transfer of glucose (and amino acids) at physiological rates of fluid absorption and concentration. In their provocative hypothesis, the primary role of Na-coupled active hexose co-transport was to provide the osmotic force for convective (absorptive) fluid flow through the paracellular pathway, and to trigger contraction of cytoskeletal proteins [75,76]. These cytoskeletal proteins, when contracted, induced dilatation of the intercellular junctions, disruption of the structure in the zonula occludens, and condensation of actomyosin in the perijunctional ring [76]. This enhances the permeability of the junction to small solutes the size of monosaccharide molecules. Thus, increasing amounts of luminal glucose follow apparently linear "diffusion" type kinetics, because the increasing glucose concentration induces corresponding increases in fluid transfer driven by the active transport mechanism. With this model, the failure of sufferers from glucose–galactose malabsorption to absorb dietarily significant amounts of free glucose–galactose liberated from carbohydrate, would thus be due to the inability to create a glucose-induced convective flow of fluid by their reduced or absent glucose/Na^+ co-transporter system, and lack of relaxation of the permeability of the intercellular junctions. Simple diffusion of hexoses across the epithelium is dismissed by Pappenheimer as being too inefficient to cause significant absorption. Thus in some respects, this mechanism explains the lack of active hexose absorption, and the reduced "permeability" of the small bowel epithelium to hexoses in individuals with glucose–galactose malabsorption.

While the evidence presented by Pappenheimer and his co-workers for the "solvent drag–junctional permeability change" hypothesis of glucose absorption is impressive, there is avoidance of experimental data on the possible role of diffusion. The work with brush border vesicles indicates that there is a "diffusive" movement component of glucose into the vesicles, which must obviously mean that the hexose diffuses across the brush border membrane proper (see above and reference 71). Other data obtained from absorption studies in rat indicate that the movement of glucose across the intestine can display linear "diffusion" kinetics in vivo even when the convective fluid flow should be minimal, viz. when glucose is absorbed across the intestine in the presence of luminal phloridzin, the inhibitor of the hexose/Na^+ co-transporter (see above and reference 26). A possible modification of the Pappenheimer model could be that convective flow is not the only mechanism in operation in vivo. Diffusion and the fluid flow washing away the absorbed hexose (maintaining the source–sink hexose gradient) could well be involved in the movement of glucose from lumen to blood.

Clearly, even although the glucose transporter can now be cloned, there are still unanswered questions about the manner in which hexoses are absorbed from dietary carbohydrates.

Absorption of Free Glucose and Galactose Compared with Those Produced from Disaccharides

The hydrolysis of maltose, sucrose and lactose by the brush border disaccharidases produces glucose, galactose and fructose. Apart from lactose, the absorption of which is rate-limited by its slow hydrolysis to its component monosaccharides, the rate of absorption of monosaccharides produced from the disaccharides should be the same from monosaccharide solutions as from solutions of disaccharides producing equimolar amounts of the free hexoses. In other words each millimole of maltose should give rise to the same glucose absorption as twice that amount of glucose. However, early experiments from Crane's laboratory on the absorption of hexoses from disaccharides by intestinal preparations of hamster jejunum in vitro, produced some surprising results [5]. Glucose liberated from sucrose was transported better than glucose produced from glucose-1-phosphate. This was so even in the presence of high concentrations of glucose oxidase in the incubating medium that should have destroyed all free glucose liberated from the disaccharide that spilled out into the medium. Similar results were obtained with other disaccharides. These results led Crane to develop the hypothesis of "kinetic advantage", namely that the hexoses liberated from disaccharides were especially close to the co-transporter, and thus had a kinetic advantage over the free hexose in the lumen that had to diffuse to the transport site through the unstirred fluid layers and glycocalyx [6]. A similar conclusion came about from the experiments of Parsons and Pritchard using frog intestine in vitro [77].

Caspary, working in Crane's laboratory, undertook further studies [78] to clarify the relationships between sucrose hydrolysis and the transport of the liberated glucose, again using hamster intestine in vitro. Addition of sucrose to the preparation, after saturating the glucose transfer process with glucose, produced an additive increase in the maximal rate of transport in excess of 90%, but the glucose derived from glucose-1-phosphate was not additive. Thus the simple formation of glucose close to the brush border membrane (due to the hydrolysis of the glucose phosphate) did not allow the glucose to enter the "kinetic advantageous system".

Further work showed that while the active transport of free glucose was dependent on the presence of Na^+ ions, that from sucrose hydrolysis was not. Confirmation that the transport mechanism for glucose produced by sucrose hydrolysis was Na^+-independent, was obtained using recordings of transfer potentials. Sucrose alone generated a transfer potential which was undiminished by the presence of glucose oxidase in the luminal fluid. The transfer potential generated by 15 mM glucose was not increased significantly by the addition of 15 mM sucrose, despite the fact that sucrose by itself generated a potential approximately 66% of the glucose transfer potential.

Caspary used these findings to produce a model where the disaccharidase was in extremely close proximity to the Na^+-independent active glucose transferring system (B system), while the classical Na^+-dependent active transfer glucose system (A system) was a little further away from the hydrolase site. In this model, sucrose hydrolysis produced a transfer potential when present on its own, because, when it was hydrolysed, the glucose which was formed saturated the B system, and the untransported glucose then had access to the Na^+-dependent electrogenic transport system. Thus, these experiments uncovered the fact that in the brush border there were at least two active transport systems for glucose: one Na^+-dependent and the other Na^+-independent. How important the latter is under physiological conditions is difficult to say, as in the normal circumstances of a meal, the intestinal lumen contains large quantities of Na^+ ions from pancreatic and intestinal secretions.

It has been suggested that the "kinetic advantage" observed under in vitro conditions is the probable explanation for the fact that hardly any glucose is found in the intestinal bulk phase fluid during the hydrolysis of luminal sucrose in human perfusion studies [79]. Cook [80] found that the rate of glucose absorption in perfusion studies of proximal jejunum in six adult Zambians was greater from a maltose solution (100 mM) than from a glucose solution (200 mM) but a number of authors in previous studies did not find any significant differences [81,82], nor any kinetic advantage of hexose absorption using other disaccharides [79,83]. Thus the importance of the kinetic advantage of hexose absorption using disaccharides is not clearly defined for in vivo studies. Recent work with glucose–galactose malabsorption patients found that the monosaccharides from lactose hydrolysis in their intestines did not have a hydrolase-related transport system to compensate for the primary defect [84].

Jones et al. [85] investigated the absorption of glucose from corn starch hydrolysates in the intubated proximal human jejunum. The absorption of glucose was compared using three isocaloric and isotonic sugar solutions containing 140 mM glucose, 70 mM maltose and a partial amylase-hydrolysate of starch (52% of its glucose content comprising glucose polymers more than 10 glucose units long). The absorption of glucose from the partial hydrolysate and the free glucose solution was similar, but it was significantly faster (+29%) from the maltose. In the partial hydrolysate the mucosal hydrolysis of the higher molecular weight oligosaccharides (glucose units >10) appeared much slower than that of the lower molecular weight fraction, probably because of their slower diffusion through the unstirred layer. As the absorption of glucose was identical between the partial hydrolysate and free glucose, the authors argued that the overall absorption obtained from the partial hydrolysate comprised a "slow" component from the large oligosaccharides, and a "fast" component from the lower oligosaccharides. Indeed, when a complete amylase hydrolysate was used (consisting predominantly of maltose, maltotriose and α-limit dextrins) glucose absorption was significantly faster (+47%) than from isocaloric free glucose. It was suggested that (1–4)- but not (1–6)-linked oligosaccharides liberated during luminal and brush border hydrolysis of dietary starch conferred a kinetic advantage on glucose absorption. Confirmation of this suggestion was obtained when the absorption of glucose from maltotriose and an oligosaccharide mixture containing predominantly four to eight (1–4)-linked glucose molecules was shown to have a kinetic advantage.

Whether this kinetic advantage demonstrated at a single concentration applies over the range of concentrations that would occur during normal digestion of a starch-containing meal cannot be assumed. Thus the importance of the finding to the nutritional/clinical situation is yet to be determined. In the case of oral tolerance tests, there was no evidence of a more rapid absorption of glucose from a partial hydrolysate of starch (caloreen) than from free glucose [86].

Adaptations Induced by Changes in Dietary Carbohydrate

Three basic mechanisms were listed above, whereby changes in the level and type of food entering the alimentary tract induced adaptations in intestinal absorptive functions. These were changes in enterocyte number, function and/or maturation. There is now a large literature about such changes in a variety of animals [87–90]. The following sections deal with studies selected to show the influence of carbohydrate on the intestinal functions previously described.

Influence of Carbohydrate on Mucosal Mass (Enterocyte Number)

When experimental animals are progressively starved, there is a progressive decrease in small intestinal enterocyte population, as exemplified by decreases in mucosal mass (dry and wet weights), enterocyte number, enterocyte column number, and villous height, especially in the jejunum [91]. Animals in this condition can then be fed specific diets or infused with carbohydrate to see what restores the mucosal mass to normal levels. Another useful preparation to assay the effects of luminal solute infusions on mucosal mass is the animal maintained on total parenteral nutrition (TPN). With this model the specific effects of the luminal solute can be examined uncomplicated by the effects of starvation on general metabolism. In experimental animals it is known that small intestinal mucosal atrophy occurs in TPN [92,93]. Thus, from studies on both starved and TPN-maintained animals the concept has arisen that luminal factors, possibly nutrients and secretions, are necessary to maintain normal small intestinal enterocyte numbers.

The number of enterocytes is important for intestinal function, because the absorptive capacity of the small bowel is partly dependent on the number present. In transport kinetic terms, the J_{max} per unit length is influenced by the number of enterocytes. However, the J_{max} per unit weight, per unit DNA (a measure of cell number), or per unit area may not be so influenced, as this is now an assessment of function per "unit" cell. Thus, different interpretations of the same absorption data can occur depending upon what "structural" basis the data is calculated. The problem of how to normalise data has bedevilled the field, and has been discussed a number of times [94–96]. In some respects it is a problem of orientation of the study. For a nutritionist the focus of attention will be on the total amount of substances absorbed, and thus transfer kinetics will be based on intestinal length, or even on whole areas, or the whole small bowel. For a cell biologist, or physiologist, the focus is finer, and the kinetics will be calculated on bases that indicate function per cell. Conflicts arise most often when the nutritional conditions employed induced large phenotypic or anatomical changes in the intestine.

The animal most used for assaying dietary factors on mucosal mass has been the rat. Infusion of dietary sugars, and even of non-metabolised, actively transferred sugars (such as 3-O-methylglucose) and non-actively absorbed sugars into the alimentary tract of TPN rats is known to prevent small bowel atrophy [97]. Thus mucosal growth is stimulated not only by luminal nutrition, but presumably also by osmotic effects of luminal solutes. The mechanism(s) by which the growth is stimulated is unclear; both local factors and enterotrophic hormones have been proposed [98]. Recently the midgut infusions of sucrose, maltose and lactose have been compared with those of glucose, galactose and fructose [99]. In TPN rats it was found that the growth threshold for disaccharide-induced mucosal adaptation was lower than that for the monosaccharides. The authors thought that the "work of hydrolysis" of the disaccharides triggered a greater growth response than the absorption of free monosaccharides. Abolishing disaccharide hydrolysis by an enzyme inhibitor apparently prevented mucosal growth. No consideration was given in the discussion as to whether the kinetic advantage of hexoses produced from the disaccharides, previously described, could play any part in the enhancement.

One obvious question about these studies is how relevant the results are in relation to the response of the human intestine to reduced or absent luminal nutrition. In obese humans who are starved for weight loss, there do not appear to be the same atrophic structural changes or reduction in the enterocyte population as in fasting rats [100], but, of course, the models are not identical because the starved obese humans can utilise their considerable fat stores, and are thus not so depleted of energy as the rats. Other studies on prolonged isolation of the human small bowel from its continuity [101], and on prolonged parenteral nutrition [102], also show a lack of atrophic changes. There is evidence, however, that luminal nutrition can play a role in maintaining enterocyte mass in human small intestine [98].

Influence of Carbohydrate Intake on Intestinal Digestive Enzymatic Functions

Since the work of Pavlov it has been known that modifications in the diet of experimental animals can change the levels of digestive enzymes. The literature is extensive, and has been reviewed a number of times [109–112]. For the purpose of this chapter it is necessary only to indicate briefly those aspects of the field that have a bearing on intestinal absorptive function.

In general, the substrate of the digestive enzyme increases the activity of the enzyme. The feeding of specific carbohydrates, or carbohydrate-rich diets, to animals is known to increase the amount of disaccharidase in the brush border, and in some situations even the specific activity. Thus, feeding increased amounts of sucrose increases sucrase activity by *de novo* synthesis [113,114]. In man, high dietary intake of sucrose or fructose, but not glucose, increases sucrase and maltase, but not lactase, activity [11,115,116]. Lactase does not appear to be regulated by lactose [117]. In adult rats, the feeding of a high starch diet increases the capacity of the intestine to hydrolyse lactose and absorb the constituent monosaccharides [118]. The level of carbohydrate in the diet can thus influence intestinal carbohydrase activity, such that increases will potentiate, and decreases reduce, its digestive/absorptive potential [119]. As the monosaccharides released by disaccharide hydrolysis act as inhibitors of the disaccharidases [2], the greater the amount of monosaccharide formed, the greater will be the inhibition of the carbohydrase, reducing its activity; this mechanism thus creates a local negative feedback system. The levels of dietary carbohydrate, because of their effects on carbohydrase levels, will thus secondarily influence intestinal transport functions for hexoses.

Dietary carbohydrate also influences the levels and activity of metabolic enzymes inside enterocytes. Studies have shown changes in glycolytic enzymes, galactose-metabolising enzymes, fructose-metabolising enzymes and in hexokinase [115,116,120–122] in man and animals, when the dietary or administered carbohydrate has been manipulated.

The field is complex. An example is the case of hexokinase, the rate-limiting enzyme of the glycolytic pathway of enterocytes. This it is present in the enterocytes in soluble and in particulate-bound activities [123,124]. The soluble hexokinase is suggested to be involved in lactate production. The binding of the enzyme is apparently controlled by diet. Glucose administration causes its solubilisation: a rise in the hexokinase activity of the particulate-free fractions prepared from the glucose-exposed enterocytes is observed [125]. Further studies [126] showed that this solubilisation occurred also with 3-*O*-methylglucose (a non-metabolised but actively transferred sugar) and with mannose passively transferred but metabolised to the same level as occurred with glucose, while sorbitol passively transferred and non-metabolised, and 2-deoxyglucose passively transferred and metabolised caused smaller increases. The mechanism causing the solubilisation of the enzyme is thus apparently not due to active transport nor to metabolism (phosphorylation) of the sugars. It was suggested that the phenomenon was due to a rise in intracellular sugar concentration, but this must be very low, because sorbitol entry into the cell is usually very low.

Perhaps the perspective of the enzymologist, who assumes that the dietary substrates must be acting directly on the enterocytes, is too limited. Changes in enzymatic distribution inside the enterocytes could be due to release of hormonal or paracrine agents, or perhaps to activation of neural arcs via receptors in the gut wall. Osmotic effects from the lumen could perhaps cause changes in osmoreceptors that would modulate neurotransmitter release around the enterocytes, and cause changes in enzyme distribution, perhaps by cytosol changes in cellular calcium. Experiments on changes in intracellular enzymes should perhaps be repeated in the presence of known neural blocking agents, or antagonists such as atropine, tetrodotoxin, hexamethonium and alpha blockers.

One of the disappointing features of the literature on intestinal enzymatic adaptation is that, despite the many published papers, the role of the observed changes in enzymes in the physiological condition of the enterocyte is usually not clearly understood, and most authors are left either to speculate as I have done above, or to ignore the physiological implications of the changes. There is much to learn about the relation between enzyme changes and enterocyte function.

Influence of Carbohydrate on Absorptive Functions

That the level of dietary carbohydrate can influence the absorption of sugars has been known for over 50 years [39,127]. Other more recent studies have examined the effects of feeding often large concentrations of specific sugars or glucalogues to fasting animals, to assess their effects on hexose absorption. Roy and Dubois [128], for example, infused the duodenum of fasted conscious rats with either the glucalogue 3-O-methylglucose, or glucose, or fructose, for 48 hours, and then tested their effects on the absorption of 3-O-methylglucose in an isolated, perfused jejunum removed from the rats. There was an increase in the glucalogue absorption induced by glucose and fructose but not by the glucalogue *per se*. It was suggested that the active transport mechanism for the glucalogue could only be induced (non-specifically) by metabolisable monosaccharide. There was no apparent relation between the levels of hexokinase and the absorptions recorded.

The problem with this type of study is that the transfer mechanism was not examined kinetically, and the only mechanism examined was that for the glucalogue, which may or may not be identical to that for the dietary hexoses. Because of this possibility (see previous sections on multiple hexose carriers), Debnam and Levin [26] investigated the effects of allowing fasted rats to drink isotonic solutions of various sugars (glucose, galactose and the glucalogue α-methylglucoside) for varying periods of time up to a maximum of 3 days. They assessed the absorption of glucose, galactose and α-methylglucoside in the various groups of rats fed the sugars, and in the 3-day-fasted rats, using their technique that allowed the active-transfer electrogenic component of absorption to be assessed using the parameters of apparent K_m and V_{max} [24]. Cleland [129] has argued that when the values for K_m and V_{max} for a mechanism exposed to changed conditions are compared. If they alter in different directions then this should constitute strong evidence that a real change in the kinetic parameters for the mechanisms has taken place. Theoretically there are nine different responses that can take place in the K_m and J_{max} (V_{max}) of a transfer mechanism exposed to a change in conditions [130]. These are shown in Table 5.3. Highly significant changes in both parameters were observed by Debnam and Levin [26], but the changes were more complex than had previously been suspected. Despite the complexity, a number of conclusions could be drawn.

In relation to the "apparent K_m" values glucose and galactose maintained their own K_m, but α-methylglucoside did not. Glucose did not maintain the apparent K_m for galactose, and only partially maintained that for α-methylglucoside. Galactose maintained the K_m for glucose and galactose, but not that for α-methylglucoside. In the case of the V_{max}, glucose and galactose feeding maintained the V_{max} at the fed level, but α-methylglucoside did not increase the

Table 5.3. The nine theoretically possible changes in K_m and J_{max} of a transfer mechanism that can occur when it is exposed to a change in conditions

Type	1	2	3	4	5	6	7	8	9
K_m	↓	↑	↓	↑	0	0	0	↑	↓
J_{max}	↓	↑	↑	↓	0	↑	↓	0	0

From Levin [91] and Cleland [129].
Changes are designated as ↑ for an increase, ↓ for a decrease and 0 for no alteration.

V_{max} for glucose above that of the fasted rats. Surprisingly, in the case of galactose, none of the fed sugars appeared able to restore its V_{max} to the fed level. With α-methylglucoside, only the feeding of glucose could maintain its level. These changes in V_{max} induced by the various feeding regimes could not be explained by changes in the morphology of the intestine, indicating that the changes occurred in the enterocytes rather than in their numbers, or in anatomical changes in the gut. Because of the long time scale over which the changes took place (at least 12 hours), and the differential changes observed in the V_{max} values, it was argued that changes in enterocyte hexokinase, which could alter the energy disposition to the active transfer mechanisms, could not be the cause of the adaptations observed. The complex changes in the kinetic parameters for the various sugars gave rise to the conclusion that more than one carrier (transfer mechanism) must exist for active hexose absorption in rat jejunum.

Since this study a number of other investigators have examined the effects of different dietary sugars as inducers of sugar transfer mechanisms [93,131–134]. The most comprehensive studies [135] involved feeding dietary rations containing 55% of either glucose, galactose, fructose, 3-O-methylglucoside or maltose to mice, and then measuring the uptake of the various sugars in vitro. The five were found to be of similar potency as inducers of aldohexose transport, although galactose and fructose may have been more potent than glucose. The results were also interpreted to indicate that multiple aldohexose transporters existed in mouse jejunum with overlapping specificities.

From another study in mice [136] the feeding of a high carbohydrate diet (55% sucrose) induced an increase in the V_{max} of active glucose transport of approximately twofold, compared with a low carbohydrate diet, which was the same as the proportional increase in the specific binding of phloridzin. This was interpreted as indicating that the sucrose diet induced an increase in glucose transporters of the individual enterocytes. Clearly, as sucrose was fed in this experiment, and as it was presumably split in the mouse gut to glucose and fructose, it is not possible to say which monosaccharide, or both, was the cause of the induction of transport sites. The exact mechanism by which the increased transport sites are induced is not known, as the mice were on the diet for 2 weeks. It could be through changes in enterocyte development [137], or direct induction of mature cells or of crypt cells, or all three.

The control of insulin secretion by ingested and absorbed glucose is well known, but study of possible influences that the hormone may have on enterocyte function has not been a very productive field until quite recently. Early studies in the rat, while showing that experimental diabetes produced functional and anatomical changes in the small intestine, were not able to demonstrate direct regulation of the metabolism of the enterocyte by insulin [138]. This led to the accepted view that the gut was an insulin-insensitive tissue. Kellet et al. [139], however, re-examined the action of insulin on enterocyte function in the rat and found that acute decreases of insulin, induced by injection of anti-insulin serum in the anaesthetised rat, dramatically reduced the production of lactate and the metabolism of glucose by the small intestine to 28% and 21% of the normal levels respectively. The activities of the mucosal enzymes pyruvate dehydrogenase and phosphofructokinase (claimed to be the rate-limiting enzyme of glycolysis in the enterocyte) were correspondingly decreased. The changes could be reversed by injecting insulin into the treated rat. The absorption of the actively transported, but not metabolised, glucalogue 3-O-methylglucose was unaffected by the insulin anti-serum blockade or by insulin injection. Glucose absorption and transport, however, was affected in the short term by insulin indirectly and by changes in enterocyte glucose metabolism. Whether this moment-to-moment control of enterocyte metabolism will be found in the unanaesthetised preparation or in man has yet to be ascertained. Ingested glucose will affect insulin levels not only by its direct action on the pancreatic cells but also by its specific action in releasing gastric inhibitory peptide (GIP) from the K cells in the duodenum [140]. The released GIP stimulates insulin secretion especially when the plasma levels of glucose are high. Glucose also releases other glucagon-like peptides from the intestine (GLP-1 and GLP-2 and derivatives) which also increase the secretion of insulin [141]. Their role in indirectly influencing and controlling enterocyte function is unknown.

Conclusion

The aim of this chapter was to review the absorption of the digestion products of starches and sugars, and how they may alter the kinetics of sugar absorption. The idea was to compare the effects of starch, disaccharides and monosaccharides. It is clear from the literature examined, however, that there are few studies that have used such a protocol. Most of the data reported have come from studies using monosaccharide or disaccharide feeding regimes, which compared the effects with normally fed or fasted intestine. Rarely has the goal been to compare the effect of monosaccharides with disaccharides or starch. Zamcheck [142] pointed out some 28 years ago that even then the nutritional literature was "replete with references to the diverse effects of diets of varied composition on growth, tissues, enzymes and diseases of man and animals", but that little was "known about the influence of these diets on the metabolism of the gut mucosa itself". He added that the "integration of the myriad phenomena occurring at the gut–lumen–mucosal cell interface was impossible in the present state of knowledge". While our information on the effects of carbohydrate diets on gut function has increased considerably since Zamcheck's review, it is a sad commentary that the echoes of his last line still resound. We have many more facts, but we are still a long way from being able to integrate them into a coherent picture.

References

1. Lipkin M (1987) Proliferation and differentiation of normal and diseased gastrointestinal cells. In: Johnson LR (ed) Physiology of the gastrointestinal tract, 2nd edn. Raven Press, New York, pp 255–289
2. Gray GM (1981) Carbohydrate absorption and malabsorption. In: Johnson LR (ed) Physiology of the gastrointestinal tract, 2nd edn. Raven Press, New York, pp 1063–1072
3. Fogel MR, Gray GR (1973) Starch hydrolysis in man: an intraluminal process not requiring membrane digestion. J Appl Physiol 35:263–267
4. Ugolev AM, Delaey P (1973) Membrane digestion – a concept of enzymic hydrolysis on cell membranes. Biochim Biophys Acta 300:105–128
5. Miller D, Crane RK (1961) The digestive function of the epithelium of the small intestine. I. An intracellular locus of disaccharide and sugar phosphate ester hydrolysis. Biochim Biophys Acta 52:281–293
6. Crane RK (1966) Structural and functional organization of an epithelial cell brush border. In: Warren KB (ed) Symposia of the International Society for Cell Biology. Academic Press, New York, pp 71–102
7. Semenza G (1968) Digestion and absorption of sugars. Mod Probl Pediatr 11:32–47
8. Crane RK (1968) Digestive–absorptive surface of the small bowel mucosa. Ann Rev Med 19:57–68
9. Sundqvist T, Magnusson KE, Sjodahl R, Stjernstrom I, Tagesson C (1980) Passage of molecules through the wall of the gastrointestinal tract. II. Application of low molecular weight polyethylene glycol and a deterministic mathematical model for determining intestinal permeability in man. Gut 21:208–214
10. Dahlqvist A, Semenza G (1985) Disaccharidases of small intestinal mucosa. J Pediatr Gastroenterol Nutr 4:857–867
11. Alpers DH (1987) Digestion and absorption of carbohydrate and proteins. In: Johnson LR (ed) Physiology of the gastrointestinal tract. 2nd edn. Raven Press, New York, pp 1469–1487
12. Crane RK (1960) Intestinal absorption of sugars. Physiol Rev. 40:789–825
13. Schultz SG, Curran PF (1970) Coupled transport of sodium and organic solutes. Physiol Rev 50:637–718
14. Kimmich GA (1981) Intestinal absorption of sugar. In: Johnson LR (ed) Physiology of the gastrointestinal tract. Raven Press, New York, pp 1035–1061
15. Hopfer U (1987) Membrane transport mechanisms for hexoses and amino acids in the small intestine. In: Johnson LR (ed) Physiology of the gastrointestinal tract, 2nd edn. Raven Press, New York, pp 1499–1526
16. Semenza G, Kessler M, Hosang M, Weber J, Schmidt U (1984) Biochemistry of the Na^+, D-glucose cotransporter of the small intestinal brush-border membrane. Biochim Biophys Acta 779:343–379
17. Hediger MA, Coady MJ, Ikeda TS, Wright E (1987) Expression cloning and cDNA sequencing of the Na^+/glucose transporter. Nature 330:379–381
18. Stevens BR, Kneer C (1988) Lanthanide-stimulated glucose and proline transport across rabbit intestinal brush-border membranes. Biochim Biophys Acta 942:205–208
19. Levin RJ (1979) Fundamental concepts of structure and function of the intestinal epithelium. In: Duthie HL, Wormsley KG (eds) Scientific basis of gastroenterology. Churchill Livingstone, Edinburgh, pp 308–337
20. Read NW, Holdsworth CD, Levin RJ (1974) Electrical measurement of intestinal absorption of glucose in man. Lancet I:624–627
21. Read NW, Barber DC, Levin RJ, Holdsworth CD (1977) Unstirred layer and kinetics of electrogenic glucose absorption in the human jejunum in situ Gut 18:865–876
22. Igarashi Y, Himukai M, Konno T (1981) In vivo recording of the glucose- and disaccharide-evoked potentials from the human jejunum in infancy. Eur J Pediatr 135:255–260
23. Luppa D, Hartenstein H, Muller F (1987) Relation between microvilli membrane potential and glucose transport capacity of rat small intestine. Biomed Biochim Acta 5:341–348
24. Debnam ES, Levin RJ (1975) An experimental method of identifying and quantifying the active transfer electrogenic component from the diffusive component during sugar absorption measured in vivo. J Physiol (Lond) 246:181–196
25. Wilson TH (1962) Intestinal absorption. Saunders, Philadelphia, pp 91–98
26. Debnam ES, Levin RJ (1976) Influence of specific dietary sugars on the jejunal mechanisms for

glucose, galactose, and α-methylglucoside absorption: evidence for multiple sugar carriers. Gut 17:92–99

27. Honegger P, Semenza G (1973) Multiplicity of carriers for free glucalogues in hamster small intestine. Biochim Biophys Acta 318:390–410

28. Thomson ABR, Gardner MLG, Atkins GL (1987) Alternate models for shared carriers or a single maturing carrier in hexose uptake into rabbit jejunum in vitro. Biochim Biophys Acta 903:229–240

29. Stevens BR, Kaunitz JD, Wright E (1984) Intestinal transport of amino acids and sugars: advance using membrane vesicles. Ann Rev Physiol 46:417–433

30. Malo C (1988) Kinetic evidence for heterogeneity in Na$^+$-D-glucose cotransport system in the normal human fetal small intestine. Biochim Biophys Acta 938:181–188

31. Harig JM, Rajendran VM, Barry A, Adams MB, Ramaswamy K (1986) Regional variations in transport of D-glucose and L-leucine in human small intestinal brush-border membrane vesicles (BBMV). Gastroenterology 90:1450

32. Syme G, Levin RJ (1980) The validity of assessing changes in intestinal absorption mechanisms for dietary sugars with non-metabolizable analogues (glucalogues). Br J Nutr 43:435–443

33. Levin RJ (1976) Digestion and absorption of carbohydrate – from embryo to adult. In: Boorman KN, Freeman BM (eds) Digestion in the fowl. British Poultry Science Ltd., Edinburgh, pp 63–116

34. Csaky TZ, Esposito G (1969) Osmotic swelling of intestinal epithelial cells during active sugar transport. Am J Physiol 217:753–755

35. Nissim JA (1964) Mechanism of intestinal absorption: the concept of a spectrum of intracellular plasma. Nature 204:148–151

36. Murer H, Hopfer V, Kinne-Saffran E, Kinne R (1974) Glucose transport in isolated brush-border and lateral-basal plasma-membrane vesicles from intestinal epithelial cells. Biochim Biophys Acta 345:170–179

37. Maenz DD, Cheeseman CI (1986) Effect of hyperglycaemia on D-glucose transport across the brush-border and basolateral membrane of rat small intestine. Biochim Biophys Acta 860:277–285

38. Karasov WH, Debnam ES (1987) Rapid adaptation of intestinal glucose transport: a brush-border or basolateral phenomenon? Am J Physiol 253:G54–61

39. Donhoffer S (1935) Über die elektive Resorption der Zucker. Arch exp Pathol Pharmakol 177:689–692

40. Barany EH, Sperber E (1942) A theoretical and experimental study of intestinal glucose absorption. Ark Zool 34:1–31

41. Fisher RB, Parsons DS (1953) Glucose movements across the wall of the rat small intestine. J Physiol (Lond) 119:210–223

42. Smyth DH (1971) Intestinal transfer mechanisms, measurements and analogies. J Clin Pathol 24 [Suppl 5]:1–9

43. Dietschy JM (1970) Difficulties in determining valid rate constants for transport and metabolic processes. Gastroenterology 58:863–874

44. Winne D (1973) Unstirred layer, source of biased Michaelis constant in membrane transport. Biochim Biophys Acta 298:27–31

45. Winne D (1977) Correction of the apparent Michaelis constant, biased by an unstirred layer, if a passive component is present. Biochim Biophys Acta (Biomembranes) 464:118–126

46. Wilson FA, Dietschy JM (1974) The intestinal unstirred layer: its surface area and effect on active transport kinetics. Biochim Biophys Acta 363:112–126

47. Dugas MC, Ramaswamy K, Crane RK (1975) An analysis of the D-glucose influx kinetics of in vitro hamster jejunum, based on considerations of the mass transfer coefficient. Biochim Biophys Acta 401:486–501

48. Parsons DS (1976) Closing summary, Appendix 2. Unstirred layer. In: Robinson JWL (ed) Intestinal ion transport. MTP Lancaster, pp 407–430

49. Yuasa H, Miyamoto Y, Iga T, Hanano M (1986) Determination of kinetic parameters of a carrier-mediated transport in the perfused intestine by a two-dimensional laminar flow model: effects of the unstirred layer. Biochim Biophys Acta 856:219–230

50. Batt RM, Peters TJ (1976) Absorption of galactose by the rat small intestine in vivo: proximal–distal kinetic gradients and a new method to express absorption per enterocyte Clin Sci Mol Med 50:499–509

51. Schedl HP, Clifton JA (1963) Kinetics of absorption in man: normals and patients with sprue. In: Schmid E et al. (eds) Proceedings of world congress on gastroenterology. S Karger, Basel, p 728

52. Levin RJ, Mitchell MA, Barber DC (1983) Comparison of jejunal and ileal absorptive functions for glucose and valine in vivo – a technique for estimating real K_m and J_{max} in the domestic fowl. Comp Biochem Physiol 74A:961–966

53. Anderson BW, Kneip JM, Levine AS, Levitt MD (1988) Physiological measurements of luminal stirring. Gastroenterology 94:A7 (abstr)

54. Darlington WA, Quastel JH (1963) Absorption of sugars from isolated surviving intestine. Arch Biochem Biophys 43:194–207

55. Kiyasu JY, Chaikoff IL (1957) On the manner of transport of absorbed fructose. J Biol Chem 224:935–939

56. Wilson TH, Vincent JN (1955) Absorption of sugars in vitro by the intestine of the golden hamster. J Biol Chem 216:851–866

57. White LW, Landau BR (1965) Sugar transport and fructose metabolism in human intestine in vivo. J Clin Invest 44:1200–1213

58. Miller M, Raig JW, Drucker WR, Woodward H (1956) The metabolism of fructose in man. Yale J Biol Med 29:335–360

59. Gracey M, Burke V, Oshin A (1972) Active intestinal transport of D-fructose. Biochim Biophys Acta 266:397–406

60. Macrae AR, Neudoerffer TA (1972) Support for the existence of an active transport mechanism of fructose in the rat. Biochim Biophys Acta 288:137–144

61. Meeuwisse GW, Melin K (1969) Glucose–galactose malabsorption – a clinical study of 6 cases. Acta Paediatr Scand [Suppl] 188:1–24

62. Andersson DEH, Nygren A (1978) Four cases of long-standing diarrhoea and colic pains cured by fructose-free diet – pathogenetic discussion. Acta Med Scand 203: 87–92

63. Rumessen JJ, Gudmand-Hoyer E (1986) Absorption capacity of fructose in healthy adults. Comparison with sucrose and its constituent monosaccharides. Gut 27:1161–1168

64. Holdsworth CD, Dawson AM (1964) The absorption of monosaccharides in man. Clin Sci 27:371–379

65. Forster H (1972) Views dissenting with the "gradient hypothesis" Intestinal sugar absorption: studies in vivo and in vitro. In: Heinz E (ed) Na-linked transport of organic solutes. Springer, Berlin Heidelberg New York, pp 134–139

66. Silk DBA, Dawson AM (1979) Intestinal absorption of carbohydrate and protein in man. In: Crane RK (ed) International review of physiology. Gastrointestinal physiology III, vol 19. University Park Press, Baltimore, p 163

67. Dowling RH (1982) Discussion after paper on functional and structural responses of the dog small intestine to resection. In: Robinson JWL et al. (eds) Mechanisms of intestinal adaptation. MTP, Lancaster, pp 409–411

68. Borgstrom B, Dahlqvist A, Lundh G, Sjovall J (1957) Studies of intestinal digestion and absorption in the human. J Clin Invest 36:1521–1536

69. Cole AS (1961) Soluble material in the gastrointestinal tract of rats under normal feeding conditions. Nature 191:502–503

70. Murakami E, Saito M, Suda M (1977) Contribution of diffusive pathway in intestinal absorption of glucose in rat under normal feeding conditions. Experientia 33:1469–1470

71. Ling KY, Im WB, Faust RG (1981) Na$^+$-independent sugar uptake by rat intestinal and renal brush border and basolateral membrane vesicles. Int J Biochem 13:693–700

72. Dawson DJ, Burrows PC, Lobley RW, Holmes RW (1987) The kinetics of monosaccharide absorption by human jejunal biopsies: evidence for active and passive processes. Digestion 38:124–132

73. Evans L, Grasset E, Heyman M, Dumontier AM, Bean J-P, Desjeux J-F (1985) Congenital selective malabsorption of glucose and galactose. J Pediatr Gastroenterol 4:878–886

74. Pappenheimer JR, Reiss KZ (1987) Contribution of solvent drag through intercellular junctions to absorption of nutrients by the small intestine of the rat. J Membrane Biol 100:123–136

75. Pappenheimer JR (1987) Physiological regulation of transepithelial impedance in the intestinal mucosa of rats and hamsters. J Membrane Biol 100:137–148

76. Madara JL, Pappenheimer JR (1987) Structural basis for physiological regulation of paracellular pathways in intestinal epithelia. J Membrane Biol 100:149–164

77. Parsons DS, Pritchard JS (1971) Relationship between disaccharide hydrolysis and sugar transport in amphibian small intestine. J Physiol (Lond) 212:299–319

78. Caspary WF (1972) Evidence for a sodium-independent transport system for glucose derived from disaccharides. In: Heinz E (ed) Na-linked transport of organic solutes. Springer, Berlin Heidelberg New York, pp 99–108

79. Gray GM, Ingelfinger FJ (1965) Intestinal absorption of sucrose in man: the site of hydrolysis and absorption. J Clin Invest 44:340–378
80. Cook GC (1973) Comparison of the absorption rates of glucose and maltose in man in vivo. Clin Sci 44:425–428
81. McMichael HB, Webb J, Dawson AM (1967) The absorption of maltose and lactose in man. Clin Sci 33:135–145
82. Gray GM, Santiago NA (1966) Disaccharide absorption in normal and diseased human intestine. Gastroenterology 51:489–498
83. Cook GC (1970) Comparison of the absorption and metabolic products of sucrose and its monosaccharides in man. Clin Sci 38:687–697
84. Beyreiss K, Hoepffner W, Scheerschmidt G, Muller F (1985) Digestion and absorption rates of lactose, glucose, galactose, and fructose in three infants with congenital glucose–galactose malabsorption: perfusion studies. J Pediatr Gastroenterol Nutr 4:887–892
85. Jones BJM, Brown BE, Loran JS et al. (1983) Glucose absorption from starch hydrolysates in the human jejunum. Gut 24:1152–1160
86. Wahlquist ML, Wilmshurst EG, Murton CR, Richardson EN (1978) The effect of chain length on glucose absorption and the related response. Am J Clin Nutr 31:1998–2001
87. Williamson RCN, Chir M (1978) Intestinal adaptation: structural, functional and cytokinetic changes. N Engl J Med 298:1393–1402
88. Williamson RCN, Chir M (1978) Intestinal adaptations: mechanisms of control. N Engl J Med 298:1444–1450
89. Williamson RCN (1982) Intestinal adaptation: factors that influence morphology. In: Polak JM et al. (eds) Structure of the gut. Glaxo Group Research Ltd., Ware, UK, pp 337–345
90. Karasov WH, Diamond J (1987) Adaptation of intestinal nutrient transport. In: Johnson LR (ed) Physiology of the gastrointestinal tract, 2nd edn. Raven Press, New York, pp 1489–1497
91. Levin RJ (1984) Intestinal adaptation to dietary change as exemplified by dietary restriction studies. In: Batt RM, Laurence TLJ (eds) Function and dysfunction of the small intestine. (Proceedings of the second George Durrant memorial symposium.) Liverpool University Press, Liverpool, pp 77–93
92. Eastwood GL (1977) Small bowel morphology and epithelial proliferation in intravenously alimented rabbits. Surgery 82:613–620
93. Richter GC, Levine GM, Shiau Y-F (1983) Effects of luminal glucose versus non-nutritive infusates on jejunal mass and absorption in the rat. Gastroenterology 85:1105–1112
94. Levin RJ (1967) Techniques, terminology and parameters in intestinal absorption. Br Med Bull 23:209–212
95. Levin RJ (1982) Assessing small intestinal function in health and disease in vivo and in vitro. Scand J Gastroenterol 17 [Suppl 74]:13–51
96. Mitchell MA, Levin RJ (1981) Amino acid absorption in jejunum and ileum in vivo – a kinetic comparison of function on surface area and regional bases. Experientia 37:265–266
97. Weser E, Tawil T, Fletcher JT (1982) Stimulation of small bowel mucosal growth by gastric infusion of different sugars in rats maintained on total parenteral nutrition. In: Robinson JWL et al. (eds) Mechanisms of intestinal adaptation. MTP, Lancaster, pp 141–149
98. Williamson RCN (1984) Adaptive intestinal hyperplasia. In: Batt RM, Laurence TJL (eds) Function and dysfunction of the small intestine. (Proceedings of the second George Durrant memorial symposium.) Liverpool University Press, Liverpool, pp 55–76
99. Weser E, Babbitt J, Hoban M, Vandeventer A (1986) Intestinal adaptation: different growth responses to disaccharides compared with monosaccharides in rat small bowel. Gastroenterology 91:1152–1527
100. Knudsen KB, Bradley EM, Lecocq FR, Bellamy HM, Welsh JD (1968) Effects of fasting and refeeding on the histology and disaccharidase activity of the human intestine. Gastroenterology 55:46–51
101. Tompkins RK, Waisman J, Watt CMH, Corlin R, Keith R (1977) Absence of mucosal atrophy in human small intestine after prolonged isolation. Gastroenterology 73:1406–1409
102. Guedon C, Shmitz J, Lerebours E, Metayer J, Audran E, Memet J, Colin R (1986) Decreased brush-border hydrolase activities without gross morphologic changes in human intestinal mucosa after prolonged parenteral nutrition of adults. Gastroenterology 90:373–378
103. Eisenthal R, Cornish-Bowden A (1974) The direct linear plot. A new graphical procedure for estimating enzyme kinetic parameters. Biochem J 139:715–720
104. Rask-Madsen J, Gudmand-Hoyer E, Krag E (1976) Electrical measurement of active sugar absorption from the human jejunum. Scand J Gastroenterol 11 [Suppl 38]:46
105. McNeish AS, Ducker DA, Warren IF, Davies IP, Harran MJ, Hughes CA (1979) The influence

of gestational age and size on the absorption of D-xylose and D-glucose from the small intestine of the human neonate. In: Development of mammalian absorptive processes. Excerpta Medica, Amsterdam, pp 267–276 (Ciba foundation symposium 70 (new series))

106. Sparso BH, Luke M, Wium E (1984) Electrogenic transport of glucose in the normal upper duodenum. I. Technique and apparent transport constants. Scand J Gastroenterol 19:561–567

107. Sparso BH, Luke M, Wium E (1984) Electrogenic transport of glucose in the normal upper duodenum. II. Unstirred water layer and estimation of real transport constants. Scand J Gastroenterol 19:568–574

108. Flourié B, Vidon N, Florent CH, Bernier JJ (1984) Effect of pectin in jejunal glucose absorption and unstirred layer thickness in normal man. Gut 25:936–941

109. Grossman MI, Greengard H, Ivy AC (1943) The effect of dietary composition on pancreatic enzymes. Am J Physiol 138:676–682

110. Knox WE, Auerbach VH, Lin ECC (1956) Enzymatic and metabolic adaptations in animals. Physiol Rev. 36:164–254

111. Spencer RP, Knox WE (1960) Comparative enzyme apparatus of the gut mucosa. Fed Proc 19:886–897

112. McCarthy DM, Nicholson JA, Kim YS (1980) Intestinal enzyme adaptation to normal diets of different composition. Am J Physiol 239:G445–G451

113. Cezard JP, Broyart JP, Cuisinier-Gleizes P, Mathieu H (1983) Sucrase–isomaltase regulation by dietary sucrose in the rat. Gastroenterology 84:18–29

114. Riby JE, Kretchmer N (1984) Effect of dietary sucrose on synthesis and degradation of intestinal sucrase. Am J Physiol 246:G757–G763

115. Rosensweig NS (1972) Dietary sugars and intestinal enzymes. J Am Diet Assoc 60:483–486

116. Rosensweig NS, Herman RH, Stifel FB (1971) Dietary regulation of small intestinal enzyme activity in man. Am J Clin Nutr 24:65–69

117. Gilat T, Russo S, Gelman-Malachi E, Aldor TAM (1981) Lactase in man: a non-adaptable enzyme. Gastroenterology 62:1125–1127

118. Tsuboi KK, Kwong LK, Yamada K, Sunshine P, Koldovsky O (1985) Nature of elevated rat intestinal carbohydrase activities after high-carbohydrate diet feeding. Am J Physiol 249:G510–G518

119. Leichter J, Goda T, Bhandari SD, Bustamente S, Koldovsky O (1984) Relation between dietary-induced increase of intestinal lactase activity and lactose digestion and absorption in adult rats. Am J Physiol 247:G729–G735

120. Espinoza J, Hritz A, Kaplan R, Clark SB, Rosensweig NS (1975) Regional variation in glycolytic adaptation to dietary sugars in rat small intestine. Am J Clin Nutr 28:453–458

121. Stifel FB, Herman RH, Rosensweig NS (1968) Dietary regulation of galactose metabolising enzymes: adaptive changes in rat jejunum. Science 162:692–693

122. Crouzoulon G (1979) Enzymes of fructose metabolism in the intestinal mucosa of the rat (*Rattus norvegicus*). Localization along the small intestine; consequences of dietary fructose. Comp Biochim Physiol 62A:789–796

123. Srivastava LM, Hubscher G (1966) Glucose metabolism in the mucosa of the small intestine. Biochem J 100:458–466

124. Anderson JW, Herman RH, Tyrrell JB, Cohn RM (1971) Hexokinase: a compartmented enzyme. Am J Clin Nutr 24:642–650

125. Mayer RJ, Shakespeare P, Hubscher G (1970) Glucose metabolism in the mucosa of the small intestine. Biochem J 116:43–48

126. Jones GM, Mayer RJ (1973) Glucose metabolism in the rat small intestine: the effect of glucose analogues on hexokinase activity. Biochem J 132:125–128

127. Westenbrink HGK (1934) Über die Anpassung der Darmresorption an die Zusammensetzung der Nahrung. Arch Neerland Physiol de l'Homme et des Animaux 19:563–583

128. Roy CC, Dubois RS (1972) Monosaccharide induction of 3-O-methyl glucose transport through rat jejunum. Proc Soc Exp Biol Med 139:883–886

129. Cleland WW (1967) The statistical analysis of enzyme kinetic data. Adv Enzymol 29:1–32

130. Levin RJ, Mitchell M (1982) Intestinal adaptations to fasting – use of corrected kinetic parameters to assess responses of jejunal and ileal absorption in vivo. In: Robinson JWL et al. (eds) Mechanisms of intestinal adaptation. MTP, Lancaster, pp 103–110

131. Bode C, Eisenhardt JM, Haberich FJ, Bode JC (1981) Influence of feeding fructose on fructose and glucose absorption in rat jejunum and ileum. Res Exp Med 179:163–168

132. Debnam ES (1985) Adaptation of hexose uptake by the rat jejunum induced by perfusion of sugars into the distal ileum. Digestion 31:25–30

133. Marrias DA, Mayer RJ (1973) Metabolism of fructose in the small intestine. I. The effect of

fructose feeding on fructose transport and metabolism in the rat small intestine. Biochim Biophys Acta 291:531–537

134. Ghishan FK, Borowitz S, Menzies C, Greene HL (1985) Adaptation of D-glucose transport by dietary carbohydrate. Nutr Res 5:221–225

135. Solberg DH, Diamond JM (1987) Comparison of different dietary sugars as inducers of intestinal sugar transporters. Am J Physiol 252:G574–G584

136. Ferraris RP, Diamond JM (1986) Use of phlorizin binding to demonstrate induction of intestinal glucose transporters. J Membrane Biol 94:77–82

137. Smith MW (1985) Expression of digestive and absorptive function in differentiating entero-cytes. Ann Rev Physiol 47:247–260

138. Levin RJ (1969) The effects of hormones on the absorptive, metabolic and digestive functions of the small intestine. J Endocrinol 45:315–348

139. Kellet GL, Jamal A, Robertson JP, Wollen N (1984) The acute regulation of glucose absorption, transport and metabolism in rat small intestine by insulin in vivo. Biochem J 219:1027–1035

140. Syke S, Morgan CM, English J, Marks V (1980) Evidence for the preferential stimulation of GIP secretion in the rat by actively transported carbohydrates and other analogues. J Endocrinol 85:201–207

141. Orskov C, Holst JJ, Knuhtsen S, Baldissera FGA, Poulsen SS, Nielsen O (1986) Glucagon-like peptides GLP-1 and GLP-2, predicted products of the glucagon gene, are secreted separately from pig small intestine but not pancreas. Endocrinology 119:1467–1475

142. Zamcheck N (1960) Dynamic interaction among body nutrition, gut mucosal metabolism and morphology and transport across the mucosa. Fed Proc 19:855–864

Chapter 6

The Role of Dietary Starches and Sugars and Their Digestive End Products on Gastrointestinal Hormone Release

T. L. Peeters

Gastrointestinal Hormones

It is often forgotten that it was the observations of Bayliss and Starling [1] concerning the effect of extracts of gastrointestinal mucosa upon gastric and pancreatic secretion that gave the strongest impetus to the development of the concept of hormonal regulation, and to the emergence of the discipline of the endocrinology. It was they who coined the term "hormone". However, at the present time endocrinologists often have limited interest in gastrointestinal hormones, which are mostly studied by a separate group of scientists. This is exemplified by the fact that the two groups publish in different journals and attend different meetings.

There are several reasons for this. The endocrine cells in the gut are not concentrated, but are diffusely distributed along the whole length of the gastrointestinal tract. This is a serious experimental disadvantage for, for example, the isolation of active substances, or the preparation of models of deficiency, so that the study of "more convenient" endocrine glands, such as the thyroid, progressed more rapidly. Furthermore the study of the physiology of the gastrointestinal hormones has led to a widening of the scope of their activities and importance. Thus, some gut hormones were discovered in neurons, not only in the gut but also in other organs and in the peripheral and central nervous system. Gut peptides with a hormonal role may apparently also be neuroendocrine agents, secreted from neurons into the blood stream, or neuromodulators affecting neurotransmission, and they may act as paracrine agents, with secretions and effects limited to neighbouring cells. It is therefore common now to speak of "regulatory peptides" rather than of "gastrointestinal hormones", and to consider them as the active agents of the "diffuse neuroendocrine system of the gut".

However, as the topic of this chapter is the release of gastrointestinal hormones by carbohydrates, the term "gastrointestinal hormones" will be used here in a restricted and somewhat old-fashioned way. The discussion will be limited to those peptides which may be defined as being released in significant amounts from the gastrointestinal tract into the circulation, and the release of which may have an endocrine regulatory function in the processes of digestion and absorption. The first part of this definition is the easiest and most useful one. It leads to the exclusion of peptides originating from the pancreas, and of those peptides released in such small amounts (enkephalins, VIP) that their release is not only difficult to measure in the laboratory, but probably also difficult to detect at distant sites. The second part of the definition has been phrased cautiously, and this is necessary, despite the fact that a hormone should of course have an endocrine role. To establish that a certain function is hormonally regulated is not easy, and therefore substances have to be included which are strictly only "candidate" hormones. However, the second part of the definition also narrows deliberately the scope of the present discussion to the processes of digestion and absorption. The intricate mechanism regulating the supply of glucose to the different organs will not be considered, although they will occasionally have to be referred to.

Based upon similarities in amino acid sequence, the regulatory peptides of the gut can be considered in groups. The gastrin group comprises gastrin and cholecystokinin (CCK); the secretin group consists of secretin, glucagon, glucagon-like peptides (GLPs), gastric inhibitory polypeptide (GIP) and vasoactive intestinal polypeptide (VIP). The enkephalins are sometimes considered as members of the gastrin group because of their similarity to part of the CCK sequence. Motilin, somatostatin and neurotensin bear no resemblance to each other, nor to the members of the gastrin and secretin groups, but pancreatic polypeptide (PP), peptide YY and neuropeptide Y are closely related to each other. However, the discussion will not include glucagon and PP because they are almost exclusively found in the pancreas; VIP, the enkephalins and neuropeptide Y because they are mainly present in neural elements; nor peptide YY and all related peptides, because of their limited interest to the present topic. Multiple molecular forms of several peptides exist corresponding to extensions of the peptide chain, but in the present context it is not important to distinguish it.

The endocrine cells of the gut have distinct morphological and cytochemical characteristics. Most have the same staining properties as the adrenaline-producing chromaffin cells of the adrenal gland, and are therefore called enterochromaffin cells. Ultrastructurally they are characterised by the presence of granules, and this has led to a classification system (G-, D-, I-, etc.) based upon shape, size, staining properties and number of granules. However, immunocyto-chemistry has allowed the definite identification of cells producing specific peptides. Usually endocrine cells are divided into "open" cells, in which part of the cell surface is in direct contact with the gut lumen, and "closed" cells, which are completely surrounded by other cells. Obviously it is more likely that secretion by "open" cells may be induced by the presence of digestive substances in the gut.

Although the endocrine cells are dispersed in the gut, there is a distinct grouping according to their secretory product. Thus endocrine cells secreting gastrin (G-cells) are mainly found in antral and duodenal mucosa; the CCK and secretin cells (respectively I-cells and S-cells) in the duodenum and jejunum; and the somatostatin cells, or D-cells, in all parts of the gut. The GIP- secreting K-cells are most prominent in duodenum and jejunum, but are also present in antrum and ileum. The M-cells that secrete motilin are found in the antrum, but are most abundant in the upper duodenum. The cells producing glucagon-like peptides are the L-cells. They are found in ileum, as are the neurotensin or N-cells.

The best-known functions and releasers of these different peptides may be summarised as follows. Gastrin is released by peptides and amino acids, and stimulates gastric acid secretion. CCK is released by digestion products of fat and by protein, and stimulates gall bladder contraction and pancreatic enzyme secretion. Secretin is released by acid, and induces the flow of water and bicarbonate from the pancreas. Somatostatin is released in small amounts post-prandially, and has many inhibitory effects. Neurotensin is released by fats, and its physiological function may well be related to effects on motility. GIP is released by sugars, and some amino acids, and stimulates insulin secretion by the pancreas. And the same applies to the glucagon-like substances. Motilin, is periodically released during fasting by an unknown mechanism, and may help to initiate inter-digestive contractions. Table 6.1 summarises some important facts of the physiology of gut hormones, but readers interested in more detail should consult some of the extensive reviews available on this topic [2].

Table 6.1.   Selected facts for some gut hormones

Peptide	Fundus	Antrum	Duodenum	Jejunum	Ileum	Colon	Pancreas	Gall bladder
Gastrin (17;G)	T H^+ secretion	+++ Peptides, amino acids	++	+				
CCK (33;I)			+++ Fatty acids, amino acids	+++	+		T Enzyme secretion	T Contraction
Secretin (27;S)			+++	+++	+		T Water, HCO_3^- secretion	
Somatostatin (14;D)	+	+	+	+	+			
Neurotensin (13;N)					+++ Fats			
GIP (42;K)		T Reduces H^+	+++ Glucose, amino acids	+++			T Insulin secretion	
GLPs (69,37;L)					+++ Glucose		T Insulin secretion	
Motilin (22;M)	T Contraction	+, T	+++ Fat, periodic during fasting	+				

The number of amino acid residues in the most prominent molecular form and the letter denoting the type of cell containing the peptides are given under the name in parenthesis. The distribution is indicated by plus signs and T denotes the target region. Under the target the best-known function is given; under the plusses the best releasers.

Release of Gastrointestinal Hormones by Carbohydrates

Gastrin

Gastrin was the second gastrointestinal hormone discovered. It has been extensively studied, perhaps because it causes an easily measured effect: the secretion of gastric acid. Gastrin is released after a meal and this release is mostly induced by the direct contact of dietary constituents with the open gastrin cells of the antrum [3]. The magnitude of the response is dependent upon the composition of the meal, but is primarily determined by its protein fraction. Indeed neither fats nor carbohydrates are able to induce gastrin release [4]. It is therefore not surprising that gastrin levels are stable during an oral glucose tolerance test [5]. In the present context gastrin needs no further discussion.

Cholecystokinin

Although the name cholecystokinin (CCK) is almost the only one now used, pancreozymin (PZ) was used as an alternative until it was discovered that both names referred to the same substance. Some authors still refer to CCK-PZ, because in this way the two most important activities of the peptide are denoted: stimulation of gall bladder contraction and stimulation of pancreatic secretion.

Because the pancreas secretes the enzyme amylase, which is of prime importance for the digestion of carbohydrates, a regulatory effect could theoretically be expected by means of CCK. However, this does not seem to be the case. Although CCK is certainly able to induce amylase secretion (the best bioassay for CCK uses the amylase release from isolated rat pancreatic acini as the biological response [6]) it is unlikely that the post-prandial rise in CCK regulates pancreatic enzyme secretion in general and amylase secretion in particular. Indeed neural factors seem to be more important [7], although a role for CCK may be derived from the finding that the CCK-antagonist proglumide abolishes the response to intraduodenal amino acids and fats in dogs [8]. Thus this possible hormonal role does not seem to be related to carbohydrate digestion, since the best releasers of CCK are fatty acids with at least nine carbon atoms [9], suggesting a role for CCK in the regulation of fat digestion through its effect on gall bladder contraction.

The matter of CCK release may not be entirely certain, though, in view of the problems posed by the radioimmunoassay of this peptide. A more recent study, which used a bioassay, showed that glucose and amino acids may also be potent CCK releasers [6]. At the same time, there is probably no true regulatory effect on the pancreas, since that is not necessary: the basal secretory rate of the pancreas seems to be adequate for the digestion of normal meals. This can be deduced from the observation that in patients with chronic pancreatitis, malabsorption only occurs when the maximal secretory rate drops below the normal basal rate [10].

CCK might also have a regulatory role via its effects on gastric emptying and satiety: administered intravenously or intracerebroventricularly, it inhibits gastric emptying and reduces food intake [11–13]. The satiety effect could be a consequence of the motility effect, as gastric distention also induces satiety. Because CCK does not cross the blood–brain barrier, different peripheral and central receptors seem to be involved. The peripheral receptors are probably present on vagal sensory afferents, because in rats vagotomy reduces the intravenous effect of CCK [14]. However, the physiological importance of the peripheral mechanism is uncertain as high doses of CCK are required to induce measurable effects on satiety [15]. Nevertheless experiments with CCK antagonists seem to indicate that CCK released into the circulation is involved in the regulation of gastric emptying [16], and it has been suggested that CCK may induce satiety via its effect upon gastric emptying [17]. In a recent study, though, it was found that although pectin reduced the gastric emptying rate and also increased the sensation of satiety, the post-prandial CCK release was unaffected [17]. Taking all the evidence together, peripheral CCK does not seem to play a role in the satiety sensation.

Secretin

Secretin was the first gastrointestinal hormone discovered. It is released upon contact of acid with the duodenal mucosa, and causes bicarbonate secretion by the pancreas. Because secretin also releases insulin [18] it has been thought that the secretin response augments the insulin response, but this effect requires very high concentrations [19]. Furthermore, intraduodenal glucose has no effect upon secretin levels [20]. Apparently the secretin response bears no relation to carbohydrate digestion. It may be noted that secretin possibly has a metabolic influence on glucose homeostasis, since secretin levels rise during prolonged fasting and are then suppressed by oral or intravenous glucose [21,22].

Somatostatin

Somatostatin was originally isolated from the hypothalamus, but it is now well known that it is present in the cell bodies of neurons of the central and peripheral nervous system, and in endocrine cells of the pancreas and gut. It has a wide range of inhibitory actions; among these most attention has been paid to the inhibition of growth hormone release from the pituitary, the action that led to its discovery, and its effect on the endocrine pancreas.

The study of its release into the circulation has been beset by technical problems associated with the radioimmunoassay of somatostatin. It is our belief that somatostatin cannot be measured in unextracted human plasma [23]. Furthermore its half-life is extremely short, so that the release pattern in the systemic circulation may be a poor reflection of local changes. This, and the fact that somatostatin-containing "endocrine" cells are usually found in close spatial association with their target cells, has led several workers to believe that somatostatin is a paracrine, rather than an endocrine agent.

Nevertheless, a mixed meal does cause a small increase in plasma somatostatin [24,25]. Fats and proteins cause a sustained increase, carbohydrates only a transient rise [26]. In dogs it has been noted that glucose has no effect [27], but measurements of plasma somatostatin during oral glucose tolerance tests have confirmed the stimulatory effect of glucose in man [28,29]. After prolonged fasting this response is even enhanced [30]. In vitro studies have shown that glucose can induce somatostatin release from both pancreas and duodenum [31]. The antrum is also a possible source, but the post-prandial rise comes mostly from the duodenum. Indeed the response to nutrients is unaffected by antrectomy, and the contribution by the pancreas can be neglected [32,33].

The amounts of somatostatin released post-prandially may be sufficient to lead to endocrine effects upon gastric acid secretion, pancreatic endocrine and exocrine secretion, and perhaps gut motility. However, it is unlikely that any of these effects will strongly influence or regulate the digestion of carbohydrates. It has also been reported that somatostatin inhibits the absorption of amino acids and of glucose in the human intestine [34], and in this way it may serve as a modulator of nutrient entry [35].

Neurotensin

Neurotensin was isolated using its vasoactive properties as a guideline. Subsequently many pharmacological actions have been described on a wide range of functions and systems. However, it is unclear which one of these is important physiologically.

Although neurotensin is also a putative neurotransmitter, certainly in the central nervous system, it is present in specific endocrine cells, the N-cells of the ileal mucosa, and its release by a meal has been well documented. However, this release is solely caused by fats. Amino acids, glucose and saline cause insignificant increases in man [36,37] and in dog [38]. These findings contradict the results found in patients with the dumping syndrome. Indeed when dumping is provoked by a glucose solution, a clear rise in neurotensin levels has been found [39]. Perhaps the rapid absorption of glucose which takes place in the duodenum prevents sufficient glucose from reaching the ileum where the neurotensin cells are located. Because neurotensin delays gastric emptying, it has also been suggested that its release may play a part in compensating for the defective neural pathways which cause the dumping [39].

GIP

GIP is the acronym for gastric inhibitory peptide, which was so named because it was identified by its inhibitory action on acid secretion. However, later studies have shown that its most important biological effect is the enhancement of insulin release under hyperglycaemic conditions.

Indeed when blood glucose is maintained at a high level, oral administration of glucose produces increases in plasma GIP and in plasma insulin, but this enhancement of insulin release is not seen when blood glucose is low [40]. These results have been confirmed by studies of the effect of glucose concentration on the GIP-stimulated insulin release from isolated pancreatic islets [41] and from the isolated perfused pancreas [42].

As suggested by Brown [43], the acronym GIP might therefore better be thought to represent "glucose-dependent insulinotropic hormone". Because GIP is released from the gut by meals but not by the intravenous administration of nutrients [44], its release seems to represent a signal from the gut to the pancreas; an inhibitory effect on acid secretion cannot be completely dismissed however. The release comes specifically from the duodenum, as has been shown by studying the effect of glucose perfusions at different sites [45]. As may be expected GIP release is reduced in patients with coeliac disease where the duodenal mucosa is damaged [46]. The release site is specific for glucose, because GIP released by fat originates mainly from the jejunum [47].

The existence of a regulatory control exerted by the gut on the pancreas had already been proposed by La Barre in 1932, who named the hypothetical regulator incretin [48]. In 1969 Unger and Eisentraut suggested the name entero-insular axis for this control loop [49]. It is now generally accepted that GIP corresponds to incretin, although it cannot be excluded that other insulinotropic factors are involved in the entero-insular axis [50]. GIP therefore represents a true regulatory mechanism which is involved not in carbohydrate digestion, but in further assimilation once the carbohydrates have entered the body. In fact, not only is GIP not released by intravenous glucose, the mere presence of monosaccharides in the intestine is also insufficient. It is only after they have been actively transported into and/or metabolised by the epithelial cells that GIP release is induced. Indeed mannose, which is not transported, and 2-deoxyglucose, which is not metabolised, have no effect upon GIP release [51,52], and hyperosmolar solutions of mannitol are also unable to release GIPO [43,44].

Nevertheless, it is clear that factors affecting the rate at which glucose or other monosaccharides become available for transport affect GIP release. Thus in patients with accelerated gastric emptying, GIP release is accelerated [39]. This may reinforce the compensatory mechanism already mentioned for neurotensin, as GIP too inhibits gastric emptying. Furthermore the composition of the diet may affect the rate at which glucose is produced. Thus the GIP response to an oral sucrose challenge was more pronounced in patients adapted to a glucose diet than in patients adapted to a starch diet, presumably because the sucrose diet induced higher levels of sucrase [55,56]. As one would expect, the response to an oral glucose load is faster than that to an oral sucrose load [57], and the presence of non-nutrient dietary fibres also delays the GIP response [58]. The α-glucosidase inhibitor acarbose, a complex oligosaccharide which has been used in diabetic patients to decrease the post-prandial rise in glucose, also reduces the GIP response, presumably because it reduces the rate at which glucose is produced [59,60].

Motilin

Motilin occupies a somewhat special position among the gastrointestinal peptides. It is released periodically during fasting [61] and this release is correlated with the occurrence of a specific motor pattern, the so-called migrating motor complex [62]. Motilin seems therefore to be unique, in that it exerts its function in the fasting state. However, it may also accelerate gastric emptying.

In man, the ingestion of a mixed meal has been reported to have no effect on motilin levels [63], to cause a small rise [64], a delayed decline [65] or a transient rise followed by a decline [66]. In the dog, mixed meals decrease motilin levels [67,68]. Because motilin levels fluctuate during fasting, and because these fluctuations are abolished by feeding, it is possible that the initial rise reported after a meal could be an artefact related to the timing of the meal. Christofides et al., who investigated post-prandial release with reference to preceding motility, concluded that a "continental breakfast" did release motilin [69]. However, the presence or absence of a rise may also be related to the composition of the meal. Indeed, oral fat has been shown to cause a rise and glucose a fall in plasma motilin [64,66]. Insufficient amounts of fat, or large amounts of glucose, could therefore prevent the initial transient rise.

The sustained decrease, on the other hand, could be related to the amount of carbohydrate present, and the rate of its conversion to glucose, but this hypothesis has not been tested. However, an equally valid, and perhaps more likely one, is that the release of other digestive hormones during the meal is responsible for the decrease in motilin. Indeed secretin, insulin and PP are good inhibitors of motilin release [70]. The extent of this decrease could help to determine the length of time the interdigestive motility pattern is disrupted. However, the available data do not suggest that this is the case. Indeed, in dogs fed equicaloric amounts of fats, albumin and sucrose, the longest disruption is observed after fats, the shortest after albumin [71]. Probably the return of the fasting pattern is less effected by changes in motilin than by changes in the true digestive hormones, as most of them are able to disrupt the interdigestive pattern [62].

Intravenous nutrients also have an effect. In man, intravenous glucose and amino acids suppressed motilin, and intravenous fat was a good releaser [64]. In the dog only the inhibitory effect of glucose and amino acids was observed, and fat had no effect [72]. Whatever the importance of intravenous nutrients, in a set of elegant experiments Mori et al. [68] clearly showed that the post-prandial decrease in motilin required the contact of nutrients with the duodenum.

Two groups should be mentioned here in whom the entry rate of glucose into the duodenum is abnormal: patients with the dumping syndrome and heavy smokers. In both groups the response to an oral glucose load is not what one would expect. In patients with the dumping syndrome no significant changes in plasma motilin were noted during dumping provoked by oral glucose [39]. In heavy smokers, on the other hand, motilin levels initially increased. This increase may be related to rapid gastric emptying and to the changes in the blood glucose response seen in such subjects [73]. It has indeed been shown that a motilin infusion accelerates the rise in blood glucose during an oral glucose tolerance test [74], and that motility effects influence the glycaemic response during such tests [75].

Glucagon-Like Peptides (GLPs)

It has been known for some time that other molecular forms of pancreatic glucagon are present in the intestine [77], but it is only recently that they have been completely identified. The original gene product contains three sequences with a high degree of homology, and its enzymatic processing gives rise to the different molecular forms. From the first of these sequences pancreatic glucagon, with 29 amino acids, is formed in the pancreas. In the small intestine the three sequences lead to glucagon-like peptides (GLPs). One of them is enteroglucagon, which corresponds to pancreatic glucagon extended at its N-terminus with 32 amino acids; another is glicentin, which contains the enteroglucagon sequence and yet another 32-amino acid extension at the N-terminal end. Both enteroglucagon and glicentin contain the first glucagon-like sequence of the glucagon gene. The two other sequences give rise to GLP-1 and GLP-2, and from GLP-1 two peptides are derived: GLP-1 7–37 and GLP-1 7–36 amide. The GLPs are produced by the L-cells, which are predominantly present in the terminal ileum [2].

GLPs are released by oral glucose [78], but as the cells producing GLPs are located in the lower parts of the small intestine, they are only released in small amounts in normal conditions. A pronounced response is seen in patients with coeliac disease or dumping syndrome [79], two conditions where abnormal concentrations of glucose are found in the ileum.

A long list of effects has been compiled for glucagon and GLPs, but probably few of them are physiologically important. Pancreatic glucagon of course causes glycogenolysis, lipolysis, gluconeogenesis and ketogenesis in the liver [80], but glicentin has only one-hundredth the potency of pancreatic glucagon [80]. Enteroglucagon has also been called oxyntomodulin, because it inhibits gastric acid secretion. Recent studies suggest that GLP-1 7–37 is a more potent insulinotropic agent than GIP. It is released in parallel with enteroglucagon after a meal and after oral glucose. High levels were found in patients with the dumping syndrome, suggesting that it mediates the hyperinsulinaemia and hypoglycaemia of this disorder [81]. It was noted in the same study that the increased enteroglucagon levels which have been described after the intake of high viscous fibre diets may be associated with high plasma levels of GLP-1 7–37 and that this could explain the improved glucose–insulin ratio observed in patients with cirrhosis taking such fibres [82].

Conclusions

Gastrointestinal hormones contribute to the regulation of the processes of digestion and absorption. However, in a recent review of the factors involved in the regulation of carbohydrate assimilation, gastrointestinal hormones were not considered to be of any importance [76]. The present review of the release pattern of the individual hormones by carbohydrates, together with what is known of their regulatory effects, mostly confirms this conclusion. In general carbohydrates are poor releasers: only GIP, GLP, somatostatin and possibly CCK are released by oral glucose; gastrin, secretin and neurotensin are unaffected; and motilin release is depressed. Of these peptides released by glucose, only one may have a regulatory effect in the processes required to bring nutrients to the *milieu intérieur*: somatostatin by regulating absorption. As mentioned above the digestion itself is apparently not regulated, as even basal levels of pancreatic secretion may be sufficient to digest carbohydrates. However, in the wider context of carbohydrate assimilation, GIP and GLPs represent an important link, and serve as mediators of the entero-insular axis.

Because of the limited regulation exerted by gastrointestinal hormones on carbohydrate digestion, not much attention has been paid to differences in their release induced by different carbohydrates. Moreover, since the oral glucose tolerance test is a standard procedure, and since the prevailing opinion is that glucose is the releaser, and not its parent molecules, studies have been almost completely limited to those on the effect of glucose. If one compares the results of these experiments with those obtained after a mixed meal, the conclusion is that rises after oral glucose are faster, confirming that carbohydrates must be hydrolysed to glucose before they affect hormone release. True comparative studies have only been performed with regard to the release of GIP – studies which confirm this rule. If the rate at which glucose appears in the duodenum is higher, because of a more rapid enzymatic breakdown or accelerated gastric emptying, the release of GIP and GLPs is faster. It can therefore safely be predicted that starches with a different rate of hydrolysis will have corresponding effects upon the rate of the gut hormone release, and, via the entero-insular axis, upon glucose metabolism.

References

1. Bayliss WM, Starling EH (1902) Mechanism of pancreatic secretion. J Physiol (Lond) 28:325–334
2. Walsh JH (1987) Gastrointestinal hormones. In: Johnson LR (ed) Physiology of the gastrointestinal tract, 2nd edn. Raven Press, New York, pp 181–253
3. Lichtenberger LM (1982) Importance of food in the regulation of gastrin release and formation. Am J Physiol 243:G429–G441
4. Richardson CT, Walsh JH, Hicks MI (1976) Studies on the mechanism of food-stimulated gastric acid secretion in normal human subjects. J Clin Invest 58:623–631
5. Djuric DS, Popovic V, Nesovic M (1984) Plasma gastrin levels during oral glucose tolerance test and insulin tolerance test in acromegaly. Horm Metab Res 16:102–103
6. Liddle R, Goldfine I, Williams J (1984) Bioassay of plasma cholecystokinin in rats: effects of food, trypsin inhibitor and alcohol. Gastroenterology 87:542–549
7. Solomon TE (1987) Control of exocrine pancreatic secretion. In: Johnson LR (ed) Physiology of the gastrointestinal tract, 2nd edn. Raven Press, New York, pp 1173–1207
8. Stubbs R, Stabile B (1985) Role of cholecystokinin in pancreatic exocrine response to intraluminal amino acids and fat. Am J Physiol 248:G347–G352
9. Meyer J, Jones R (1974) Canine pancreatic responses to intestinally perfused fat and products of fat digestion. Am J Physiol 226:1178–1187
10. DiMagno EP, Go VLW, Summerskill WHJ (1973) Relations between pancreatic enzyme outputs and malabsorption in severe pancreatic insufficiency. N Engl J Med 288:813–815
11. Debas H, Farooq O, Grossman M (1975) Inhibition of gastric emptying is a physiological action of cholecystokinin. Gastroenterology 68:1211–1217.
12. Gibbs J, Smith G (1977) Cholecystokinin and satiety in rats and rhesus monkeys. Am J Clin Nutr 30:758–761
13. Della Ferra M, Baile C (1979) Cholecystokinin octapeptide: continuous picomole injections into the cerebral ventricles of sheep suppress feeding. Science 206:471–473
14. Smith G, Jerome C, Cushin B, Eterno R, Simansky K (1981) Abdominal vagotomy blocks the satiety effect of cholecystokinin in the rat. Science 213:1036–1037
15. Pappas TN, Mellendez RL, Strah KM, Debas HT (1985) Cholecystokinin is not a peripheral satiety signal in the dog. Am J Physiol 249:G733–G738
16. Shillabeer G, Davison J (1984) The cholecystokinin antagonist, proglumide, increases food intake in the rat. Regul Pept 8:171–176
17. Lorenzo C, Williams CM, Hajnal F, Valenzuela JE (1988) Pectin delays gastric emptying and increases satiety in obese subjects. Gastroenterology 95:1211–1215
18. Hubel K (1972) Secretin: a long progress note. Gastroenterology 62:318–341
19. Fahrenkrug J, Schaffalitzky De Muckadell O, Kuhl C (1978) Effect of secretin on basal and glucose-stimulated insulin secretion in man. Diabetologia 14(4):229–234
20. Boden G, Essa N, Owen OE, Reichle FA (1974) Effects of intraduodenal administration of HCl and glucose on circulating immunoreactive secretin and insulin concentrations. J Clin Invest 53:1185–1193
21. Oektedalen O, Opstad P, Schaffalitzky De Muckadell OB (1982) Secretin – A new stress hormone? Reg Pept 14:213–219
22. Oektedalen O, Opstad PK, Jorde R, Shaffalitzky De Muckadell OB (1984) Responses of vasoactive intestinal polypeptide, secretin, and human pancreatic polypeptide to glucose during fasting. Scand J Gastroenterol 19:59–64
23. Peeters TL, Depraetere Y, Vantrappen G (1981) Radioimmunoassay for somatostatin using a simple extraction method. Clin Chem 27:888–891
24. Colturi T, Unger R, Feldman M (1984) Role of circulating somatostatin in regulation of gastric

acid secretion, gastrin release and islet cell function. Studies in healthy subjects and duodenal ulcer patients. J Clin Invest 74:417–423

25. Peeters TL, Vantrappen G, Janssens J (1982) Control of gut motility. In Bloom SR et al. (eds) Systemic role of regulatory peptides. Schettauer Verlag, Stuttgart, pp 195–210

26. Penman E, Wass J, Medbak S et al. (1981) Response of circulating immunoreactive somatostatin to nutritional stimuli in normal subjects. Gastroenterology 81:692–699

27. Chayvialle JA, Miyata M, Rayford P, Thompson J (1980) Effects of test meal, intragastric nutrients, and intraduodenal bile on plasma concentrations of immunoreactive somatostatin and vasoactive intestinal peptide in dogs. Gastroenterology 79:844–852

28. Itoh M, Hirooka Y, Nihei N (1983) Response of plasma somatostatin-like immunoreactivity (SLI) to a 75 g oral glucose tolerance test in normal subjects and patients with impaired glucose tolerance. Acta Endocrinol 104:468–474

29. Wass JAH, Penman E, Dryburgh JR et al. (1980) Circulating somatostatin after food and glucose in man. Clin Endocrinol 12:569–574

30. Verillo A, de Teresa A, Martino C, di Chiara G, Verillo L (1988) Somatostatin response to glucose before and after prolonged fasting in lean and obese non-diabetic subjects. Regul Peptides 21:185–196

31. Schauder P, McIntosh C, Arends J, Arnold R, Frerichs H, Creutzfeldt W (1976) Somatostatin and insulin release from isolated rat pancreatic islets stimulated by glucose. FEBS Lett 68:225–227

32. Taborsky GJ, Eisinck JW (1984) Contribution of the pancreas to circulating somatostatin-like immunoreactivity in normal dog. J Clin Invest 73:216–223

33. Glaser V, Valtysson G, Fajans SS, Vlinik AI, Cho K, Thompson N (1981) Gastrointestinal/pancreatic hormone concentrations in the portal venous system of 9 patients with organic hyperinsulinism. Metabolism 30:1001–1010

34. Krejs GJ, Browne R, Raskin P (1980) Effect of intravenous somatostatin on jejunal absorption of glucose, amino acids, water and electrolytes. Gastroenterology 78:26–31

35. Schusdziarra V (1980) Somatostatin – a regulatory modulator connecting nutrient entry and metabolism. Horm Metab Res 12:563–577

36. Rosell S, Rokaeus A (1979) The effect of ingestion of amino acids, glucose and fat on circulating neurotensin-like immunoreactivity (NTLI) in man. Acta Physiol Scand 107:263–267

37. Flaten O, Hanssen L (1982) Concentration of neurotensin in human plasma after glucose, meals and lipids. Acta Physiol Scand 114:311–313

38. Go V, Demol P, Reed N, Koch M, DiMagno E (1981) Role of nutrients in the gastrointestinal release of immunoreactive neurotensin. Peptides 2[Suppl 2]:267–269

39. Lawaetz O, Blackburn AM, Bloom SR, Aritas Y, Ralphs DNL (1983) Gut hormone profile and gastric emptying in the dumping syndrome. Scand J Gastroenterol 18:73–80

40. Anderson D, Elahi D, Brown J, Tobin J, Andres R (1978) Oral glucose augmentation of insulin secretion. Interactions of gastric inhibitory polypeptide with ambient glucose and insulin levels. J Clin Invest 62:152–161

41. Schauder P, Brown J, Frerichs H, Creutzfeldt W (1975) Gastric inhibitory polypeptide: effect on glucose-induced insulin release from isolated rat pancreatic islets in vitro. Diabetologia 11:483–484

42. Pederson R, Brown J (1978) Interaction of gastric inhibitory polypeptide, glucose, and arginine on insulin and glucagon secretion from the perfused rat pancreas. Endocrinology 103:610–615

43. Brown J, Dryburgh J, Ross S, Dupre J (1975) Identification and actions of gastric inhibitory polypeptide. Recent Prog Horm Res 31:487–532

44. Cataland S, Crockett S, Brown J, Mazzaferri E (1974) Gastric inhibitory polypeptide (GIP) stimulation by oral glucose in man. J Clin Endocrinol Metab 39:223–228

45. Thomas F, Shook D, O'Dorisio T et al. (1977) Localization of gastric inhibitory polypeptide release by intestinal glucose perfusion in man. Gastroenterology 72:49–54

46. Creutzfeldt W, Ebert R, Arnold R, Frerichs H, Brown J (1976) Gastric inhibitory polypeptide (GIP), gastrin and insulin: response to test meal in coeliac disease and after duodenopancreatectomy. Diabetologia 12:279–286

47. Schattenmann G, Ebert R, Siewert R, Creutzfeldt W (1984) Different response of gastric inhibitory polypeptide to glucose and fat from duodenum and jejunum. Scand J Gastroenterol 19:260–266

48. La Barre J (1932) Sur les possibilités d'un traitement du diabete par l'incretine. Bull Acad Roy Med Belg 12:620–634

49. Unger RH, Eisentraut AM (1969) Enteroinsular axis. Arch Intern Med 123:261–266

50. Creutzfeldt W, Ebert R (1985) New developments in the incretin concept. Diabetologia 28:565–573
51. Ebert R, Creutzfeldt W (1980) Decreased GIP-secretion through impairment of absorption. Front Horm Res 7:192–201
52. Sykes S, Morgan LM, English J, Marks V (1980) Evidence for preferential stimulation of gastric inhibitory polypeptide secretion in the rat by actively transported carbohydrates and their analogues. J Endocrinol 85:201–207
53. Martin EW, Sirinek KR, Crockett SE et al. (1975) Release of gastric inhibitory polypeptide: comparison of hyperosmolar carbohydrate solutions in stimuli. Surg Forum 26:381–382
54. O'Dorisio TM, Spaeth JT, Martin EW et al. (1978) Mannitol and glucose: effects on gastric acid secretion and endogenous gastric inhibitory polypeptide (GIP). Am J Dig Dis 23:1079–1083
55. Reiser S, Hondler HB, Gardner LB, Hallfrish JG, Michaelis IV OE, Prather ES (1979) Isocaloric exchange of dietary starch and sucrose in humans. Am J Clin Nutr 32:2206–2213
56. O'Dorisio TM, Cataland S (1981) Effect of diet on GIP release. In: Bloom SR, Polak JM (eds) Gut hormones. Churchill Livingstone, Edinburgh, pp 269–272
57. Reynolds JC, Falko JM, O'Dorisio TM, Cataland S (1980) The role of gastric inhibitory polypeptide (GIP) in the hyperinsulin response to oral sucrose. Clin Res 232A
58. Jenkins DJA (1980) Influence of fiber and guar-supplemented food on insulin secretion and glucose tolerance. In: Creutzfeldt W (ed) Frontiers of hormone research. S Karger, Basel, pp 202–217
59. Folsch UR, Ebert R, Creutzfeldt W (1981) Response of serum levels of gastric inhibitory polypeptide and insulin to sucrose ingestion during long-term application of acarbose. Scand J Gastroenterol 16:629–632
60. Ruppin H, Hagel J, Feuerbach W, et al. (1988) Fate and effects of the α-glucosidase inhibitor acarbose in humans. An intestinal slow-marker perfusion study. Gastroenterology 95:93–99
61. Peeters TL, Vantrappen G, Janssens J (1980) Fasting motilin levels are related to the interdigestive motility complex. Gastroenterology 79:716–719
62. Vantrappen G, Janssens J, Peeters TL (1981) The migrating motor complex. Med Clin N Am 65:1311–1329
63. Collins SM, Lewis TD, Fox JET, Track N, Meghji MM, Daniel EE (1981) Changes in plasma motilin concentration in response to manipulation of intragastric and intraduodenal contents in man. Can J Physiol Pharmacol 59:188–194
64. Christofides ND, Bloom SR, Besterman HS, Adrian TE, Ghatei MA (1979) Release of motilin by oral and intravenous nutrients in man. Gut 20:102–106
65. Brown JC, Dryburgh JR (1982) Radioimmunoassay of motilin. Methods Enzymol 84:359–368
66. Imura H, Seino Y, Mori K, Itoh Z, Yanaihara N (1980) Plasma motilin levels in normal subjects and patients with diabetes mellitus and certain other diseases. Fasting levels and responses to food and glucose. Endocrinol Jpn 27:151–155
67. Lee KY, Kim MS, Chey WY (1980) Effects of a meal and gut hormones on plasma motilin and duodenal motility in dog. Am J Physiol 238:G280–G283
68. Mori K, Seino Y, Yanaihara N, Imura H (1981) Role of the duodenum in motilin release. Regul Pept 1:271–277
69. Christofides ND, Bloom SR, Vantrappen G, Janssens J, Peeters TL, Hellemans J (1981) Postprandial release of motilin in relation to the interdigestive motor complex in man. Biomed Res 2:67–68
70. Vantrappen G, Peeters TL (1989) Motilin. In: Maklouf G (ed) Handbook of physiology. Endocrinology of the gastrointestinal tract. (In press)
71. De Wever I, Eeckhout C, Hellemans J, Vantrappen G (1978) Disruptive effect of test meals on the interdigestive complex in dogs. Am J Physiol 235:661–665
72. Mori K, Seino Y, Itoh Z, Yanaihara N, Imura H (1981) Motilin release by intravenous infusion of nutrients and somatostatin in conscious dogs. Regul Pept 1:265–270
73. Hanson M, Almer LO, Ekman R, Janzon L, Trell E (1987) Motilin response to a glucose load aberrant in smokers. Scand J Gastroenterol 22:809–812
74. Long RG, Christofides ND, Fitzpatrick ML, O'Shaughnessy DJ, Bloom SR (1982) Effects of intravenous somatostatin and motilin on the blood glucose and hormonal response to oral glucose. Eur J Clin Invest 12:331–336
75. Thompson DG, Wingate DL, Thomas M, Harrison D (1982) Gastric emptying as a determinant of the oral glucose tolerance test. Gastroenterology 82:51–55
76. Alpers DH (1987) Digestion and absorption of carbohydrates and proteins. In Johnson LR (ed) Physiology of the gastrointestinal tract, 2nd edn. Raven Press, New York, pp 1469–1487

77. Sasaki H, Rubalcava B, Baetens D et al. (1975) Identification of glucagon in the gastrointestinal tract. J Clin Invest 56:135–145
78. Unger R, Ohneda A, Valverde I, Eisentraut A, Exton J (1968) Characterisation of the responses of circulating glucagon-like immunoreactivity to intraduodenal and intravenous administration of glucose. J Clin Invest 47:48–65
79. Unger R, Orci L (1976) Physiology and pathophysiology of glucagon. Physiol Rev 56:778–826
80. Jarrouse C, Bataille D, Jeanrenaud B (1984) A pure enteroglucagon, oxyntomodulin (glucagon 37), stimulates insulin release in perfused rat pancreas. Endocrinology 115:102–105
81. Jenkins DJA, Thorne MJ, Taylor RH (1984) Slowly digested carbohydrate food improves impaired carbohydrate tolerance in patients with cirrhosis. Clin Sci 66:649–657
82. Kreyman B, Williams G, Ghatei MA, Bloom SR (1987) Glucagon-like peptide-1 7–36: a physiological incretin in man. Lancet II:1320–1304

Commentary

Flatt: Ingested glucose has been known for more than 20 years [1,2] to stimulate the release of glucagon-like immunoreactive peptides (referred to by some as enteroglucagon) from the intestine. However their nature, and thus their correct terminology, has only recently been established by application of the methods of recombinant DNA technology [3]. It has been shown that mammalian progluca-gon is processed in the intestine to proglucagon 1–69 (glicentin), proglucagon 33–69 (oxyntomodulin), GLP-1 (glucagon-like peptide-1, proglucagon 72–108) and GLP-2 (proglucagon 126–158). GLP-1 is further processed to GLP-1 7–37 and/or GLP-1 7–36 amide [4,5].

Recent studies have shown that the release of GLP-1 7–36 immunoreactivity was stimulated in human volunteers by oral glucose or a mixed meal [5], and from the isolated pig ileum by luminal glucose or vascular perfusion of gastrin-releasing peptide [6]. Effects of other dietary sugars and starches have yet to be examined, but GLP-1 7–36 has been established as a potent glucose-dependent stimulator of insulin secretion from the isolated perfused rat or porcine pancreas [4,7]. It has also been shown to stimulate insulin release directly from human insulinoma cells [8], and to increase insulin secretion at physiological concent-rations when infused in man [5]. Since plasma concentrations of GLP-1 7–36 in the latter experiments were no greater than 50 pmol/l, this also serves to illustrate that measurement of low levels of peptides by radioimmunoassay in blood taken from a site remote from secretion and/or biological action does not necessarily rule out an important physiological role. However, it remains to be established whether GLP-1 7–36 is as important as GIP in the physiology and pathophysio-logy of insulin secretion [9].

References

1. Samols E, Tyler J, Megyesi L, Marks V (1966) Immunochemical glucagon in human pancreas, gut and plasma. Lancet II:727–729
2. Samols E, Marks V (1967) Nouvelles conceptions sur la signification fonctionnelle du glucagon (pancréatique et extrapancréatique). Journ Annu Diabetol Hôtel Dieu 7:43–66
3. Conlon JM (1988) Proglucagon-derived peptides: nomenclature, biosynthetic relationships and physiological roles. Diabetologia 31:563–566
4. Holst JJ, Ørskov C, Vagn Nielsen O, Schwartz TW (1987) Truncated glucagon-like peptide 1: an insulin-releasing hormone from the distal gut. FEBS Lett 211:169–174

5. Kreymann B, Williams G, Ghatei MA, Bloom SR (1987) Glucagon-like peptide-1 7–36: a physiological incretin in man. Lancet II:1300–1304
6. Ørskov C, Holst JJ, Knuhtsen S, Baldiserra FGA, Poulsen SS, Vagn Nielsen O (1986) Glucagon-like peptides GLP-1 and GLP-2, predicted products of the glucagon gene, are secreted separately from pig small intestine but not pancreas. Endocrinology 119:1467–1475
7. Mojsov S, Weir GC, Habener JF (1987) Insulinotropin: glucagon-like peptide 1 (7–37) co-encoded in the glucagon gene is a potent stimulator of insulin release in the perfused rat pancreas. J Clin Invest 79:616–619
8. Flatt PR, Shibier O, Hampton SM, Marks V (1989) Effects of glucagon-like peptides on insulin release by human insulinoma cells and clonal RINm5F cells. J Molec Endocrinol (submitted)
9. Morhan LM, Flatt PR, Marks V (1988) Nutrient regulation of the enteroinsular axis and insulin secretion. Nutr Res Rev 1 (in press)

Marks: Only actively absorbed sugars, i.e. glucose and galactose, stimulate the release of GIP; other sugars, e.g. fructose and xylose, do not [1,2]. Nor do lactose and sucrose, unless they have been hydrolysed into their monosaccharides.

References

1. Sykes S, Morgan ML, English J, Marks V (1980) Evidence for preferential stimulation of gastric inhibitory polypeptide secretion in the rat by actively transported carbohydrates and their analogues. J Endocrinol 85:201–207
2. Salminen S, Salminen E, Marks V (1982) The effects of xylitol on the secretion of insulin and gastrin inhibitory polypeptiudes. Diabetologia 18:480–482

Flatt: GIP is also released by amino acids. The significance of this is that co-ingestion of these nutrients with carbohydrate elicits a greater GIP response which in augmenting insulin release moderates the glycaemic excursion following the absorption of glucose.

Marks: GIP has many properties, apart from its ability to stimulate the release of insulin, that qualify it for serious consideration as the "obesity hormone". Its secretion is stimulated by absorption of long chain, but not short chain fatty acids, and their incorporation into chylomicrons. GIP itself activates lipoprotein lipase in adipose tissue, and serves therefore as a "clearing factor" favouring the uptake and storage of any dietary fat in adipocytes. GIP secretion stimulated by fat, but not GI secreted in response to actively absorbed sugars, is inhibited by endogenous insulin, and possibly C-peptide, which serves therefore as a negative feedback control.

High plasma GIP levels are found in many examples of human and animal obesity, especially those of a hereditary form. The GIP content of the gut is increased by high fat feeding and the blood GIP response to fat and carbohydrate feeding is increased by such dietary measures.

I believe that by directly favouring the deposition of fat in adipocytes, and increasing fat synthesis from carbohydrates via its insulin stimulatory effects, GIP is well qualified to enable fat to be stored during periods of food surplus, but imposes no metabolic burden upon the body during periods of food deficiency. This topic has been discussed in detail elsewhere [1].

Reference

1. Marks V (1988) GIP: the obesity hormone. In: James WPT, Parker SW (eds) Current approaches: obesity. Duphar Medical Relations, Southampton, pp 13–20

Riou: It has been shown recently that VIP inhibits glucose oxidation and stimulates lipolysis in rat enterocytes in a dose-dependent manner, with an apparent K_m of 10^{-11}M [1]. This finding raises the possibility that regulatory peptides present in the gut might regulate the metabolic pathway of nutrients in the enterocytes, and therefore play an important physiological role locally.

Reference

1. Vidal H, Conte B, Beylot M, Riou J-P (1988) Inhibition of glucose oxidation by vasoactive intestinal peptide in isolated rat enterocytes. J Biol Chem 263:9206–9211

Chapter 7

The Metabolism of Sugars and Starches

V. Marks and P. Flatt

Introduction

The dietary carbohydrates, whether ingested as pre-formed sugars or as starches, enter the body proper almost exclusively as their constituent monosaccharides. In the case of starches, including "animal starch" or glycogen, this is always, and exclusively, as glucose. The sugars are more variable in the composition. Apart from glucose and fructose themselves – which occur naturally as constituents of many fruits, as well as of honey and some vegetables, but are derived, in the case of fructose, mainly from sucrose – the only other dietary monosaccharide of major importance is galactose. This occurs in combination with glucose in the form of lactose, and also free, in association with glucose, as a lactose hydrolysate. Of lesser dietary importance are certain pentoses and polyols derived from fruits, or by addition to foods, where they serve as sweeteners or fillers.

All of the dietary monosaccharides and their corresponding polyols are inter-convertible within the body through a number of metabolic pathways of varying complexity [1]. This is the reason for the apparent paradox that, whilst none of the sugars, including glucose, is an essential ingredient of the diet, the presence in the tissues of each of the three absorbed monosaccharides (glucose, galatose and fructose) is essential for a normal, healthy life (Fig. 7.1).

Glucose is ordinarily one of the main metabolic fuels of the body, but can be replaced in this function by fatty acids and ketones and, in the diet, by fructose. This happens, for example, in sufferers from the very rare disease glucose/galactose malabsorption [2,3]. Glucose is however, irreplaceable as a constituent of some structural cellular components, and as a precursor of certain other vital metabolites. The same is true of galactose, which is an essential constituent of many structural cellular components. Fructose is a metabolic fuel used by spermatozoa, and may, therefore, be looked upon as essential for continuation of the species, if not for the individual. Xylitol is a key component in one of the main metabolic pathways by which glucose is metabolised within the body [4]. Sorbitol is the intermediary whereby glucose is turned into fructose in the seminal vesicles.

Myoinositol is an almost universal constituent of animal cells, and was once

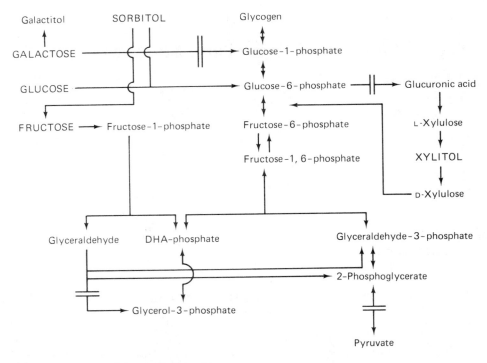

Fig. 7.1. Pathways of the metabolism of sugars.

thought to be derived primarily from the food. This led to it erroneously being called a vitamin. This it may be in some species, but certainly not in human beings, who can, if necessary, synthesise all the myoinositol they require by cyclisation of glucose-6-phosphate [5]. Clements and Diethelm [6], for example, showed that the normal human kidney synthesises about 4 g of myoinositol each day, compared to a daily dietary intake of about 1 g [7]. Nevertheless, the possibility remains that conditions of increased need, possibly because of accelerated destruction in the body or loss in the urine, might give rise to a state of inositol deficiency. Thus far, however, the existence of myoinositol deficiency in man has not been established, and its status as a vitamin must be considered quashed.

Metabolism: General Considerations

In general, metabolism can be considered either at the whole body or at the cellular level. Metabolism at the whole body level involves complex interactions between different organs of the body under a variety of different conditions. It depends upon transport across membranes, and into and out of cells, and it is generally more relevant to pathophysiology than is metabolism at the cellular, subcellular, or even molecular levels. Without knowledge of these lower levels, however, no proper understanding of the overall metabolic economy of the body is possible. Much, indeed most, of the entire intermediary metabolism of the main dietary sugars leading to the production of energy necessary for the maintenance of life, is common to all of them [1,8] (see Fig. 7.1). The main energy-providing metabolic pathways followed by the carbohydrates, namely the glycolytic and pentose phosphate shunt pathways and the tricarboxylic acid cycle, utilise the same enzymes and co-enzymes, many of which consist in part of one or other of the B group of vitamins, regardless of the sugar from which they originated. The intracellular pathways followed by intermediate metabolites, the enzymes and co-enzymes involved, and the nature and control of their regulators, are dealt with extensively in most textbooks of biochemistry. Only those points of difference between the various dietary sugars will be considered in detail here.

Glucose Metabolism

Glucose will be used illustratively, since it not only comprises some 70%–90% of all carbohydrates that we eat, but also because, of all the dietary sugars, its overall metabolism is the most tightly controlled by hormonal and other regulators. The main hormonal regulators, in particular, are glucose-specific, their effects often being exerted primarily and most strongly at the whole body and cellular transport level rather than at the subcellular and molecular levels.

Glucose is by far the most plentiful dietary carbohydrate; it is also the only one that occurs in significant amounts free in the body, that is to say not as part of a larger molecule. It is scarcely surprising that its metabolism has been studied more thoroughly than that of all the other sugars put together, although, as has been pointed out, it is strictly speaking no more vital than they are.

Starch

The metabolism of starch, once it has been absorbed into the body, is really that of glucose; any differences in the pathophysiological effects of different starches are due entirely to differences in their behaviour in the lumen of the gut before digestion and absorption are complete. These differences can manifest themselves in many ways, but possibly the most important, from the viewpoint of the overall energy economy of the body, is the anatomical site at which absorption actually takes place. Thus endocrine cells responsible for the secretion of entero-insular hormones such as GIP and GLP-1 are not uniformly distributed in the small intestine, and the nature of the hormonal response will vary accordingly [9,10]. Once it is inside the body, the metabolism of glucose, like that of all carbohydrates, is determined by intracellular enzymes the specificities and capacities of which are primarily genetically determined, but the activities of which are modulated by hormones and other regulators of metabolic control [1,8].

The most important exception to this general rule is what is known as non-enzymatic, post-translational glycosylation of proteins, often referred to more concisely as glycation, which has been incriminated in the causation of some of the so-called medical "complications" of diabetes, as well as of the ageing process [11–13]. This aspect of metabolism, which concerns all of the dietary monosaccharides to some extent, and is different for each one of them, will be considered in detail later.

Glucose: Whole Body Economy

In the fed state much, though not all, of the glucose in the circulation is derived directly from the food. The remainder, and in the fasting state all of it, is derived either from the breakdown of pre-formed glycogen in the liver or by new glucose molecules formed in the liver from circulating intermediate metabolites [14].

Dietary glucose, whether pre-formed or derived from starches or disaccharides, enters the body through the portal vein. It traverses the liver where a proportion of it is removed, not necessarily in its first passage, and converted by phosphorylation into glucose-6-phosphate and eventually into glycogen. Insulin facilitates, and glucagon inhibits, the conversion of glucose-6-phosphate into glycogen. Glucose not removed by the liver may be broken down by extrahepatic tissues to trioses.

The absorption of glucose is associated with an increase in insulin release and a decrease in glucagon release by the pancreas, but the secretion and the actions of both of these hormones are to some extent modulated by other hormones released from the gut [9,10.15].

The glucose not taken up by the liver is assimilated in peripheral tissues, especially skeletal muscle, under the influence of insulin [16]. This hormone has the almost unique property, shared only with the insulin-like growth factors and exercise, of allowing glucose to be transported actively into muscle and fat cells, where it is immediately converted into glucose-6-phosphate by hexokinase. This rapid removal of glucose from within the cells permits more of it to enter the cells. The cells of the brain, erythron (haemopoietic system: bone marrow and red cells), liver and endocrine pancreas are unique in that they do not require insulin in order to permit entry of glucose from the extracellular fluid. This has been wrongly construed by many authors to mean that insulin does not influence glucose metabolism within these cells.

The concentration of insulin necessary to permit entry of glucose into insulin-dependent cells is highly variable, not only between persons but also in the same individual from one occasion to another, and perhaps, most important of all, between one cell type and another. It is profoundly altered by either subacute or chronic changes in the quantity or quality of the diet, and also by rigorous athletic training, which alters both the number and affinity of insulin receptors, and makes the tissues more sensitive to the hypoglycaemic actions of insulin [17].

Although it is believed that a high dietary fat intake decreases insulin sensitivity, we have shown this is not always the case [18]. A possible explanation for the apparently detrimental effect of high fat feeding on glucose tolerance observed in the past is the concomitant decrease in carbohydrate, i.e. glucose, intake that generally accompanies the institution of high fat feeding.

In the absence of insulin, or when, for any reason, the peripheral cells become resistant to its actions, cellular glucose uptake is markedly reduced, as is its metabolism to non-glucose intermediates or glucose end products such as glycogen and fat. Consequently glucose accumulates in the blood, which gives rise to hyperglycaemia. Even in healthy subjects, and despite a massive net increase in glucose uptake by the liver and its storage as glycogen, neither the production of glucose by gluconeogenesis nor the breakdown of pre-formed glycogen by glycogenolysis are completely halted [14]. Hyperglycaemia produced by carbohydrate ingestion, which is always comparatively modest in normal healthy subjects, is consequently aggravated in persons with defective insulin secretion and/or insulin insensitivity, because not only are their peripheral tissues incapable of assimilating glucose, but their livers are also incapable of suppressing their own output of glucose properly [19]. This failure to suppress the hepatic contribution to the body glucose pool compounds the hyperglycaemia caused by the under-utilisation of dietary glucose by diabetic subjects. It also explains the small but larger than usual rise in blood glucose concentration that follows the ingestion of fructose, galactose and other gluconeogenic substrates by diabetic subjects with impaired insulin secretion.

Galactose

Galactose shares the same active transport system as glucose in the enterocytes of the small intestine. It enters the portal venous blood at a rate not much slower than that of glucose but, because of its greater uptake by the liver, very little unaltered galactose enters the peripheral circulation. Consequently, peripheral blood galactose levels rarely rise above 1.0 mmol/l in normal healthy subjects, even after large doses (100 g) of lactose [20]. If, however, galactose is ingested without glucose, higher plasma concentrations are obtained. Alcohol specifically inhibits galactose uptake and metabolism by the liver. It can, therefore, produce a modest but transient galactosaemia when taken together with either galactose or lactose [21]. Whether the galactosaemia so produced is ever sufficient to cause adverse effects is not known with certainty.

Galactose, after its absorption from the gut, enters the liver cells, where it is converted by the galactose-specific enzyme galactokinase into galactose-1-phosphate. This is then rapidly and completely converted, by means of a two-stage transformation involving galactose-1-phosphate uridyl transferase and uridyl diphosphate glucose-4-epimerase, into glucose-1-phosphate [21]. This in turn is converted into glycogen in exactly the same way as glucose-1-phosphate formed from glucose itself. Theoretically the glucose-1-phosphate produced from galactose in the liver should be able to enter the glycolytic pathway of the liver cells directly, but there is little evidence that this happens to any significant extent under normal circumstances. In this respect, therefore, the metabolic fate of galactose differs from that of fructose, where substantial amounts of its metabolic products do seem to enter the glycolytic pathway directly, without first being converted into glycogen [22–25].

All of the enzymes involved specifically with galactose metabolism, including galactokinase, occur widely in mammalian tissues [21]. They are, however, especially plentiful in liver which is, to all intents and purposes, the exclusive site of exogenous (dietary) galactose metabolism [21]. Nevertheless, even in the complete absence of galactose from the diet, sufficient galactose can be made by reversal of the normal galactose–glucose interconversion to enable all of the essential structural elements of body cells and connective tissues that contain galactose and galactosamine as constituents of their glycoprotein and mucopolysaccharides, to be replaced or made *de novo* as, and when, required.

Levels of galactose in peripheral blood rarely rise above 1.0 mmol/l. When for any reason they do, certain tissues, and notably the lens of the eye, can extract it from the blood and, by means of the ubiquitous enzyme aldehyde reductase, convert it into galactitol (also called dulcitol). There are no enzymes present in mammalism tissues for which this polyol can serve as a substrate [21]. Consequently it is not metabolised but instead accumulates in the tissues, where it can cause cellular disruption due to an increase in intracellular osmolarity. In the lens of the eye this disruption manifests itself as a cataract.

Cataracts have long been recognised as a complication of diabetes mellitus, and of each of the two major inborn errors of galactose metabolism caused by galactose-1-phosphate uridyl transferase and galactokinase deficiencies. The possibility also exists that cataracts might occur as a complication of chronic liver disease in people consuming more than modest amounts of galactose, such as might happen if they drank large amounts of milk or alcoholic drinks with mixers sweetened with hydrolysed lactose.

Classical galactosaemia, which is due to galactose-1-phosphate uridyl transferase deficiency, is a very serious disease and manifests itself in the first few days of life. It may have a fatal outcome, or there will be mental retardation, unless it is treated promptly by total exclusion from the diet of galactose taken in the form of lactose in human or animal milk. Substitution of soya milk, supplemented by sucrose, glucose and/or fructose as an energy source, leads to rapid restoration of health and normal growth and development. The intolerance to galactose (and lactose) persists throughout life. The disease itself is very rare, with a prevalence at birth of about 1 in 60 000. The heterozygotes have only half the normal amount of galactose-1-phosphate uridyl transferase activity, and constitute about 1% of the population [21]. As far as is known they do not manifest any clinical abnormalities, but this may merely reflect the insensitivity of the methods currently used for their detection – exactly as happened, until recently, with carriers of the hereditary fructose intolerance (*HFI*) gene.

Galactokinase deficiency produces little in the way of systemic disturbance, even in homozygotes, but is associated with galactosaemia and galactosuria following the ingestion of milk and, unless prevented by exclusion of milk from the diet, manifests itself by a high incidence of cataracts in children as well as in adults [21]. Though seemingly rare in Europeans, the gene frequency may be higher in some populations with a high incidence of early cataracts. The long-term effects of heterozygosity for galactokinase deficiency upon whole body galactose economy are still unknown.

Fructose

In man, fructose enters the portal venous blood largely unchanged following its absorption from the small intestine. This is not true for all species, in some of which it undergoes almost complete fructolysis within the enterocytes whence it is discharged into the portal circulation either as glucose or as lactate [24]. When given in equimolar amounts, fructose is absorbed more rapidly after ingestion of sucrose than after glucose and fructose (see Chap. 8).

Fructose is avidly removed from the blood after a single passage through the liver. Consequently, little of it gains access to the systemic circulation even after the ingestion of large amounts of either fructose or sucrose, even when it is ingested in solution rather than as a constituent of solid food. Within the liver, fructose undergoes phosphorylation to fructose-1-phosphate under the influence of the fructose-specific enzyme fructokinase. Although fructose can serve as a substrate for hexokinase, its affinity for this enzyme is so low, especially in the presence of glucose, that very little is ordinarily phosphorylated by this mechanism either in the liver or elsewhere. Exceptionally however, in the congenital absence of fructokinase due to a specific inborn error of metabolism, hexokinase provides almost the only way that dietary fructose can gain access to the body metabolic pool. Even so, in this condition much of the ingested fructose is excreted unchanged in the urine with few, if any, ill effects.

Fructose and the Liver

Fructokinase is present in large amounts in the liver where it converts fructose into fructose-1-phosphate. In doing so, it consumes large amounts of ATP and inorganic phosphate, which are normally regenerated as the fructose-1-phosphate formed is split into glyceraldehyde and dihydroxyacetone phosphate. Cleavage of fructose-1-phosphate is effected by the hepatic type of aldolase (aldolase B) that occurs in abundance only in the liver [22]. Hepatic aldolase differs from the aldolase found in muscle and most other tissues in having an equal affinity for fructose-1-phosphate and fructose-1,6-diphosphate; the latter is an important intermediary in the glycolytic pathway of all tissues and the gluconeogenic pathways of liver and kidney [24–26].

Dihydroxyacetone phosphate, one of the two metabolites formed by cleavage of fructose-1-phosphate, is an intermediary metabolite in both the glycolytic and gluconeogenic pathways. Glyceraldehyde, the other metabolite formed by fructose-1-phosphate cleavage, is not an intermediary in either pathway, but it can serve as a substrate for any one of a number of enzymes found mainly or exclusively in the liver which convert it into glycolytic intermediary metabolites. Probably most of the glyceraldehyde formed by fructolysis is converted, through the agency of triokinase, to glyceraldehyde-3-phosphate which can then, in turn, condense stoichiometrically with dihydroxyacetone phosphate to produce fructose-1,6-diphosphate and ultimately glycogen from which glucose can be formed by glycogenolysis [1].

Alternative metabolic pathways that glyceraldehyde can follow include its conversion through the combined action of glyceraldehyde dehydrogenase and glycerate kinase into glycerate-2-phosphate, or via glycerol and the agency of the liver-specific enzyme glycerol kinase into glycerol-3-phosphate. This can be used either for the esterification of fatty acids and their conversion into triglycerides, or for transport of cytosolic reducing equivalents into mitochondria. Both 2-phosphoglycerate and glycerol-3-phosphate are, like dihydroxyacetone phosphate, intermediary metabolites in the glycolytic and gluconeogenic pathways. Which one of the two pathways they actually follow at any particular time is determined by many factors, of which the prevailing nutritional and hormonal status of the individual are probably the most important.

The consumption of either fructose or sucrose in substantial amounts on a single occasion produces a larger rise in peripheral blood pyruvate and lactate levels than the ingestion of an isoenergetic amount of glucose alone [23]. This suggests that at least some of the intermediates produced from fructose enter the glycolytic rather than the gluconeogenic pathway of the liver. However, only when fructose is given intravenously together with alcohol, which inhibits gluconeogenesis from pyruvate and lactate, do these metabolites accumulate in the blood in amounts sufficient to constitute a hazard to health [25]. Fructose can increase the rate of alcohol metabolism by 10%, and under exceptional circumstances by as much as 50%. Moreover, because neither fructose alone nor the combination stimulates insulin release, there is no risk of rebound hypoglycaemia occurring [27,28] as there is sometimes when glucose or sucrose and alcohol are taken together on an empty stomach.

Fructose and Adipose Tissue

Although fructose is ordinarily only a very poor substrate for hexokinase, the situation changes in the absence of insulin, since fructose, unlike glucose, does not need insulin in order to gain access to the interior of muscle and fat cells. This is important in respect of the different roles they play in the control of fat metabolism when, for any reason, insulin action is reduced.

The long chain fatty acids stored in the body as triglycerides are obtained largely from the fat in the diet. They can, however, also be synthesised in the body from intermediate metabolites produced via the "glycolytic pathway" from carbohydrates, in combination with the "reducing equivalents" obtained from the pentose phosphate [1]. These routes are probably not major contributors to total body fatty acid content under ordinary dietary conditions, but may become so when carbohydrates are fed in excess of daily energy requirements [29].

Fatty acids taken as food are not stored in the body as such, but must either be metabolised directly to carbon dioxide and water through the carboxylic acid cycle, or be immediately converted into, and stored as, triglycerides. Indeed, it is in the form of triglycerides that most fatty acids enter the body proper through the thoracic duct. Fats that are composed of long chain fatty acids and have undergone partial hydrolysis in the intestinal lumen before being absorbed are re-synthesised into triglycerides in the enterocytes of the intestinal mucosa, and packaged, along with apolipoproteins B and A and phospholipids, into chylomic-rons before being transported to the tissues. There they are split by lipoprotein lipase back into free fatty acids, which can be taken up by the cells and utilised.

Both the conversion of fatty acids into carbon dioxide and water and their storage as triglycerides can occur in virtually all of the tissues of the body, but formation and deposition of triglycerides is especially important in the adipocytes [30]. These cells are not, as was once thought, metabolically inert, but in a state of constant flux. Under the influence of the intracellular enzyme hormone-depen-dent lipase, the triglycerides they contain are constantly broken down into glycerol and free fatty acids, both of which diffuse out of the adipocyte into the extracellular fluid, and from there into the body as a whole.

All of the glycerol produced by lipolysis in adipose tissue does indeed behave in this way, since adipocytes, unlike liver cells, do not possess the enzyme glycerol kinase which is necessary to effect the phosphorylation of free glycerol and bring it back into the metabolic pool. Much of the fatty acid produced is, however, available for re-esterification by glycerol-3-phosphate formed as an intermediate in the glycolytic pathway. Glucose will not enter adipocytes from the blood, however, unless insulin is also present in the extracellular fluid (ECF) at a concentration above a certain threshold. The exact threshold level is variable, but depends largely upon the number and affinity of insulin receptors on the surface of the adipocytes. When the insulin level in the ECF falls below this threshold, glucose cannot enter the cells, and consequently glycolysis decreases and insufficient glycerol-3-phosphate is available for re-esterification of the free fatty acids produced by lipolysis, especially since lipolysis is increased in the absence of insulin and in the presence of lipolytic hormones such as glucagon. Under these circumstances the free fatty acids, like the glycerol formed with them, enter the blood stream in greatly increased amounts. From there they can be extracted by certain tissues, such as heart and striated muscle, for use as fuel, or by the liver to be converted either into ketone bodies such as acetoacetate and β-hydroxybuty-rate, or back into triglycerides for re-export as very low density lipoprotein (VLDL).

Fructose, as mentioned above, can enter peripheral tissues, including adipo-cytes, when ECF insulin levels are below the threshold level for the entry of glucose. Consequently it can serve as a substrate for hexokinase, which converts it into fructose-6-phosphate; this in turn can undergo glycolysis to produce glycerol-3-phosphate. The glycerol-3-phosphate becomes available for re-esteri-fication of free fatty acids produced by intra-adipocyte lipolysis, and reduces their entry into the blood and their delivery to the liver.

The non-dependence of fructose upon the presence of insulin in the ECF for its entry into cells, and consequently its availability for phosphorylation into glycolytic pathway intermediates, may explain why fructose given intravenously or even by mouth can reduce ketosis in untreated insulin-dependent diabetics, whereas glucose itself is either ineffective or detrimental [23]. It must, however, be borne in mind that in diabetic, as in healthy subjects, peripheral blood fructose levels seldom, if ever, rise above 1 mmol/l, even after the largest doses of fructose that can be tolerated by mouth, so that whilst this difference in metabolism between fructose and glucose is interesting, its pathophysiological relevance is probably small.

Fructose Metabolism and Hyperuricaemia

One important consequence of the extremely rapid removal of fructose from the blood and its conversion into fructose-1-phosphate in the liver, is an increase in ATP turnover and intrahepatic depletion of inorganic phosphate. This can, under certain circumstances, lead to a rise in body urate production [24,25]. Other mechanisms may also be involved, which between them can cause a small but significant rise in plasma urate levels in normal healthy subjects given a single large dose of fructose or sucrose, or following the institution of high fructose and/ or sucrose feeding [31–33]. This is, however, a far from universal finding [34,35], and may reflect unrecognised environmental or genetic influences [31].

Though seldom of clinical importance in its own right, the increased production of uric acid can, especially when combined with other conditions such as chronic alcohol abuse which themselves reduce renal urate excretion, predispose to the development of gout. Recently Radda and his colleagues [31] have shown that some cases of clinical gout can be explained largely, if not entirely, on the basis of increased urate production as a manifestation of heterozygosity for the *HFI* gene.

Although homozygotes for the *HFI* gene have been recognised for more than 30 years by virtue of their profound and characteristic clinical disease [24,26], the heterozygotes were, until very recently, thought to be free from any metabolic disturbances. Using ^{31}P magnetic spectroscopy, Oberhaensli et al. [31] showed that fructose-1-phosphate accumulates in the liver of obligatory heterozygotes for *HFI* after ingestion of substantial, but not unrealistically large amounts of fructose or sucrose by mouth. They also showed that the heterozygotes had a greater rise in plasma urate than control subjects treated similarly, and that a high proportion of them suffered from clinical gout. The claims made some 20 years ago [32,33] that the frequency and severity of gouty attacks can sometimes be diminished by eating a fructose-restricted diet would seem, therefore, to gain some support from these observations, which relate to possibly 0.4% of the population in Britain and 1.2% in Switzerland and other central European countries where *HFI* gene frequency is high.

The contentious questions of whether fructose and sucrose ingestion specifically predisposes to the development of hypertriglyceridaemia [22], and its relevance to human health, are discussed elsewhere in this volume and will not be considered here.

Polyols

Sorbitol was formerly used quite extensively as a sweetener and filler in foods specifically designated as suitable for use by diabetic patients. The possible reason for this was that it did not qualify as a "carbohydrate" according to food labelling regulations then in vogue, but it was also used on the grounds that it produces a smaller rise in blood glucose concentration than most other nutritive energy-providing sweeteners.

Sorbitol is absorbed more slowly from the gut than any of the monosaccharides, and it is extracted from the portal blood in its entirety by the liver, where it is converted by the enzyme sorbitol dehydrogenase into fructose before being phosphorylated by fructokinase into fructose-1-phosphate [24]. The metabolism of dietary sorbitol once it has been absorbed is, therefore, largely that of fructose, save that any sorbitol that does escape metabolism by the liver probably cannot be utilised by adipose and other tissues in the same way as fructose, since it does not gain access to the interior of the cells and nor do they possess large amounts of sorbitol dehydrogenase.

Sorbitol can be formed from glucose within the body by the action of aldose reductase; this is the mechanism whereby the fructose in semen is manufactured. This reaction is also suspected to play a role in the pathogenesis of diabetic cataracts, in a manner analogous to that of galactitol in galactosaemia, but the evidence is less convincing despite intensive research involving the use of aldose reductase inhibitors [36]. These agents may, however, have a useful part to play in the alleviation of diabetic neuropathy, which is also suspected of being due to sorbitol accumulation and does seem to respond to treatment with them [36].

Xylitol [37] is the physiological polyol corresponding to the pentose xylulose. This is an important intermediary in the glucuronic acid pathway of glucose metabolism which, in most mammalian species, is concerned with the synthesis of vitamin C. The discovery of the glucuronic–xylulose metabolic pathway was a result of investigations into the cause of the benign inborn error of metabolism pentosuria. In this condition, large amounts of L-xylulose are excreted in the urine, due to a deficiency of the enzyme L-xylulose reductase which ordinarily converts it into xylitol. Xylitol itself undergoes further metabolism, eventually ending up as glucose-6-phosphate.

Xylitol is sweeter and fresher tasting than sorbitol, which it resembles in being only slowly absorbed from the gut and producing only a very small rise in blood glucose concentration [38]. It is being increasingly used in the manufacture of sweets and other confectionery because it is non-cariogenic.

Dietary xylitol is metabolised by the same metabolic pathway as the xylitol produced as an intermediary in the metabolism of glucose. It has been given with apparent safety to many people by mouth in large amounts over prolonged periods, without noticeable adverse effects upon several biochemical markers of toxicity [39,40]. Xylitol can also be tolerated by most people when given in large amounts intravenously. It has, however, been incriminated as the cause of calcium oxalate crystal deposition in the kidneys, with subsequent kidney failure, in some patients given massive doses of xylitol intravenously. The causal relationship has, though, been disputed [37,41]. Nothing similar has ever been observed in normal healthy subjects given xylitol by mouth in unphysiologically large amounts over prolonged periods [37].

Carbohydrate Metabolism and Vitamins

Many of the enzymes involved in the metabolism of carbohydrates utilise vitamin B metabolites as essential co-enzymes. In doing so they may increase the normal

rate of metabolic degradation of the vitamins, and increase the dietary requirement for them. There is, however, very little experimental or indeed epidemiological evidence to substantiate this hypothesis at least so far as man is concerned
[42].

All of the steps in the intracellular metabolism of dietary carbohydrates that
are catalysed by co-enzymes derived from vitamin B are common to pathways
followed by all carbohydrates (e.g. glycolytic pathway, pentose phosphate shunt
and tricarboxylic acid cycle). None of the specific processes involved in the
metabolism of one or other of the dietary sugars and their interconversion is
known to make any special or unusual demand upon vitamin-B-derived co-
enzymes. Consequently vitamin requirements are unlikely to be any different,
regardless of whether the dietary carbohydrate is supplied as starches or sugars.
The concern expressed by some popularist writers, therefore, that diets containing large amounts of sugars might predispose to vitamin deficiencies, whilst those
rich in more complex carbohydrates do not, is unjustified on the basis of any
differences in the metabolism of the different classes of dietary carbohydrate,
although it could possibly be so on other grounds.

Starches are seldom eaten in their pure form, and what are generally referred
to as starchy foods are really mixtures of anutrients* and nutrients, including
trace elements, vitamins and toxins of one sort or another. Added sugars,
especially sucrose, on the other hand, are often taken in a highly purified form,
uncontaminated by other chemical sustances, whether nutrient or anutrient.

If the total amount of food habitually eaten is determined solely by its energy
content it is conceivable, but no more than that, that diets providing a very large
proportion of their energy as pure carbohydrates could, regardless of their
chemical nature or class, lead to vitamin or trace element deficiency. A key
question in this regard concerns the conceptual basis of requirements and
whether they are related to energy expenditure. However, in experimental
animals fed bizarre diets this does undoubtedly occur. Indeed, some of the lesions
formerly attributed to the sugars themselves [43] were probably due to unrecognised trace nutrient deficiency. There are, however, no grounds for believing that
this situation is common in human nutrition, or that it has anything specifically to
do with "added" sugar content. It does, however, occur in people eating
outrageously bizarre diets of all kinds, including those in which added sugars have
been specifically excluded [44].

Non-enzymatic Glycosylation

The ability of sugars to react non-enzymatically with proteins so as to produce
glycosylated products has been known to food chemists for many years, but was
recognised as important by physiological chemists only within the last ten years or
so [45–47].

Sugars in body fluids in general, and in blood in particular, can react with free
amino groups on proteins to produce sequentially (i) a readily chemically

---

*A component of food with neither nutritional value nor toxic activity.

reversible glycosylated product (a Schiff base), which then undergoes internal rearrangement to produce (ii) a much more stable, but still at least potentially reversible Amadori-type glycosylation product, which ultimately is converted into (iii) a chemically irreversible compound referred to as an advanced glycosylation end product (AGE). AGE can interact with other proteins to produce a variety of polymers, all of which are suspected of being detrimental to the continuing good health of the organism [11–13].

A large body of information has accumulated in respect of the interaction between glucose and diverse proteins, but much less is known about interaction with other sugars of biological importance, such as fructose and galactose, that can, and do, occur in the blood from time to time.

Both aldose and ketose sugars can react with proteins to form Schiff bases, but only when they are present in the reaction mixture in the chain form [46]. Sugars vary widely in the proportion of their molecules that are in the chain form at any one time. Glucose has an unusually low proportion of its molecules (0.002%) in the chain form, which provides the free carbonyl group for participation in the glycosylation reaction. In the case of fructose, 0.7% of the molecules are in the chain form. Because of the much lower concentration of fructose than of glucose in blood and extracellular fluid, the contribution made by fructose to the amount of non-enzymatic glycosylated proteins present in the blood at any one time, or of AGE in the body as a whole, has received little or no attention in the past. However, because of its more than 50-fold greater reactivity than glucose in glycosylation processes on a molar basis, dismissal of fructose as trivial in this respect may be misplaced.

Galactose too has a higher, though not precisely known proportion of its molecules in the chain configuration than glucose [46], and does participate in the non-enzymatic glycosylation reaction [48,49] leading to AGE. This is especially likely to occur in subjects who, for any reason, have higher than usual blood galactose levels. This too has received scant attention in the pathophysiological and medical literature up to now, as indeed has participation in the non-enzymatically mediated glycosylation process of all the so-called rare sugars that occur to a limited extent in many fruits and vegetables. Although their concentrations in blood may never reach levels comparable with those even of fructose or galactose, their possible participation in post-translational glycosylation processes cannot be dismissed without investigation [50]. Since AGE are now thought to play a central role not only in the pathogenesis of the somatic complications of diabetes but in the very process of ageing itself, this deficiency in our knowledge is in urgent need of rectification.

References

1. Coleman JE (1980) Metabolic interrelationships between carbohydrates, lipids and proteins. In: Bondy PK, Rosenberg LE (eds) Metabolic control and disease Saunders, Philadelphia, pp 161–274
2. Lindquist B, Meevwisse G (1962) Chronic diarrhoea caused by monosaccharide malabsorption. Acute Paediatr 51:674–686
3. Elsas LJ, Hillman RE, Patterson JH, Rosenberg L (1970) Renal and intestinal hexose transport in familial glucose–galactose malabsorption. J Clin Invest 49:576–581

4. Hiatt HH (1972) Pentosuria. In: Stanburg JB et al. (eds) The metabolic basis of inherited disease, 3rd ed. McGraw Hill, New York, pp 119–130
5. Holub BJ (1986) Metabolism and function of myo-inositol and inositol phospholipids. Ann Rev Nutr 6:563–597
6. Clements RS, Diethelm AG (1979) The metabolism of myoinositol by the human kidney. J Lab Clin Med 93:210–219
7. Clements RS, Reynertson R (1977) Myoinositol metabolism in diabetes mellitus. Effect of insulin treatment. Diabetes 26:215–221
8. Beitner R (ed) (1986) Regulation of carbohydrate metabolism, vols 1 and 2. CRC Press, Boca Raton
9. Morgan LM, Marks V (1981) The gastrointestinal hormones. In: Pennington GW, Naik S (eds) Hormone analysis: methodology and clinical interpretation, vol II. CRC Press, Boca Raton, pp 63–100
10. Morgan LM, Flatt PR, Marks V (1988) Nutrient regulation of the enteroinsular axis and insulin secretion. Nutr Res Rev 1:79–97
11. Cerami A, Vlassara H, Brownlee M (1987) Glucose and aging. Sci Am 256:82–88
12. Brownlee M, Cerami A, Vlassara H (1988) Advanced glycosylation end products in tissue and the biochemical basis of diabetic complications. N Engl J Med 318:1315–1321
13. Monnier VM, Sell DR, Abdul-Karim FW, Emancipator SN (1988) Collagen browning and cross-linking are increased in chronic experimental hyperglycaemia. Relevance to diabetes and aging. Diabetes 37:867–872
14. Jackson RA, Roshama RD, Hawa MI, Lim BM, Disivio L (1986) Impact of glucose ingestion on hepatic and peripheral glucose metabolism in man: an analysis based on simultaneous use of the forearm and double isotope techniques. J Clin Endocrin Metab 63:541–549
15. Hartmann H, Ebert R, Creutzfeldt W (1986) Insulin dependent inhibition of hepatic glycogenolysis by gastric inhibitory polypeptide in perfused rat liver. Diabetologia 29:112–114
16. Daniel PM, Love ER, Pratt OE (1974) Insulin-stimulated entry of glucose into muscle in vivo as a major factor in the regulation of blood glucose. J Physiol (Lond) 247:273–288
17. Krotkiewski MK, Björntorp P, Holm G et al. (1984) Effects of physical training on insulin, connecting peptide (C-peptide), gastric inhibitory polypeptide (GIP) and pancreatic polypeptide (PP) levels in obese subjects. Int J Obes 8:193–199
18. Morgan LM, Hampton SM, Tredger J, Cramb R, Marks V (1988) Modifications of gastric inhibitory polypeptide (GIP) secretion in man by a high-fat diet. Br J Nutr 59:373–380
19. Rizzia R, Mandarino L, Gerich J (1981) Dose response characteristics for the effects of insulin on production and utilisation of glucose in man. Am J Physiol 240:E630
20. Haworth JC, Ford JD, Robinson TJ (1965) Peripheral and portal vein blood sugar after lactose and galactose feedings. Clin Sci 29:83–92
21. Segal S (1983) Disorders of galactose metabolism. In: Stanburg JB et al. The metabolic basis of inherited disease, 5th ed. McGraw Hill, New York, pp 167–191
22. Shafrir E (1986) Effect of sucrose and fructose on carbohydrate and lipid metabolism and the resulting consequences. In Beitner R (ed) Regulation of carbohydrate metabolism, vol II. CRC Press, Boca Raton, pp 95–140
23. Miller M, Craig JW, Drucker WR, Woodward H (1956) The metabolism of fructose in man. Yale J Biol Med 29:335–360
24. Froesch ER (1972) Essential fructosuria and hereditary fructose intolerance. In: Stanburg JB et al. (eds) The metabolic basis of inherited disease, 3rd ed. McGraw Hill, New York, pp 131–148
25. Woods HF (1986) Pathogenic mechanisms of disorders of fructose metabolism. In: Burman D et al. (eds) Inherited disorders of carbohydrate metabolism. MTP, Lancaster, pp 191–203
26. Baerlocher U, Gitzelmann R, Steinmaan B (1980) Chemical and genetic disorders in fructose metabolism. In: Burman D et al. (eds) Inherited disorders of carbohydrate metabolism. MTP, Lancaster, pp 163–190
27. O'Keefe SJD, Marks V (1977) Lunchtime gin and tonic: a cause of reactive hypoglycaemia. Lancet I:1286–1288
28. Marks V, Wright J (1980) Alcohol-provoked reactive hypoglycaemia. In: Andreani D et al. (eds) Current views on hypoglycaemia and glucagon. Academic Press, London, pp 283–295
29. Bjorntorp P, Sjostrom L (1978) Carbohydrate storage in man: Speculations and some quantitative considerations. Metabolism 27:1853–1859
30. Bjorntorp P (1978) The fat cell: a clinical view. In: Bray G (ed) Recent advances in obesity research II. Newman, London, pp 153–168
31. Oberhaensli RD, Rajagopalan B, Taylor DJ et al. (1987) Study of hereditary fructose intolerance by use of ^{31}P magnetic resonance spectroscopy. Lancet II:931–934

32. Perheentupa J, Raivio K (1967) Fructose-induced hyperuricaemia. Lancet II:528–531
33. Stirpe F, della Corte E, Bonetti E, Abbondanza A, Abbati A, de Stefano F (1970) Fructose induced hyperuricaemia. Lancet II:1310–1311
34. Grigoresco G, Rizkalla SW, Halfon P et al. (1988) Lack of detectable deleterious effects on metabolic control of daily fructose ingestion for 2 months in NIDDM patients. Diabetes Care 11:546–550
35. Bantle JP, Laine DC, Thomas JW (1986) Metabolic effects of dietary fructose and sucrose in types I and II diabetic subjects. JAMA 256:3241–3246
36. Dvornik D (1987) Aldose Reductase Inhibition. McGraw Hill, New York
37. Mäkinen KK (1978) Biochemical principals of the use of xylitol in medicine and nutrition with special consideration of dental aspects. Birkauser Verlag, Basel
38. Salminen S, Salminen E, Marks V (1982) The effects of xylitol on the secretion of insulin and gastric inhibitory polypeptide in man and rats. Diabetologia 22:480–482
39. Huttunen JK, Mäkinen KK, Scheinin A (1975) Turka sugar studies XI. Effects of sucrose, fructose and xylitol diets on glucose, lipid and urate metabolism. Acta Odontol Scand 33:239–245
40. Mäkinen KK, Scheinin A (1975) Turka sugar studies. XIII. Effect of the diet on certain clinico-chemical values of serum. Acta Odontol Scand 33:265–276
41. Salminen E, Salminen S, Marks V, Bridges JW (1983) Urinary excretion of orally administered oxalic acid in xylitol fed mice. In: Hayes AW et al. (eds) Developments in the science and practice of toxicology. Elsevier, Amsterdam, pp 333–336
42. Hollenbeck CB, Leklem JE, Riddle MC, Connor WE (1983) The composition and nutritional adequacy of subject-selected high carbohydrate low fat diets in insulin dependent diabetes mellitus. Am J Clin Nutr 38:41–51
43. Thornberg JM, Eckert CD (1984) Protection against sucrose-induced retinal capillary damage in the Wistar rat. J Nutr 114:1070–1075
44. Labib M, Gama R, Wright J, Marks V (1988) Dietary maladvice as a cause of hypothyroidism and short structure. Br Med J 298:232–233
45. Cerami A (1985) Accumulation of advanced glycosylation products on proteins and nucleic acids: role in aging. In: Eaton JW et al. (eds) Cellular and molecular aspects of aging: the red cell as a model. Alan R Liss, New York, pp 79–88
46. Winterhalter KH (1985) Non-enzymatic glycosylation of proteins. In: Eaton JW et al. (eds) Cellular and molecular aspects of aging: the red cell as a model. Alan R Liss, New York, pp 169–120
47. Bunn HF, Gabbay KJ, Gallop PM (1978) The glycosylation of hemoglobin. Relevance to diabetes mellitus. Science 200:21–27
48. Urbanowski JC, Cohenford MA, Dain JA (1982) Nonenzymatic galactosylation of human albumin. J Biol Chem 257:111–115
49. Urbanowski JC, Cohenford MA, Levy HL, Crawford JD, Dain JA (1982) Nonenzymatically galactosylated serum albumin in a galactosemic infant. N Engl J Med 306:84–86
50. Burden AC (1984) Fructose and misleading glycosylation data. Lancet II:986

Commentary

Riou: There is much evidence that a substantial amount of glycogen is synthesised in the liver by the indirect pathway [1], in which glucose is transformed in peripheral tissues into lactate and pyruvate which are recycled back to the liver and processed in the gluconeogenic pathway to form glycogen. Actually it is difficult to state whether in normal feeding this pathway represents 20% or 80% of the glucose-to-glycogen pathway. Nevertheless it is difficult to reject these new data completely. If the hypothesis is true it will mean that peripheral tissues are very important in the metabolic transformation of glucose to glycogen in the liver. It seems that when the glucose load is reasonably small the indirect pathway is predominant; and when the load is large the direct pathway (glucose to glucose-6-

phosphate to GIP to glycogen) is predominant. This could be important in the physiology and metabolism of different sugars.

Reference

1. Katz J, McGarry JD (1984) The glucose paradox: is glucose a substrate for liver metabolism? J Clin Invest 74:1901–1909

Southgate: The comments regarding the metabolic differences between the various sugars and the claims for differences in the nutritional properties of diets (such as they have been demonstrated) that are rich in sugars compared with starches are, as the authors state, probably due to other dietary compositional factors. It is true, however, that the consumption of products which contain "pure" starch as the major component are rare. The alterations in nutrient density of diets rich in sucrose may or may not be important, the key question being the conceptual basis of requirements and whether they are related to energy expenditure. If nutrient requirements are truly nutrient density specific, then changes in nutrient density would be expected to be nutritionally important.

Levin: While it is true that fructose is found in high concentrations in human semen it is not the exclusive source of metabolic fuel for the spermatozoa. Normal spermatozoa in the seminiferous tubules/epididymis are in fluids devoid of fructose. It is added at ejaculation from the seminal vesicle secretion. Thus in the male, fructose is not essential for sperm survival. In the female vagina the sperm rapidly leave the male fluids to swim into cervical/uterine/fallopian tube fluids where there is no fructose: there they utilise glucose. There is no evidence that dietary fructose has any role to play in the seminal vesicle fructose which comes from cell metabolism of the seminal vesicle. Dietary fructose is rapidly metabolised by the liver and there is thus usually no rise in blood fructose.

Macdonald: The metabolic effects of fructose depend on the sex of the consumer. In men and post-menopausal women a raised sucrose intake will increase the fasting level of triglyceride in serum, whereas this is not the case in pre-menopausal women [1].

Reference

1. Macdonald I (1976) Sex differences in the metabolic response to dietary carbohydrate. In: Berdanier C (ed) Carbohydrate metabolism. Hemisphere Publishing Corporation, Washington, pp 211–222

Chapter 8

Glycaemic Responses to Sugars and Starches

V. Marks

Introduction

It was recognised long before the era of insulin treatment or of clinical blood sugar measurements, that meals consisting mainly of carbohydrates led to a prompt and rapid increase in urinary glucose excretion by patients with diabetes mellitus, whereas meals consisting mainly of fat and protein did not. By excluding carbohydrates from the diet the life of sufferers from diabetes could sometimes be prolonged by a few years. Fructose alone amongst the carbohydrates appeared not to produce any adverse clinical effects and for some time before the discovery of insulin, and for a short time after, it was used in the treatment of diabetes [1].

Even after the introduction of insulin therapy, however, dietary carbohydrates were considered anathema for diabetics, and their intake was rigorously controlled. Indeed, until comparatively recently restriction of carbohydrate was considered more important than all the other dietary considerations put together.

For more than half a century it has been taken as axiomatic that blood sugar levels rise less after starchy foods than after sugars, and this is the basis for the recommendation to diabetics that all simple sugars should be excluded from their diet. By a similar logic, individuals said to have "reactive hypoglycaemia", also known simply as hypoglycaemia, resulting from too rapid an early rise in blood glucose concentration in response to eating a meal, were advised to avoid sugar and other rapidly absorbed carbohydrates.

It is, however, difficult to see upon what evidence this widely held belief was based since, as long ago as 1913, Jacobsen [2] had shown that, in normal healthy subjects, capillary blood glucose levels rose just as much after meals consisting largely of starchy foods (e.g. bread and potatoes) as they did after glucose. This was confirmed by MacLean and de Wesselow [3] who, in 1921, wrote "the rapid splitting of the starches and disaccharides is very noticeable; the curve obtained with cane-sugar and potatoes showing quite as rapid a rise as with glucose. Indeed, potatoes seem to be very suitable for testing the diabetic patient since they are looked on by most diabetics with less suspicion than an actual dose of sugar". Bread too gave almost identical blood sugar curves to those produced by glucose.

This early work, which was confirmed many times in diabetics [4] as well as in normal subjects, seems largely to have been forgotten. Admittedly it was poorly controlled and performed on only a small number of subjects. Moreover, it is open to criticism that the blood sugar methods available were not specific for glucose and, although glucose is much the most important sugar present in the circulation, it is not the only one. Other sugars, notably fructose, can appear in the blood after ingestion of foods containing them [5]. Indeed, the fall in blood glucose levels that invariably accompanies the ingestion of fructose-containing foods by patients with hereditary fructose intolerance was overlooked by Chambers and Pratt [6] in the first patient with this condition to have been described in the literature, because the methods they used to measure blood sugar also measured fructose as though it were glucose [7].

That the presence of other sugars in blood might not be as clinically irrelevant as has always been supposed, is suggested by the fact that in vivo they are non-enzymatically incorporated into proteins just as actively as glucose [9], and possibly even more so [8]. Since post-translational glycosylation of proteins is now thought to account for many of the adverse effects of prolonged hyperglycaemia [10] it is appropriate, with the emphasis in the dietary management of diabetes having shifted from the use of low to high carbohydrate diets, to examine critically the effects upon blood sugar levels of dietary carbohydrates which differ from one another, not only in their physical form but also chemically. Before doing so, however, it may be instructive to consider the factors that can affect the glycaemic responses to meals (including drinks), the methods available for measuring them and the reasons for doing so.

Because there are large inter-species differences in the way that sugars and starches are handled, both in the gut and after absorption, I have confined this commentary as far as possible to studies on human beings, and have distinguished between those carried out on normal healthy volunteers and those on patients with one or other of the two main types of diabetes, i.e. non-insulin-dependent diabetes mellitus (NIDDM) and insulin-dependent diabetes mellitus (IDDM).

Reasons for Measuring Blood Sugar Concentrations

Apart from scientific curiosity, the main reasons for measuring blood glucose concentrations are that when they are high they may be associated causally with either the acute or chronic manifestations of diabetes, and when they are low they may cause brain malfunction (neuroglycopenia). They are influenced by a large number of factors, of which the presence or absence of disease, especially diabetes (which can be looked upon as a disorder of insulin action) is overwhelmingly the most important. It has proved quite impossible to extrapolate from data obtained from studies of carbohydrate metabolism in normal healthy subjects, and their glycaemic responses to various foods, to what is likely to happen in diabetic or other patients. The converse is equally true.

Factors Affecting Glycaemic Responses to Meals

The most important factors, apart from disease, that affect the magnitude and time course of the glycaemic response to the ingestion of carbohydrates in solution or to carbohydrate-containing meals are listed below, though probably not in order of importance.

1. The specificity of the method used for measuring the blood sugar.
2. Whether arterial or venous blood is sampled.
3. Whether long-term or only acute responses are taken into account.
4. The chemical nature of the carbohydrate under study, and especially the monosaccharides into which it is split by hydrolysis in the gut.
5. Whether it is taken alone or with other nutrients.
6. Its physical form and texture.
7. The "dose" or amount of food eaten at one sitting.
8. Whether it is raw, cooked or processed in some other way.

Also important are:

9. The energy content and composition of the habitual diet.
10. Whether healthy subjects or those with some abnormality of carbohydrate metabolism and/or intestinal function are being investigated.

Only some of these factors can be considered in detail here.

Method of Blood Sugar Measurement

Clinically useful methods for measuring blood sugar concentrations first became available in the early 1920s, but it was not until much later that methods specific for glucose became popular. They are still not used by everyone.

Fructose, galactose, and many poorly identified substances known collectively as "saccharoids", are detected and measured as though glucose by the older reductiometric blood sugar techniques. Though neither fructose nor galactose appear ordinarily to make a major contribution to total blood sugar, they may do so under some rare circumstances. Consequently, comparisons of glycaemic responses to foods that do not utilise glucose-specific methodologies must be viewed with suspicion. There is, however, so little information available with regard to blood fructose and galactose levels following ordinary feeding, and obtained using modern, specific and sensitive methods of measurement, that no further consideration will be given to this potentially important topic here.

Site of Collection of Blood Sample

In normal healthy subjects there is virtually no difference between fasting arterial, capillary and venous blood glucose levels, reflecting the almost complete absence of glucose uptake by the forearm in the post-absorptive and fasting states [11,12].

The situation changes dramatically and immediately when food containing absorbable carbohydrates is ingested [12,13]. In arterial blood the glucose concentration rises rapidly, but soon falls to just above fasting levels, where it remains until absorption is complete. In venous blood, on the other hand, the rise in glucose concentration occurs more slowly, and to a smaller extent. It generally returns to, or commonly below, the fasting level within 2 hours of ingestion of the meal – long before its absorption from the gut is complete. The resulting arterior-venous (A-V) blood glucose difference is a measure of the uptake of glucose by the tissues that occurs under the influence of insulin secreted in response to ingestion of the meal, and is a useful indication of the extent of glucose assimilation by all the tissues of the body apart from the liver.

The glucose concentration in capillary blood and in blood from the distended veins of the warmed hand, so-called arterialised blood, is similar to that of arterial blood throughout the absorptive phase [13]. Consequently, glucose measurements made on capillary or arterialised venous blood give a far more meaningful indication of the true glycaemic response to food than can be obtained by the use of mixed forearm venous blood.

In general the more, and particularly the earlier, insulin is secreted in response to a meal, and the more insulin-sensitive the tissues, the greater the A-V glucose difference. In healthy young subjects with good glucose tolerance the A-V glucose difference can be 3.0 mmol/l or more and can exceed the actual rise in arterial blood glucose concentration. This can give rise to the apparent absurdity, if venous blood alone is sampled, of a fall in blood glucose concentration to below fasting levels during a period of active glucose absorption [13].

This fact was well known to clinical investigators long before its cause had been elucidated [3,7,14]. It appears, however, to have escaped the attention of many modern authors, who have continued not only to use venous rather than arterialised, or capillary, blood in situations where it is wholly inappropriate, but also to draw erroneous conclusions from their results. In comparing glycaemic responses to foods and, more especially, meals of different composition and different insulin stimulatory properties, totally different results will be recorded depending on whether arterial or venous blood is sampled.

It has long been known [3,12,15,16] that in normal healthy persons the extent, and even duration, of the rise in blood glucose concentration, especially when measured in venous blood, bears little relationship to the size of the glucose load once it exceeds approximately 50 g. This is due to a combination of saturation of the absorptive mechanism for glucose, on the one hand, and an increase in the rate of glucose assimilation, under the influence of insulin, on the other. The situation is totally different in people with diabetes, especially of the insulin-dependent variety, where the blood glucose concentration continues to rise in response to an increase in the size of the glucose load, only reaching a plateau when the rate of renal excretion of glucose approximates that of glucose absorption [12]. People with NIDDM exhibit a pattern mid-way between the two extremes of normal and total insulin deficiency, as do individuals with other disorders of glucose homeostasis or gastrointestinal malfunction.

Because the perturbations in blood glucose concentration produced by carbohydrate feeding in normal healthy subjects are ordinarily very small, quite small changes in absolute terms can result in seemingly large differences between the size of response to different food when expressed in percentage terms. The same considerations do not apply in diabetes.

Methods of Comparison

Traditionally, glycaemic responses to different nutrients or meals have been compared by plotting mean blood sugar concentrations against time for varying periods, usually 2 or occasionally 3 hours, following ingestion of a test meal. Since the time course and shapes of curves so obtained are seldom the same with meals of different composition and texture, valid comparisons are difficult, if not impossible to make, especially if, as often happens when venous blood is sampled, the glucose concentration is below the pre-glucose ingestion baseline for much of the time.

Short-Term Comparisons

The most popular method of comparison in recent years has been expression of the glycaemic response or the area under the curve (AUC) during the 2, or more commonly 3 hours after ingestion of the test meal [18–20]. The result can be expressed either in absolute terms, i.e. mmol glucose/hour [18], or relative to the AUC produced by an equivalent amount of glucose or other "standard material" such as white bread [20].

The relative AUC has been called the "glycaemic index". For some time following its initial description and widespread application to foods by Jenkins and co-workers, the glycaemic index enjoyed considerable popularity as a tool for the dietetic management of diabetes and hypoglycaemia, but it is, at best, a crude instrument of questionable value [20,21]. Reproducibility of the glycaemic index of foods between centres is low, partly, I suspect, because of inconsistencies in the sampling site, some authors using mixed venous, rather than arterialised or capillary blood as originally described by Jenkins and co-workers [19].

This is not the whole story, however, since even oral glucose tolerance tests are notoriously non-reproducible, not only between individuals, but in the same individual from one occasion to another [23].

The shape of the blood glucose curve and the AUC are determined by the balance between the rate at which glucose enters the body glucose pool and the rate at which it leaves it to be assimilated by the tissues. Factors affecting the rate of entry of dietary carbohydrates into the body glucose pool are many, varied and complex, but are simplified by the fact that, regardless of their chemical composition at the time of ingestion, carbohydrates are absorbed only as their constituent monosaccharides. Fructose and galactose, whether taken as monosaccharides or as sucrose and lactose respectively, make relatively little contribution to the size of the glucose pool in healthy subjects and can, therefore, be largely discounted.

The rate of gastric emptying is an important determinant of carbohydrate absorption which commences in the duodenum and, except in the case of the so-called resistant starches and dietary fibres, is normally completed in the jejunum.

Ordinary meals comprise a mixture of liquid and solid components, and it has become clear that different mechanisms are involved in emptying liquids and solids from the stomach [24]. Any comparison of the glycaemic effects of nutrients that does not take this into account is consequently suspect. Yet only very rarely in the past have sugars been tested for their glycaemic effect when taken as constituents of solid foods rather than as solutions, on, usually, an empty stomach.

Small particles are expelled from the stomach more quickly than larger ones, though more slowly than liquids, and in ways that are still poorly understood. The physical properties of food are therefore important [25–27] in determining not only the rate of access of digestive enzymes to the insoluble nutrients within the food [28], but also the rate at which it is delivered to the duodenum. Energy density [29], and also the chemical composition of the food, in some ways determines the rate of gastric emptying and consequently rate of delivery of digested food to the absorptive surfaces in the small intestine. Hyperglycaemia itself, produced by intravenous glucose for example, slows gastric emptying, possibly through a mechanism involving the release of insulin [30], and this could account for some of the differences that have been observed in gastric emptying between normal and diabetic subjects in some of whom the normal discrimination between solid and liquid gastric emptying is lost.

Gastric emptying is difficult to measure accurately, and delayed emptying is more often inferred from delayed appearance of nutrients (or markers) in the blood than demonstrated by direct measurement. Delayed gastric emptying has, for example, been invoked to explain the smaller rises in blood glucose concentration that occur when the soluble fibre, guar gum, is taken with glucose or a carbohydrate-containing meal, than when carbohydrates are taken alone. Direct measurement of gastric emptying, however, shows this not to be the case [30,31]: the explanation for the smaller rise in blood glucose concentration lies in its retarded absorption in the small intestine. This is possibly a direct specific effect of the guar upon glucose absorption, since it appears not to affect the absorption of fats [32] or alcohol [33] to the same extent, if at all.

Long-Term Comparisons

In recent years clinicians interested in disorders of glucose homeostasis have become more concerned about long-term changes in blood glucose concentration than in the short-term changes that follow ingestion of a single meal. Until the discovery and application of post-translational glycosylation of proteins by glucose and other sugars, including fructose, galactose and mannose, there was no valid method of assessing average blood glucose levels over prolonged periods. It is now possible, however, by measuring glycolated blood proteins, such as haemoglobin A_{1c} and glycolated albumin, to obtain a useful assessment of the prevailing blood glucose concentration during the preceding few months. This is thought to correlate better with the risks of sustaining "diabetic complications" than do single or even multiple blood glucose measurements [10].

Chemical Composition

The main forms in which digestible carbohydrates are ingested, in decreasing order of quantitative importance, are starch, sucrose, lactose, fructose and glucose. Only the last two do not undergo digestion before they are absorbed, but whereas lactose and sucrose are single chemical entities with known composition and predictable behaviour within the gut, this is not true of starch, which is really a class of polysaccharides rather than a single substance. Starch is the main storage carbohydrate of most plants and is composed of the straight-chain glucose polymer, amylose, and the branched polymer, amylopectin. The ratio of amylose to amylopectin varies widely between the starches and may be one determinant of their individual rates of digestibility [33]. Not only are different enzymes involved in hydrolysis of each of the two constituent parts, but hydrolysis of amylopectin is generally incomplete whereas amylose forms highly organised and insoluble aggregates, and cleavage sites become inaccessible to α-amylase. Starch, unlike the simple sugars, is insoluble in water, although it can be suspended as a sol. It exists in plants and foods as granules which vary in shape and size and in the substances with which they are associated. Starch is uncooked wheat flour, for example, is mixed with gluten, and this reduces its digestibility to such an extent [34] that up to 15% of the granules survive digestion in the small intestine, only to be delivered into the colon where they can undergo fermentation into smaller absorbable compounds. Food processing, especially toasting or excessive heating followed by storage, can convert available starch into a form that cannot be completely digested by human enzymes [35].

Cooking and other forms of processing radically alter the rate and extent of digestion of "whole foods" and even pure starch preparations, which are generally very poorly absorbed unless they are cooked before ingestion [7,36]. The glycaemic responses to plant foods depend, therefore, just as much, and often more, upon how they were treated before they were eaten as upon their composition [7,37–40]. The physical form is also very important [37]. The variation in the glycaemic response to rice, for example, depends much more upon [25] whether it is eaten ground or whole grain than on whether it is polished (white) or intact (brown). The same applies to white beans, potatoes and most other vegetable and prepared foods that have been investigated. These differences can, in part, be explained on the basis of the different rates at which these foods leave the stomach, but, more importantly, the variations reflect the rate at which they are solubilised by intraluminal enzymes. Indeed, Jenkins and co-workers have shown a significant, though far from perfect, correlation between the in vitro digestibility of foods using human post-prandial jejunal juice and their glycaemic potential in vivo [28].

The effect of physical form alone, independent of digestion, was investigated by Haber and co-workers using whole apples, apple purée and apple juice [26], all of which contain their available dietary carbohydrate only in the form of sugar, namely glucose (15.4%), fructose (63.8%) and sucrose (24.6%). The apple juice was processed to remove the soluble fibre, pectin, before consumption. The three foods produced equally rapid and large rises in blood glucose concentration of similar duration, but, whereas the juice and puréed apples both produced a "rebound" hypoglycaemia, whole apples, which also produced the smallest rise in plasma insulin levels, did not. Plasma glucose was measured in mixed venous blood, and consequently the significance of the "reactive hypoglycaemia" that was observed only with the juice and purée is difficult to assess. It may merely have indicated the presence of a large A-V glucose difference at a time when active absorption of carbohydrate was still occurring [13].

Other Food Constituents

The presence in the gut of other food constituents profoundly alters the rate of absorption and hormonal response and, consequently, the glycaemic responses to carbohydrate-containing foods [30,32,41,42]. Although most attention has been given to the effects of the various non-digestible carbohydrates, collectively known as "dietary fibre", upon this aspect of glucose homeostasis – due mainly to powerful advocacy – the modifying effects of fat and proteins is equally or more profound, albeit through a totally different mechanism. It is probably because of this latter fact that the concept of the glycaemic index of individual foods has little or no relevance to the real-life situation, in which mixtures rather than single foods are eaten as meals. Fats and proteins reduce the glycaemic response to carbohydrate mainly by increasing insulin secretion in response to hyperglycae-mia, but possibly also by delaying gastric emptying. Dietary fibres, on the other hand, work either by providing a mechanical barrier to the digestive juices acting upon the starch granules contained within the fibrous foods, or, in the case of soluble fibres such as guar and pectin, by interfering with the absorption of glucose produced by digestion of the starches within the lumen of the small intestine. Such interference is either mechanical or, as seems increasingly likely, produced by a physico-chemical mechanism, possibly involving an interaction between the soluble fibre and the glucoreceptor or the transporters on the surface of the enterocytes.

Glycaemic Response to Sugars and Starches in Solution

Among the various dietary sugars it has long been known that, given in equal amounts in solutions of equal molarity, glucose is the most hyperglycaemic and fructose the least [3,43,44]. Sucrose, lactose and galactose are, in descending order, intermediate between the two extremes. Pre-formed maltose is absorbed from the intestine even more rapidly than glucose, though the maximum rise in blood glucose level it produces is no greater [45,46]. This enhanced glycaemic effect is not seen when maltose is formed as an intermediate in the intraluminal hydrolysis of starch [46], and the explanation is far from clear. Something similar does occur with fructose, which is absorbed more rapidly and causes a faster rise in peripheral venous blood fructose levels when given as sucrose than when given in an equimolar mixture with glucose [47]. Even under these optimum conditions, however, the rise in plasma fructose levels in peripheral venous blood measured by specific enzymatic methods or preferably by gas–liquid chromatography (GLC), rarely exceeds 0.5 mmol/l [48]. Glucose levels, in contrast, commonly rise ten times as much, even in healthy subjects. Fructose taken alone in solution produces a small and delayed rise in blood glucose concentration, rarely exceeding 1 mmol/l in normal subjects [49], and only two to three times as much in patients with IDDM [50]. Fructose should not, therefore, be used to reverse hypoglycaemia should it occur in the course of insulin treatment for IDDM or NIDDM [51].

When taken with glucose, as high fructose corn syrup for example, fructose has no modulating effect upon the hyperglycaemia produced, which reflects only the amount of glucose rather than that of the total carbohydrate present in the mixture [52].

Galactose taken alone produces a consistently small but significant rise in blood glucose concentration, which is rarely greater than 1.0 mmol/l in normal healthy subjects [53,54]. The rise in plasma galactose in response to an oral galactose load does, however, vary enormously between individuals. In one series of 17 healthy volunteers [54] the maximum rise in plasma galactose in response to a dose of 0.5 g/kg varied 20-fold, between 0.23 and 4.56 mmol/l. Though fascinating, and still poorly understood, this finding has little relevance to real-life situations since, except under experimental conditions, galactose is never taken orally without glucose, either as lactose or as pre-formed lactose hydrolysate. Under these conditions the galactosaemic response to galactose is virtually abolished, except in people with severe liver disease or with one of the rare inborn errors of galactose metabolism. The cause and mechanism are unknown.

Alcohol delays removal of galactose from the blood, permitting its accumulation in the circulation after ingestion, even in the presence of glucose, to such an extent that the blood galactose becomes easily measurable.

Uncooked starches, whatever their origin, are slowly and possibly incompletely hydrolysed in the gut lumen, and produce minimal increases in blood glucose concentration following their ingestion as a suspension in water on an empty stomach (unpublished). When taken boiled, on the other hand, they produce a rise in venous blood glucose concentration indistinguishable from that of an equivalent amount of glucose [7,36]. Indeed, measurement of the glycaemic response to boiled starch has long been used by clinical biochemists as a diagnostic test for chronic pancreatic insufficiency.

Glycaemic Responses to Sugars in Solution Versus Solid Food

It has long been assumed, and recently reiterated [51], that sugars taken as confectionery, e.g. sweets and chocolates, are just as hyperglycaemic as glucose in solution, or pure glucose tablets taken with water. Brodows et al. [51], for example, suggested that a 2.5-oz. (70-g) milk chocolate bar containing 20 g glucose would be just as effective in combating insulin-induced hypoglycaemia as 20 g glucose in solution. In an elegant study, marred only by the fact that they used mixed, rather than arterialised venous blood for the blood glucose measurements, Shively et al. [55] showed that this assumption was totally unfounded. None of a number of different confectionery snacks, all containing at least 90% of their total of 25-g carbohydrate content in the form of sugars, produced anything like the same rate of rise in blood glucose concentration as pure glucose. The hyperglycaemic effect of the sugar-containing confectionery was, on the other hand, almost identical to that produced by potato chips (crisps), in which over 80% of the carbohydrate content was starch. Both the sugar-containing confectionery and the starch-containing potato chips contained varying amounts of fat and protein, which themselves can modify glycaemic responses to dietary carbohydrates.

Effects of Non-carbohydrate Constituents on the Glycaemic Response to Foods

Jacobsen pointed out over 70 years ago [2] that bread with butter on produced a much smaller rise in blood glucose concentration than bread alone. The attenuating effect of co-ingested fat and, to a lesser extent, proteins upon the glycaemic effect of carbohydrate-containing food was attributed mainly to delayed gastric emptying. It is now known to be at least equally due to increased glucose assimilation, under the influence of the increased amounts of insulin secreted in response to a mixed meal, despite the lower arterial blood glucose concentrations [13]. This moderating effect of fat and protein does not occur in untreated patients with IDDM and is reduced in patients with NIDDM, in whom impaired insulin release and increased insulin resistance are common.

Blood Glucose Responses to Meals

Most of the dietary sugars ingested, and virtually all of the starches, are taken as solid foods, either in their native form (as whole vegetables and cereals) or as processed foods. These are generally eaten with other foods as a mixed meal. Under such circumstances the acute glycaemic response seen when various carbohydrate-containing foods are eaten singly has little relevance and makes no contribution to the achievement of glycaemic stability in patients with IDDM or NIDDM [21,22].

Acute and Chronic Glycaemic Responses to Sugary and Starchy Diets

Normal Subjects

Mann and Truswell [18] studied a group of metabolically normal, middle-aged men under rigidly controlled conditions, in which 70 g sucrose was substituted for either 70 g potatoes or 70 g rice (expressed as starch) in the midday and evening meals respectively. The glycaemic responses, expressed as AUC, were almost identical for the rice and the sucrose-substituted meals, but only two-thirds of that of the potato-containing meal.

Bantle et al. [56] studied ten healthy volunteers who, on each of five separate occasions, ate a traditional (American) mixed breakfast supplemented with 42 g carbohydrate as fructose, glucose, sucrose, wheat starch (as pancakes) or potato starch (as fried potato and potato starch) so as to provide 700 kcal (2940 kJ: 33% as fat, 18% as protein and 49% as carbohydrate). The venous blood glucose responses, expressed as AUC, were, in descending order of magnitude, glucose, wheat, potato, sucrose and fructose. The AUC for glucose was roughly five times as great as that for the fructose-substituted meal, but only 20% greater than for the wheat/starch-containing meal. This in turn was almost twice that of the sucrose-substituted meal.

NIDDM Patients

In contrast to the paucity of reports recording short-term blood glucose responses in normal healthy subjects to meals in which sugars were substituted for starches in mixed meals, there is a plethora of such reports relating to diabetics.

In a study already referred to, Bantle et al. [56] found the same order of glycaemic responses to carbohydrate-substituted breakfasts in ten patients with NIDDM and 12 patients with IDDM as in healthy control individuals. The absolute and relative values of the AUCs were, however, vastly different.

Slama et al. [57] measured mixed venous plasma glucose levels in 12 patients with NIDDM and six with IDDM after thay had consumed, on each of two occasions, a mixed lunch providing 517 kcal (2171 kJ: 49% carbohydrate, 25% fat, 26% protein). On one occasion 20 g of the 70 g total carbohydrate, supplied as rice starch, was substituted by sucrose. The blood glucose curves obtained in the NIDDM patients were superimposable, and their AUCs did not differ significantly from one another on the two occasions. In the IDDM patients, who throughout the period of observation were infused with insulin by an artificial pancreas, the amount of insulin required to maintain euglycaemia was the same after the sucrose-substituted as after the regular meal.

Hollenbeck et al. [58] compared blood glucose responses in 12 NIDDM patients to mixed meals which contained widely varying proportions of complex and simple carbohydrates but each provided 1000 kcal (4200 kJ). Meals that provided 80% of their carbohydrate energy as starches produced larger AUCs than those providing 80% of their carbohydrates as sugars, and also produced significantly more glycosuria.

Bornet et al. [59] observed the rise in venous blood glucose concentration in 21 NIDDM patients after they had consumed one of three types of breakfast providing 400 kcal (1680 kJ). The standard breakfast provided 51 g carbohydrate as white bread, 15 g of which was replaced either by sucrose or by honey. The blood glucose curves were superimposable, regardless of whether the patients were, or were not, well controlled.

Peterson and co-workers [60] studied 11 NIDDM and 12 IDDM patients who were given diets of similar composition, except that in one, 45 g of complex carbohydrate in the form of starch was replaced by 45 g sucrose each day over two consecutive periods, each lasting 6 weeks. All measurements of glycaemic control, including the size of the blood glucose response to individual meals, mean daily blood sugar levels, fasting blood sugar levels, during the course of the two dietary periods, and the level of glycosylated haemoglobin at the end of each of them were indistinguishable. The authors commented that the amount of sugar (45 g/day) provided by the sucrose diet in their study was below the national average [60].

This was not the case, however, in the study performed by Bantle et al. [61], in which 12 subjects with NIDDM and 12 with IDDM participated. The protocol called for the volunteers to eat one of three isocaloric diets for 8 days, followed first by one and then by the second alternative in random order. All three diets provided 55% of their energy as carbohydrate, 15% as protein and 30% as fat. In one of the diets 21% of total energy was provided as fructose, in another 23% came from sucrose, and in the last all of the carbohydrate energy came from starch. In this study too, all measures of glycaemic control were comparable between the starch and sucrose-substituted diets, especially in the NIDDM patients. Peak plasma glucose levels and AUC were generally lower after fructose-substituted meals, but mean glycosylated albumin levels at the conclusion of each period were not significantly different between the three diets.

An even more vigorous test of the effect of substituting sucrose for starch in the diet of diabetic patients was carried out by Abraira and Derler [62]. They fed 17 NIDDM volunteers, confined within a special diagnostic treatment unit over a 40-day period, with diets which provided 50% of their energy as carbohydrate, 15% as protein and 35% as fat. After assessing each subject's daily energy requirements during the first 5 days, the volunteers were fed a diet which contained either 220 g sucrose or its equivalent weight of starch. All measurements of glycaemic control, including post-prandial mixed venous blood glucose levels and glycosylated haemoglobin levels, were indistinguishable between the two groups of matched subjects throughout the period of observation.

Reactive Hypoglycaemia

The pioneers of blood sugar measurement generally remarked upon the fall in blood glucose to below fasting levels that characteristically occurred whenever a large amount of glucose was consumed by normal people on an empty stomach [3,7,14,16]. The hypoglycaemic rebound is much more pronounced if venous rather than arterial blood is sampled, and in a small minority of subjects may be accompanied by neuroglycopenia. This is especially likely to occur with larger doses of glucose and in people who, for some reason or other, have accelerated gastric emptying. However, since it does not occur except in very rare circumstances following ingestion of mixed meals, regardless of their refined sugar content, the concept that dietary sugars, especially in the form of white sugar, uniquely predispose to hypoglycaemia – and for which there has never been any documentary evidence – must now be regarded as finally disproved [24,63,66].

Comment

The public perception of sugars, especially "added sugar", as more detrimental to health than starches as a source of dietary carbohydrate, and which has been promoted by a number of writers of "popular" books and magazine articles for reasons that are not always clear, appears to have little basis in fact. The issue has been confused by reference to "empty calories", which is a deliberately and unjustifiably derogatory term applied to nutrients such as sucrose and alcohol which supply dietary energy without any of the other nutrients and anutrients associated with the ingestion of unrefined and unprocessed starchy foods.

In so far as the glycaemic responses to sugars and starches are concerned, they depend much more upon the nature of the monosaccharides from which the compounds are constructed, the presence of other nutrients (such as fats, proteins and dietary fibre) and anutrients, and the physical form in which they are ingested, than upon whether they are or are not defined as sugars or starches.

Summary and Conclusions

There is now a wealth of experimental evidence to support the conclusion reached by earlier investigators: that although it may be possible to demonstrate different blood sugar responses to different dietary sugars and starches under laboratory conditions, these responses are of little relevance under real-life conditions when meals and diets are considered. The importance attached to the nature and source of dietary carbohydrates, at least in so far as their effects upon blood glucose concentrations are concerned, may have distracted attention away from other more important nutritional considerations relating to intestinal motility and hormonal responses to foods and their disposition within the body.

A possible explanation for the comparative lack of importance of the chemical nature of dietary carbohydrates in the overall metabolic economy of the body, except in the presence of certain genetic traits, is that, apart from rare cases of malabsorption, all dietary carbohydrates eventually enter the body, where, through various hormone and substrate-controlled enzyme systems, they are interconvertible.

References

1. Moorhouse JA, Kark RM (1957) Fructose and diabetes. Am J Med 23:46–58
2. Jacobsen AThB (1913) Untersuchungen uber den Einfluss verschiedener Nahrungsmittle auf den Bludzucker bei normalen, zuckerkranken und graviden Personen. Biochem Z 56:471–494
3. MacLean H, de Wesselow OLV (1920) The estimation of sugar tolerance. Q J Med 14:103–119
4. Wishnofsky M, Kane AP (1935) The effect of equivalent amounts of dextrose and starch on glycemia and glucosuria in diabetics. Am J Med Sci 189:545–550
5. Bantle JP, Laine DC, Thomas JW (1986) Metabolic effects of dietary fructose and sucrose in Types I and II diabetic subjects. JAMA 23:3241–3246
6. Chambers RA, Pratt RTC (1956) Idiosyncrasy to fructose. Lancet II:340
7. Folin O, Berglund H (1922) Some new observations and interpretations with reference to transportational, retention and exretion of carbohydrates. J Biol Chem 51:213–269
8. Urbanowski JC, Cohenford MA, Dain JA (1982) Non-enzymatic galactosylation of human serum albumin: in vitro preparation. J Biol Chem 257:111–115
9. Urbanowski JC, Cohenford MA, Levy HL, Crawford JD, Dain JA (1982) Non-enzymatically galactosylated serum albumin in a galactosemic infant. N Engl J Med 306:84–86
10. Brownlee M, Cerami A, Vlassara H (1988) Advanced glycosylation end products in tissue and the biochemical basis of diabetic complications. N Engl J Med 318:1315–1321

11. Whichelow MJ, Wigglesworth A, Cox BD, Butterfield WJH, Abrams ME (1967) Critical analysis of blood sugar measurements in diabetes detection and diagnosis. Diabetes 16:219–226
12. Peters J, Van Slyke KK (1946) Quantitative clinical chemistry: interpretation, vol 1. Bailliere, London
13. Jackson RA, Blix PM, Matthews JA, Morgan LM, Rubenstein AH Nabarro JDN (1983) Comparison of peripheral glucose uptake after oral glucose loading and a mixed meal. Metabolism 32:706–710
14. Foster GL (1923) Some comparisons of blood sugar concentrations in venous blood and in finger blood. J Biol Chem 55:291–307
15. Bailey CV (1919) Studies on alimentary hyperglycemia and glycosuria. Arch Intern Med 23:455–483
16. Hansen KM (1923) Investigations on the blood sugar in man. Conditions of oscillations, rise and distribution. Acta Med Scand 1[Suppl IV]
17. Otto H, Bleyer G, Pennartz M, Sabin G, Schauberger G, Spaethe R (1973) Kohlenhydrataustausch nach biologischen Aquivalenten. In: Otto H, Spaethe R (eds) Diatetic bei Diabetes Mellitus. Verlag Hans Huber, Bern, pp 40–50
18. Mann JI, Truswell AS (1972) Effects of isocaloric exchange of dietary sucrose and starch on fasting serum lipids, postprandial insulin secretion and alimentary lipaemia in human subjects. Br J Nutr 271:395–405
19. Jenkins DJA, Wolever TMS, Taylor RH et al. (1980) Rate of digestion of foods and postprandial glycaemia in normal and diabetic subjects. Br Med J 281:578–585
20. Jenkins DJA, Wolever TMS, Jenkins AL, Josse RG, Wong GS (1984) The glycaemic response to carbohydrate foods. Lancet II:388–392
21. Laine DC, Thomas WMS, Levitt MD, Bantle JP (1987) Comparison of predictive capabilities of diabetic exchange lists and glycemic index of foods. Diabetes Care 10:387–394
22. Coulston AM, Hollenbeck CB, Swislocki ALM, Reaven GM (1987) Effect of source of dietary carbohydrate on plasma glucose and insulin responses to mixed meals in subjects with NIDDM. Diabetes Care 10:395–400
23. McDonald GW, Fisher GF, Burnham C (1965) Reproducibility of the oral glucose tolerance test. Diabetes 14:473–480
24. Schusdziarra V, Dangel G, Klier M, Henricks I, Pfeiffer EF (1981) Effect of solid and liquid carbohydrates upon postprandial pancreatic endocrine function. J Clin Endocrin 53:16–20
25. O'Dea K, Nestel PJ, Antonoff L (1980) Physical factors influencing post-prandial glucose and insulin responses to starch. Am J Clin Nutr 33:760–765
26. Haber GB, Heaton KW, Murphy D, Burroughs LF (1977) Depletion and distribution of dietary fibre. Effects on satiety, plasma-glucose and serum-insulin. Lancet II:679–682
27. Arends J, Ahrens K (1987) Physical factors influencing the blood glucose response to different breads in Type II diabetic patients. Klin Wochenschr 65:469–474
28. Jenkins DJA, Ghafari H, Wolever TMS et al. (1982) Relationship between rate of digestion of foods and post-prandial glycaemia. Diabetologia 22:450–455
29. Hunt JN, Cash R, Newland P (1975) Energy density of food, gastric emptying and obesity. Lancet II:905–906
30. Morgan LM, Tredger JA, Hampton SM, French AP, Peake JCF, Marks V (1988) The effect of dietary modification and hyperglycaemia on gastric emptying and gastric inhibitory polypeptide (GIP) secretion. Br J Nutr 60:29–37
31. Rainbird AL, Low AG (1986) Effect of guar gum on gastric emptying in growing pigs. Br J Nutr 55:87–98
32. Morgan LM, Goulder TJ, Tsiolakis D, Marks V, Alberti KGMM (1979) The effect of unabsorbable carbohydrate on gut hormones. Diabetologia 17:85–89
33. Tredger J, Wright J, Marks V (1979) Use of guar gum to prevent alcohol induced reactive hypoglycaemia. Br J Alcohol Alcohol 16:135–140
33. Goddard MS, Young G, Marcus R (1984) The effect of amylose content on insulin and glucose response to ingested rice. Am J Clin Nutr 39:388–392
34. Anderson IH, Levine AS, Levitt MD (1981) Incomplete absorption of the carbohydrate in all-purpose wheat flour. N Engl J Med 304:891–892
35. Engyst H, Wiggins HS, Cummings JH (1982) Determination of the non-starch polysaccharides in plant foods by gas-liquid chromatography of constituent sugars as alditol acetates. Analyst 107:308–318
36. Collings P, Williams C, MacDonald I (1981) Effects of cooking on serum glucose and insulin responses to starch. Br Med J 282:1032

37. Collier G, O'Dea K (1982) Effect of physical form of carbohydrate on the postprandial glucose, insulin and gastric inhibitory peptide responses in type 2 diabetes. Am J Clin Nutr 36:10–14
38. Brand JC, Nicholson PI, Thorburn AW. Truswell AS (1985) Food processing and the glycemic index 1–3 Am J Clin Nutr 42:1192–1196
39. Golay A, Coulston A, Hollenbeck CB, Kaiser LL, Wursch P, Reaven CM (1986) Comparison of metabolic effects of white beans processed into two different physical forms. Diabetes Care 9:260–266
40. Wolever TMS, Jenkins DJA, Thompson LU, Wong GS, Josse RG (1987) Effect of canning on the blood glucose response to beans in patients with type 2 diabetes. Hum Nutr Clin Nutr 41C:135–140
41. Estrich D, Ravnik A, Schlierf G, Fukayama GAB, Kinsell L (1967) Effects of co-ingestion of fat and protein upon carbohydrate-induced hyperglycemia. Diabetes 16:232–237
42. Collier G, McLean A, O'Dea K (1984) Effect of co-ingestion of fat on the metabolic responses to slowly and rapidly absorbed carbohydrates. Diabetologia 26:50–54
43. Field DW (1919) Blood sugar curves with glucose, lactose, maltose, mannitose and cane-sugar. Proc Soc Exp Biol Med 17:29–30
44. Swan DC, Davidson P, Albrink MJ (1966) Effect of simple and complex carbohydrates on plasma non-esterified fatty acids, plasma-sugar and plasma-insulin during oral carbohydrate tolerance tests. Lancet I:60–63
45. Dodds C, Fairweather FA, Miller AL, Rose CFM (1959) Blood-sugar response of normal adults to dextrose, sucrose and liquid glucose. Lancet II:485–488
46. Szepesi B, Michaelis OE (1986) Disaccharide effect – comparison of metabolic effects of the intake of disaccharides and their monosaccharide constituents. Prog Biochem Pharmacol 21:192–218
47. McDonald I, Turner LJ (1968) Serum fructose levels after sucrose or its constituent monosaccharides. Lancet I:841–843
48. Gerard J, Jandrain B, Pirnay F et al. (1986) Utilisation of oral sucrose load during exercise in humans. Effect of the α-glucosidase inhibitor acarbose. Diabetes 35:1294–1301
49. MacDonald I, Keyser A, Pacy D (1978) Some effects in man, of varying the load of glucose sucrose, fructose or sorbitol on various metabolites in blood. Am J Clin Nutr 31:1305–1311
50. Crapo PA, Kolterman OG, Okefsky JM (1980) Effects of oral fructose in normal, diabetic and impaired glucose tolerance subjects. Diabetes Care 3:575–581
51. Brodows RG, Williams C, Amatruda JM (1984) Treatment of insulin reaction in diabetics. JAMA 252:3378–3381
52. Akgun S, Ertel NH (1981) Plasma glucose and insulin after fructose and high-fructose corn syrup meals in subjects with non-insulin-dependent diabetes mellitus. Diabetes Care 4:464–467
53. Morgan LM, Wright JW, Marks V (1979) The effect of oral galactose on GIP and insulin secretion in man. Diabetologia 16:235–239
54. Williams CA, Phillips T, MacDonald I (1983) The influence of glucose on serum galactose levels in man. Metabolism 32:250–256
55. Shively CA, Apgar LL, Tarka SM (1986) Postprandial glucose and insulin responses to various snacks of equivalent carbohydrate content in normal subjects. Am J Clin Nutr 43:335–342
56. Bantle JP, Laine DC, Castle GW et al. (1983) Postprandial glucose and insulin responses to meals containing different carbohydrates in normal and diabetic subjects. N Engl J Med 309:7–12
57. Slama G, Haardt MJ, Jean-Joseph P (1984) Sucrose taken during mixed meal has no additional hyperglycaemic action over isocaloric amounts of starch in well-controlled diabetics. Lancet II:122–126
58. Hollenbeck CB, Coulston AM, Donner CC, Williams RA, Reaven GM (1985) The effects of variations in percent of naturally occurring complex and simple carbohydrates on plasma glucose and insulin response in individuals with non-insulin-dependent diabetes mellitus. Diabetes 34:151–155
59. Bornet F, Haardt MJ, Costagliola D, Blayo A, Slama G (1985) Sucrose or honey at breakfast have no additional acute hyperglycaemic effect over an isoglucidic amount of bread in type 2 diabetic patients. Diabetologia 28:213–217
60. Peterson DB, Lambert J, Gerring S (1986) Sucrose in the diet of diabetic patients – just another carbohydrate? Diabetologia 29:216–220
61. Bantle JP, Laine DC, Thomas JW (1986) Metabolic effects of dietary fructose and sucrose in types I and II diabetic subjects. JAMA 256:3241–3246
62. Abraira C, Derler J (1988) Large variations of sucrose in constant carbohydrate diets in type II diabetes. Am J Med 84:193–200
63. Charles MA, Hofeldt F, Shackleford A (1981) Comparison of oral glucose tolerance tests and

mixed meals in patients with apparent idiopathic postabsorptive hypogylcaemia. Absence of hypoglycemia after meals. Diabetes 30:465–470

64. Buss RW, Kansal PC, Roddam RF, Pino J, Boshell BR (1982) Mixed meal tolerance test and reactive hypoglycemia. Horm Metab Res 14:281–283

65. Hogan MJ, Service FJ, Sharbrough FW, Gerich JE (1983) Oral glucose tolerance test compared with a mixed meal in the diagnosis of reactive hypoglycaemia. A caveat on stimulation. Mayo Clin Proc 58:491–596

66. Lefebvre P, Andreani K, Marks V, Creutzfeldt W (1988) Statement on postprandial or reactive hypoglycaemia. Diabetes Care 11:439

Commentary

Levin: Glucose can be absorbed into the blood stream by the duodenum, jejunum and ileum. "Absorbed" in this sense means the translocation of the hexose from the lumen, across or between the enterocytes into the tissue fluid beneath the enterocyte layer. From there it will diffuse into the blood or lymph streams. Glucose so absorbed is thus in the blood stream, and the rest of the body can use it, and will not recognise that it has been absorbed from any special part of the gut. Other glucose molecules probably enter the endocrine cells at the crypts from the lumen and activate the cells to secrete the various hormones. It is highly unlikely that this "absorbed" glucose passes into the blood stream for subsequent metabolic use. It certainly will not be a metabolically significant quantity in relation to the joules it could provide, but it will of course be fundamental to the hormones it releases. In this sense the "absorption" of glucose into the crypt cells is not really true "absorption", for the translocation into the blood or lymph is not required. One other possible way that the endocrine cells can be influenced by the "absorbed" glucose from the "metabolically active load" may be the entry of already-absorbed glucose diffusing from the tissue fluid into the "back door", as it were, of the cells.

The point of all this is that while the site of the absorbed glucose load will not matter to the rest of the body, the hormones that the glucose can affect at each site (duodenal, jejunal and ileal) will obviously be important. We should distinguish between the effects of glucose as a hormone releaser and as a simple provider of metabolic fuel at each intestinal site.

Würsch: Mixed-meal tolerance tests have produced very contradictory results due to the introduction of a number of uncontrolled variables in the meals: type of starchy food (source, processing, form), amount of protein and fat, and interference from dietary fibre. Another important variable which has been recently stressed is the wide intra- and interindividual variation in glycaemic response to any given load [1]. Nevertheless, positive correlation between the glycaemic response to the single food, and the response when it is eaten in a mixed meal, has often been reported (for more detail, see Chap. 2, reference 1).

Reference

1. Coulston AM, Hollenbeck CB, Swislocki LM, Reaven GM (1987) Effect of source of dietary carbohydrate on plasma glucose and insulin responses to mixed meals in subjects with NIDDM. Diabetes Care 10:195–200

Flourié: Sorbitol does not require insulin in its metabolism. It has been proposed as food for diabetics, and it is commonly used in several countries, such as Germany. Do you think there are definitive data on the rationale or on the beneficial effect of sorbitol in the treatment of diabetes?

Author's reply: It is a misconception that sorbitol does not require insulin for its metabolism in vivo. Without insulin human beings cannot live for more than a very short time. Gluconeogenesis in the liver produces glucose which is released into the blood stream from anything and everything capable of serving as a substrate. Sorbitol is converted stoichiometrically into fructose and eventually into glucose. In the non-insulin-dependent diabetic, i.e. someone who has some insulin but insufficient insulin action to maintain normal glucose homeostasis, sorbitol is no less hyperglycaemic than fructose into which it is converted in the liver. It is, however, less palatable than fructose and there is consequently no justification for its continued use as a sweetener in foods prepared specially for diabetics. Indeed it is extremely doubtful that any foods should be prepared specially for diabetics, since evidence that they are better for glycaemic control than ordinary foods when consumed appropriately is completely lacking.

Würsch: I agree about the low relevance of different blood sugar responses to different sugars and starches in real life when meals and diets are considered under the usual Western conditions. But what about high carbohydrate, low fat diets, in particular for diabetics and in hyperlipidaemia? Improvement of diabetic control was observed mainly with diets rich in dietary fibre; but in most cases dietary fibre interfered with starch digestion. The beneficial effect of slow digestion of starch was elegantly showed with the use of Acarbose. After several weeks of treatment, an improvement in glucose tolerance and a decrease in Hb_{A1} was observed [1]. This subject, which is very much discussed by the scientific community and the media, should be emphasised more.

Reference

1. Sailer D, Röder G (1980) Treatment of non-insulin dependent diabetic adults with a new glycoside hydrolase inhibitor (Bay g 5421). Arzneimittel Forsch 30:2182–2185

Riou: It should be emphasised that liver is about four to five times more sensitive to insulin than is muscle. Therefore, soon after an oral glucose load the first metabolic parameter which is going to be decreased by insulin is hepatic glucose production: increase in glucose uptake by muscle tissue occurs only at 30 µU/ml insulin. It should also be remembered that glucose transport in muscle is crucially dependent on glucose concentration. Therefore part of the increase in glucose utilisation is related to a glucose effect *per se*, independently of the insulin action which indeed plays the major role.

Reference

1. Rizza RA, Mandarino LJ, Gerich JE (1981) Dose response characteristics for effects of insulin on production and utilisation of glucose in man. Am J Physiol 240:E630–E639

Chapter 9

The Relationship Between Lipid Metabolism and Dietary Starches and Sugars

I. Macdonald

The knowledge that dietary carbohydrate can be converted into lipid by the metabolic processes of the body was first demonstrated in 1852 [1], and the existence of a difference between various carbohydrates in this respect was shown in animals in 1913 [2], and in man in 1919 [3].

Two fairly recently identified clinical conditions, namely coronary heart disease (CHD) with its raised plasma lipid levels, and kwashiorkor with its excessively fatty liver, have also put the spotlight on dietary carbohydrates as being possible aetiological factors, and the question has been raised as to whether all dietary carbohydrates are equal in their lipogenic effects in these conditions.

Most of the recent work has considered whether fructose in the diet *per se*, or as part of the sucrose molecule, has a metabolic effect different from that of glucose or its polymers, such as dietary starch, and there has been a tendency, not entirely justified, to assume that dietary starch would result in metabolic consequences not dissimilar to those found when glucose is consumed.

Plasma Lipids

It is important to bear in mind that the concentration of any substance in the plasma (or blood or serum) is the result of a balance between input and output; and that, by and large, plasma is not a site of carbohydrate or lipid metabolism. The time a sample of plasma is taken is important when lipid analysis is to be carried out; for example, after a fatty meal the triglyceride level (as chylomicrons) of plasma is high. For this reason many studies are carried out on samples taken after a 12–14-hour fast, when no chylomicrons are normally present.

Immediate Effects

One mechanism whereby dietary carbohydrate might influence plasma lipid levels would be for the carbohydrate to be converted to lipid and transported as such. Another mechanism would be that the dietary carbohydrate has no direct effect on the plasma lipid but exerts its effect indirectly, through influencing a compound that in turn affects plasma lipid levels.

Thirty years ago it was reported that glucose given hourly to men lowered the serum triglyceride levels, with less marked falls in phospholipid and cholesterol concentrations [4], and similar experiments in rats were shown to cause a decrease in low and high density lipoproteins [5]. Even when the glucose is given intravenously the fall in serum triglyceride still occurs [6], thereby eliminating an absorption factor. If insulin is added to the glucose infusion, the post-prandial lipaemia is further decreased [7]; the lipaemia following a meal containing 60 g fat in men could be abolished by adding 100–250 g glucose to the meal [8]. On the other hand, a single oral dose of fructose given to healthy human subjects after a fat-containing meal increased the post-prandial hypertriglyceridaemia [9], and after a liquid meal containing fat the degree of lipaemia was less when the meal contained glucose than when it contained an isocaloric amount of sucrose [10]. Confirmation of this is seen in non-human primates: no fall in serum triglyceride level occurred in the males after intravenous fructose, but a fall was seen after intravenous glucose [11].

The most likely explanation of these findings is that there is increased removal of the plasma triglycerides due to increased lipoprotein lipase activity, and it is known that this enzyme is stimulated by insulin [12]. Thus, the fall in serum triglyceride concentration after ingesting glucose could be due indirectly to the insulin it releases, while the absence of a fall after fructose and the reduced fall after sucrose are due to fructose being non-insulinogenic [13].

An acute load of carbohydrate in fasting man has resulted in either no alterations in plasma triglyceride levels [14], or else a fall which is independent of the type of carbohydrate consumed [13].

Long-Term Effects

Triglycerides

Chronic or long-term ingestion of very high carbohydrate diets in man raises the level of triglyceride in fasting serum [15] – a rise that over the weeks tends to return to pre-high-carbohydrate levels [9]. The triglycerides present in fasting plasma, or in plasma after ingestion of a fat-free meal, are those produced by the body (endogenous), and are considered to be of greater prognostic value in disease than the plasma triglycerides found during the absorptive phase after a meal. The triglyceride level in plasma after a 12–14-hour fast may equal the highest found during a normal 24-hour period, as the intermittent intake of carbohydrate will tend to lower the level [16].

The ingestion of a high sucrose diet by rats results in an increase in the fasting plasma triglyceride fraction when compared with other carbohydrates [17,18], and this is associated with increased production of very low density lipoprotein (VLDL) [19]. An overall increase in plasma lipids has been found in rats on high sucrose diets, with little change occurring when starch replaced the sucrose [20] in rats weaned on various dietary carbohydrates. Sucrose is associated with higher plasma triglyceride levels than is starch or glucose, and the raised plasma triglyceride level usually found in pregnancy is accentuated, in rats, by sucrose [21].

Increasing the proportion of non-sucrose carbohydrate over a 7-day period in both normal and hyperlipidaemic men increases plasma triglyceride levels by 41% [22], but ultimately the mechanisms of peripheral tissue clearance "catch up" and the plasma triglyceride levels return to normal [23]. The defect in carbohydrate and triglyceride metabolism of the hyperlipidaemic subjects may be at the level of triglyceride synthesis, triglyceride clearance, or both [24].

In experiments in man the replacement of starch by sucrose in low fat diets was associated with a marked increase in fasting serum triglyceride levels [25]. In patients with carbohydrate-induced hyperlipidaemia, feeding with sucrose aggravated the hypertriglyceridaemia, whereas with starch there was a fall in serum triglyceride level [26,27]; but in a study on two hypertriglyceridaemic patients the substitution of glucose for starch in the diet did not result in any alteration in plasma triglyceride level [28]. It has been suggested that the primary responsibility for the raised plasma triglyceride after sucrose ingestion lies with an inadequate removal of the triglyceride from the plasma [29].

In two studies where the reverse procedure was adopted, namely the withdrawal of sucrose from the diet and its replacement by glucose or its polymers, the result was a fall in plasma triglyceride concentration [30,31], especially in those whose pre-diet level of plasma triglyceride was raised. The feature of sucrose that distinguishes it from other disaccharides is the pressure of fructose in the molecule. As it is not possible for man to consume large quantities of fructose because of the intestinal hurry it produces, experiments were carried out in healthy young men given diets high in carbohydrate and containing either fructose plus starch or glucose plus starch. It was found that the fructose-containing diet raised the fasting plasma triglyceride level, whereas the glucose plus starch mixture did not [32].

Sucrose, certainly in large quantities in the diet, therefore seems to be associated with an increase in the level of triglyceride in fasting plasma when compared with the effect of starch, and it seems this response occurs in normoglyceridaemic as well as hyperglyceridaemic persons. The role of dietary glucose is not so clear-cut, but probably in normal persons the triglyceride response is no different from that after starch. In the hypertriglyceridaemic patient glucose will raise the fasting plasma triglyceride level, but not so markedly as sucrose.

There are many people in the world who exist on a high carbohydrate diet and do not develop hypertriglyceridaemia [33,34], and there are several possible explanations for this:

1. *Effect of sex of the consumer.* The fasting plasma triglyceride response to dietary carbohydrate of pre-menopausal women is different from that of men and post-menopausal women, in that both men and post-menopausal women have increased plasma triglyceride levels after a high sucrose diet whereas younger women have a similar endogenous triglyceride response to both wheat starch and sucrose [35]. Further evidence of this sex difference in response to sucrose is that the fasting serum triglyceride is directly related to the serum level of fructose after a sucrose load in men, but not in young women [36]. In baboons the rate at which fructose is metabolised appears to be greater in males than in sexually mature females, while no difference between the sexes in the rate of metabolism of

glucose was detected [37]. It has been suggested that females clear plasma triglycerides more rapidly than males, and that there is an equal increase in lipogenesis from sucrose in both sexes [38]. The hormonal factor (or factors) responsible for this sex difference has not been identified, though from available evidence it seems it is not androgens.

2. *Effect of type of dietary fat.* It is well recognised that the type of fat in the diet can modify the plasma cholesterol concentration, but as the plasma triglyceride is the lipid fraction which is, in the main, carbohydrate-related, what effect does dietary fat have on the triglyceride response to dietary carbohydrates? Plasma triglyceride levels are reduced with an increase in unsaturated fat in the diet [39], and in a short-term study the increase in plasma triglyceride associated with dietary sucrose was absent when the diet contained sunflower seed oil, but was not prevented when cream replaced the polyunsaturated fat [40]. A synergistic effect of sucrose and animal fat on plasma triglyceride levels of hyperlipoprotei-naemic patients has been shown whereas, in contrast, starch in place of sucrose is not so uniformly hyperlipidaemic [41]. Similar findings have been reported in normolipidaemic men [42]. In a comparison between cream and sunflower seed oil given with glucose, fructose or starch, it was shown that, irrespective of the carbohydrate, sunflower seed oil always resulted in a marked fall in the level of fasting serum triglyceride, the difference between fructose and starch being masked by the polyunsaturated fat [43].

Thus it seems that in terms of influence on fasting plasma triglyceride levels, polyunsaturated fats may be of more consequence than the type of dietary carbohydrate.

3. *Sensitivity of the consumer.* Not everyone responds to dietary carbohydrate to the same degree. There are individuals whose plasma triglyceride response is more marked, and they have been defined as those persons whose fasting plasma triglyceride level is over 150 mg/100 ml, and whose response to sucrose load is 2.5–4 times normal [44]. In studies on patients with type IV hyperlipoprotein-aemia, giving diets containing sucrose as opposed to starch increased fasting plasma triglycerides several-fold [45]. In view of this finding it has been suggested that starch appears to be an acceptable source of carbohydrate for carbohydrate-sensitive individuals, but that sucrose is not [46].

4. *Adaptation.* There are a few animal studies that suggest that the increase in plasma triglycerides seen after exposure to a high sucrose diet returns to normal after 4–6 weeks [47,48]. Similarly, in man, a few studies suggest that metabolic adaptation can occur after several weeks [49,50].

5. *A disaccharide effect.* It has been suggested that some of the effects of sucrose on plasma triglycerides are due to the fact that sucrose is a disaccharide, as it has been found that the mean 24-hour triglyceride level is higher during the ingestion of sucrose than during the ingestion of glucose plus fructose [52]. This may be due to the greater release of an intestinal hormone by the disaccharide [52].

Plasma Cholesterol

There are reports in man and animals that would lead one to suppose that dietary carbohydrates are an important influence in determining plasma cholesterol level.

However, before drawing any conclusions with regard to effects on plasma cholesterol, it must be remembered that dietary protein can change the effect of dietary carbohydrate on plasma cholesterol, and that the endocrine state may modify the response, as may the reduction in dietary cholesterol which often accompanies an experimental diet high in carbohydrate. An alteration in the type and amount of dietary fat, or a loss in body weight also alters the level of plasma cholesterol. It seems that the effect of dietary carbohydrate *per se* on total plasma cholesterol is minor compared with its effect on plasma triglycerides.

Of more interest is the effect of dietary carbohydrate on high density lipoprotein (HDL) cholesterol in man. All dietary carbohydrates seem to reduce the level of HDL cholesterol [53] and the HDL cholesterol:total cholesterol ratio in the plasma is reduced to a greater extent by sucrose than by glucose [54].

Adipose Tissue

Lawes and Gilbert in 1852 [1] were the first to show that dietary carbohydrate can be converted by the body to adipose tissue, though this conversion is an energy-requiring process (10%–15% of consumed glucose energy) [55]. The adipocyte can convert glucose to fatty acid, the energy cost of this process being reduced by fasting [56] and a high fat diet [57] and increased by a high carbohydrate diet [56], probably via insulin [58].

Fructose can pass into the adipocyte but its transport across the cell membrane is hindered by the carrier having a relatively high K_m, so that only at high levels of fructose do significant quantities enter the fat cell [59]. In the intact animal very little fructose enters the adipose tissue as such, most of it having been converted in the liver to triglyceride which then enters the adipocyte [60], though the amount so formed is unknown.

When dietary sucrose (54% w/w) is compared with starch in rats there is an increase in fat deposition in the epididymal and peri-renal areas in the sucrose-fed animals. Insulin sensitivity of the adipose tissue was decreased in the sucrose-fed rats [61]. When rats were given drinking water containing 32% sucrose their energy intake increased while on a standard chow diet compared with animals with no sucrose in the drinking water, and though the body weights of the two groups were the same there was a greater percentage of body fat in those drinking the sucrose water [62]. In studies on rats with either sucrose or glucose in their diet, no difference in body weight gain was observed between the two high carbohydrate regimens, although, as in the previous study, more carcass fat per unit of gross energy consumed was deposited by animals on the sucrose diet than by those on the isoenergetic glucose diet [63]. This study also showed that, on a reduced energy intake, the rate of weight loss was significantly greater on the glucose diet, even though the amounts of glucose and sucrose consumed were isoenergetic. Earlier studies had reported that the weight gain in rats was greater when the dietary carbohydrate was sucrose than when it was glucose or fructose [64] or partially hydrolysed starch [65]. When either sucrose or glucose was added to a diet based on cooked wheat flour, rats fared better on the sucrose-supplemented regime, though in this case a Maillard-type reaction between the glucose and the lysine in the cooked diet may have been partly responsible for the difference [66]. In baboons there was a greater increase in body weight per unit of energy consumed when sucrose, compared with glucose syrup, was the dietary carbohydrate source [67].

One possible explanation for the difference between the effects of sucrose and other dietary carbohydrates on body weight is that the thermic effect of the ingested carbohydrate varies between the carbohydrates. In rats, giving a sucrose solution led to a 10-fold increase in lipid synthesis and to an increase in the size of the brown adipose tissue [68]. The acute thermic response in rats to various carbohydrates showed that after sucrose the dietary-induced thermogenesis was significantly greater than that after glucose [69]. In view of these apparently contradictory findings it would be unwise to conclude that an increase in thermogenesis is the cause of an increase in the fat stores seen in rats on sucrose diets.

There are only a few studies in man comparing the effect of body weight (and, by inference, on adipose tissue) of various dietary carbohydrates. In one study a liquid diet containing 85% of the energy as either sucrose or glucose was given by gastric tube at a level of 90% of the calculated energy requirement of each individual. The rate of body weight loss was greater with the glucose than with the sucrose diet in men but not in women [70]. In a similar study in overweight men and women and lasting 28 days, again the rate of weight loss was greater on the glucose diet [71]. In healthy volunteers on a diet containing normal amounts of refined carbohydrate with very little complex carbohydrate, there was increased energy intake, but no change in body weight was reported [72].

The sugar consumption of the overweight tends to be lower than that in normal weight persons [73]. In teenagers the sucrose consumption in the upper tertile of body weight was similar to that in the lower tertile [74]. The hypothesis that sucrose ingestion results in a reactive hypoglycaemia, and that this elicits hunger and hence obesity [75], has to be reconciled with the observation that sugar intake in obese individuals is lower than or equivalent to that of lean persons [44].

Liver Lipid

The peasants of the Strasbourg region have known for centuries that an excessive intake of corn by their geese will produce *foie gras*. Benedict and Lee [76], taking advantage of this knowledge, measured the respiratory quotient of these animals and found that it was 1.4, from which they concluded that dietary starch was being converted to fat. In more recent times the long-held metabolic view that much of the glucose absorbed from the intestine is removed in a first pass through the liver has been questioned, and by implication glucose *per se* may not be an important substrate for liver metabolism. It seems that glucose is not converted to glycogen or fatty acid directly, but via a C_3 molecule that is formed in tissues other than the liver [77], and that this C_3 unit is then recycled to the liver. It has been reported that in normal man over two-thirds of the glucose absorbed escapes splanchnic removal, and that peripheral tissues quantitatively play the dominant role in glucose disposal [78]. So far as is known, this does not apply to fructose.

There are conflicting views on the effects of sucrose ingestion on hepatic lipogenesis. Some reports state that in rats, hepatic lipogenesis is increased by all sugars, sucrose and maltose producing an equal effect [79,80]. Others have reported that sucrose induced greater liver lipid deposition in rats than did glucose [81,82], and that fructose was similar to sucrose in this respect [83]. Compared with glucose, a diet of 73% sucrose given to female rats for 2–3 weeks caused increased liver triglyceride levels [85], and in rabbits sucrose gave rise to more fat in the liver than did other dietary carbohydrates [87]. Dietary sucrose in the rat seemed to increase glucose-6-phosphate dehydrogenase and malic enzyme activity in the liver to a greater extent than did glucose or fructose [86]. In other studies it has been shown that, after a 48-hour fast, re-feeding of sucrose had a greater lipogenic effect than starch [87], and that sucrose given with beef tallow or safflower oil to rats resulted in greater hepatic lipid infiltration than when starch replaced the sucrose [88].

The presence of large quantities of fat in the liver is a feature of protein-energy malnutrition in children in whom the energy intake may approach normal but protein intake is inadequate. The fat in the liver under these abnormal circumstances may represent an extension of the depot fat and is found in malnourished children where the staple is either starch-based or sucrose-based.

Conclusion

The main effect of dietary carbohydrate on lipid metabolism seems to be its effect on triglycerides, whether in plasma, adipose tissue or liver. Dietary sucrose appears to be more effective in this respect than is dietary starch.

References

1. Lawes JB, Gilbert JH (1852) Composition of foods in relation to respiration and feeding of animals. Br Assoc Adv Sci ref 323
2. Togel O, Brezina E, Durig A (1913) Ueber die kohlenhydratsfarende Wirkung des Alkohols. Biochem Z 1:296–345
3. Higgins HL (1919) The rapidity with which alcohol and some sugars may serve as a nutrient. Am J Physiol 41:258–265
4. Havel RJ (1957) Early effects of fasting and of carbohydrate ingestion or lipids and lipoproteins of serum in man. J Clin Invest 36:855–859
5. Bragdon JH, Havel RJ, Gordon RS (1957) Effects of carbohydrate feeding on serum lipids and lipoproteins of the rat. Am J Physiol 189:63–67
6. Perry WF, Corbett BN (1964) Changes in plasma triglyceride concentration following the intravenous administration of glucose. Can J Physiol Pharmacol 42:353–356
7. Krut LH, Barsky RF (1964) Effect of enhanced glucose utilization on post-prandial lipemia in ischaemic heart disease. Lancet II:1136–1138
8. Albrink MJ, Fitzgerald JR, Man EB (1958) Reduction of alimentary lipemia by glucose. Metabolism 7:162–171
9. Antonis A, Bersohn I (1961) The influence of diet on serum triglycerides in South African white and Bantu prisoners. Lancet I:3–9

10. Mann JI, Truswell AS, Pimstone BL (1971) The different effects of oral sucrose and glucose on alimentary lipaemia. Clin Sci 41:123–129
11. Jourdan MH (1972) The effect of a sucrose-enriched diet on the metabolism of intravenously administered fructose in baboons. Nutr Metab 14:28–37
12. Kessler J (1963) Effect of diabetes and insulin on the activity of myocardial and adipose tissue lipoprotein lipase of rats. J Clin Invest 42:362–367
13. Macdonald I, Keyser A, Pacy D (1978) Some effects in man of varying the load of glucose, sucrose, fructose or sorbitol on various metabolites in blood. Am J Clin Nutr 31:1305–1311
14. Ferlito S, Maugeri D, Lo Porno R, Calafato M (1978) Effetti di un carico orale di glucosio di fruttosio and di saccarosio sui livelli glicemici, insulinemici and lipidemici in soggetti normali. Arch Sci Med 135:447–460
15. Ahrens EH, Hirsch S, Oettle K, Farquhar JW, Stein Y (1961) Carbohydrate-induced and fat-induced lipemia Trans Assoc Am Physicians 74:134–146
16. Schlierf G, Reinheimer W, Stossberg V (1971) Diurnal patterns of plasma triglycerides and free fatty acids in normal subjects and in patients with endogenous (type IV) hyperlipoproteinemia. Nutr Metab 13:80–91
17. Bruckdorfer KR, Kang SS, Khan IH, Bourne AR, Yudkin J (1974) Diurnal changes in the concentrations of plasma lipids, sugars, insulin and corticosterone in rats fed diets containing various carbohydrates. Horm Metab Res 6:99–106
18. Sheorain VS, Mattock MB, Subrahmanyam D (1980) Mechanism of carbohydrate-induced hypertriglyceridemia:plasma lipid metabolism in rats. Metabolism 29:924–929
19. Kamuzi T, Vranic M, Steiner G (1985) Changes in very low density lipoprotein particle size and production in response to sucrose feeding and hyperinsulinaemia. Endocrinology 117:1145–1150
20. Urbanowicz M, Chalcarz W, Jelone B, Czanocinska J (1985) Plasma lipid and protein responses in rats fed diets containing pectin and varying levels of sucrose and starch. Nutr Rep Int 32:649–658
21. Bourne AR, Richardson DP, Bruckdorfer KR, Yudkin J (1975) Some effects of different dietary carbohydrates on pregnancy and lactation in rats. Nutr Metab 19:73–90
22. Ginsberg H, Olefsky JM, Kimmerling G, Crapo P, Reaven GM (1976) Induction of hyperglycaemia by a low fat diet. J Clin Endocrinol Metab 42:729–735
23. Cahlin E, Jonsson J, Persson B et al. (1973) Sucrose feeding in man. Effects on substrate incorporation into hepatic triglycerides and phospholipids in vitro and on removal of intravenous fat in patients with hyperlipoproteinaemia. Scand J Clin Lab Invest 32:21–33
24. Kimmerling G, Javorski WC, Olefsky JM, Reaven GM (1976) Locating the site of insulin resistance in patients with non-ketotic diabetes mellitus. Diabetes 25:673–678
25. Macdonald I, Braithwaite DM (1964) The influence of dietary carbohydrates on the lipid patterns in serum and adipose tissue. Clin Sci 27:23–26
26. Kuo PT, Bassett DR (1963) Dietary sugar in the production of hyperglyceridemia. Ann Intern Med 62:1199–1212
27. Kaufmann NA, Poznanski R, Blondheim SA, Stein Y (1966) Changes in serum lipid levels of hyperlipemic patients following the feeding of starch, sucrose and glucose. Am J Clin Nutr 18:261–269
28. Porte D, Bierman EL, Bagdade JD (1966) Substitution of dietary starches for dextrose in hyperlipemic subjects. Proc Soc Exp Biol Med 12:814–816
29. Bolzano K, Sanhofer F, Sailer S, Braunsteiner H (1972) The effect of oral administration of sucrose on the turnover rate of plasma-free fatty acids on the esterification rate of plasma-free fatty acids to plasma triglycerides in normal subjects, patients with primary endogenous hypertriglyceridemia, and patients with well controlled diabetes mellitus. Horm Metab Res 4:439–446
30. Rifkind BM, Lawson DH, Gale M (1966) Effects of short-term sucrose restriction on serum lipid levels. Lancet II:1379–1381
31. Roberts AM (1971) Some effects of a sucrose-free diet on serum lipid levels. Proc Nutr Soc 30:71–72
32. Macdonald I (1966) Influence of fructose and glucose on serum lipid levels in men and pre- and post-menopausal women. Am J Clin Nutr 18:369–372
33. Higginson J, Pepler WJ (1954) Fat intake, serum cholesterol concentration, and atherosclerosis in South African Bantu. J Clin Invest 33:1366–1371
34. Schwartz MJ, Rosenzweig B, Toor M, Lewitus Z (1963) Lipid metabolism and arteriosclerotic heart disease in Israelites of Bedouin, Yemenite and European origins. Am J Cardiol 12:157–168
35. Macdonald I (1966) The lipid response of post-menopausal women to dietary carbohydrates. Am J Clin Nutr 18:86–90

36. Crossley JN (1967) Sucrose tolerance and fasting serum glyceride concentrations in young men and women. Proc Nutr Soc 26:iii–iv

37. Jourdan MH (1972) The effect of a sucrose-enriched diet on the metabolism of intravenously administered fructose in baboons. Nutr Metab 14:28–37

38. Kekki M, Nikkila EA (1971) Plasma triglyceride turnover during use of oral contraceptives. Metabolism 20:878–879

39. Englebert H (1966) Mechanisms involved in the reduction of serum triglycerides in man upon adding unsaturated fats to the normal diet. Metabolism 15:796–807

40. Macdonald I (1967) Interrelationship between the influence of dietary carbohydrates and fats on fasting serum lipids. Am J Clin Nutr 20:345–351

41. Antar MA, Little JA, Lucas P, Bucilley GC, Csima A (1970) Interrelationship between the kind of dietary carbohydrate and fat in hyperlipoproteinemic patients. 3. Synergistic effect of sucrose and animal fat on serum lipids. Atherosclerosis 11:191–201

42. Nestel PJ, Carroll KF, Havenstein N (1970) Plasma triglyceride response to carbohydrates, fats and calorie intake. Metabolism 19:1–18

43. Macdonald I (1972) Relationship between dietary carbohydrates and fats in their influence on serum lipid levels. Clin Sci 43:265–274.

44. Glinsmann WH, Irausquin NH, Park YK (1986) Evaluation of health aspects of sugars contained in carbohydrate sweeteners. J Nutr 116(11S):590

45. Reiser S, Hallfrisch J, Michaelis OE, Lazar FL, Martin RE, Prather ES (1979) Isocaloric exchange of dietary starch and sucrose in humans. I. Effects on levels of fasting blood lipids. Am J Clin Nutr 32:1659–1669

46. Reiser S (1982) Metabolic risk factors associated with heart disease and diabetes in carbohydrate-sensitive humans when consuming sucrose as compared to starch. In: Reiser S (ed) Metabolic effects of utilizable dietary carbohydrates. Marcel Dekker, New York, pp 239–259

47. Vrana A, Fabry S, Kazdova L (1976) Effects of dietary fructose on serum triglyceride concentrations in the rat. Nutr Rep Int 14:593–596

48. Lombardo YB, Chicco A, Mocchiutti N, de Rodi MA, Nusimovich B, Gutman R (1983) Effects of sucrose diet on insulin secretion in vivo and in vitro and on triglyceride storage and mobilisation of the heart of rats. Horm Metab Res 15:69–76

49. Rath A, Masek J, Kujalova V, Slabochova Z (1974) Effect of a high sugar intake on some metabolic and regulatory indicators in young men. Nahrung 18:343–353

50. Wille LE, Vellar OD, Hermansen L (1973) Changes in serum pre-beta-lipoprotein following the feeding of sucrose. J Oslo City Hosp 23:141–156

51. Thompson RG, Hayford JT, Hendrix JA (1979) Triglyceride concentrations: the disaccharide effect. Science 206:838–839

52. Szepesi B, Michaelis OE (1986) "Disaccharide effect" – comparison of metabolic effects of the intake of disaccharides and of their monosaccharide constituents. Prog Biochem Pharmacol 21:192–218

53. Schonfeld G, Weidman SW, Witzum JL, Bowen RM (1976) Alterations in levels and interrelationships of plasma apolipoproteins induced by diet. Metabolism 25:261–275

54. Macdonald I (1978) The effects of dietary carbohydrates on high density lipoprotein levels in serum. Nutr Rep Int 17:663–668

55. Baldwin RL (1968) Estimation of theoretical calorific relationship as a teaching technique. J Dairy Sci 51:104–111

56. Jansen GR, Hutchinson CF, Zanett NE (1966) Studies on lipogenesis in vivo. Biochem J 99:323–332

57. Hausberger FY, Milstein SW (1955) Dietary effects on lipogenesis in adipose tissue. J Biol Chem 214:483–488

58. Hausberger FY (1958) Action of insulin and cortisone on adipose tissue. Diabetes 7:211–217

59. Froesch ER (1972) Fructose metabolism in adipose tissue. Acta Med Scand [Suppl] 542:37–42

60. Maruhama Y, Macdonald I (1973) Incorporation of orally administered glucose-u-^{14}C and fructose-U-^{14}C into the triglycerides of liver, plasma and adipose tissue of rats. Metabolism 22:1205–1215

61. Reiser S, Hallfrisch J (1977) Insulin sensitivity and adipose tissue weight of rats fed starch or sucrose diets ad libitum or in meals. J Nutr 107:147–155

62. Kanarek RB, Marks-Kaufmann R (1979) Developmental aspects of sucrose-induced obesity in rats. Physiol Behav 23:881–885

63. Macdonald I, Grenby TH, Fisher MA, Williams CA (1981) Differences between sucrose and glucose diets in their effects on the rate of body weight change in rats. J Nutr 111:1543–1577

64. Allen RJL, Leahy JS (1966) Some effects of dietary dextrose, fructose, liquid glucose and sucrose in the adult male rat. Br J Nutr 200:339–347
65. Grenby TH, Leer CJ (1974) Reduction in "smooth surface" caries and fat accumulation in rats when sucrose in the drinking water is replaced by glucose syrup. Caries Res 8:368–372
66. Landes DR, Miller J (1976) Effect of substituting glucose for sucrose in baked wheat flour-based diets in growth and liver composition of rats. Cereal Chem 3:678–682
67. Brook M, Noel P (1969) Influence of dietary liquid glucose, sucrose and fructose on body fat formation. Nature 222:562–563
68. Granneman JG, Campbell RG (1984) Effects of sucrose feeding and denervation on lipogenesis in brown adipose tissue. Metabolism 33:257–261
69. Sharief NN, Macdonald I (1982) Different aspects of various carbohydrates on the metabolic rate in rats. Ann Nutr Metab 26:66–72
70. Macdonald I, Taylor J (1973) Differences in body weight loss on diets containing either sucrose or glucose syrup. Guy's Hosp Rep 122:155–159
71. Macdonald I (1986) Dietary carbohydrate and energy balance. Prog Biochem Pharmacol 21:181–191
72. Heaton KW, Emmett PD, Henry CL, Thornton JR, Manhire A, Hartog M (1983) Not just fibre. The nutritional consequences of refined carbohydrate foods. Hum Nutr Clin Nutr 37c:31–35
73. West KM (1978) In: Epidemiology of diabetes and its vascular lesions. Elsevier, New York, pp 224–274.
74. Walker ARP (1975) Are high compared with low consumers of sugar more prone to obesity, diabetes and coronary heart disease? S Afr J Sci 71:201–205
75. Geiselman PJ, Novin D (1982) The role of carbohydrates in appetite, hunger and obesity. Appetite 3:203–223
76. Benedict FG, Lee RC (1937) Lipogenesis in the animal body. Carnegie Institute, Washington Publication No. 489
77. Katz J, McGarry JD (1984) The glucose paradox – is glucose a substrate for liver metabolism? J Clin Invest 74:1901–1909
78. Katz LD, Gluckman G, Rapoport S, Ferrannini E, Defronzo RA (1983) Splanchnic and peripheral disposal of oral glucose in man. Diabetes 32:675–679.
79. Michaelis OE, Nace CS, Szepesi B (1975) Demonstration of a specific metabolic effect of dietary disaccharides in the rat. J Nutr 105:1186–1191
80. Naismith DA, Rana IA (1974) Sucrose and hyperlipidaemia. II. The relationship between the roles of digestion and absorption of different carbohydrates and their effects on enzymes of tissue lipogenesis. Nutr Metab 16:285–294.
81. Marshall MW, Wormack MJ (1954) Influence of carbohydrates, nitrogen source, and prior state of nutrition balance and liver composition in adult rat. J Nutr 52:51–64.
82. Litwack G, Hankes LV, Elvehjem CA (1952) Effect of factors other than choline on liver fat deposition. Proc Assoc Exp Biol Med 1:441–445
83. Harper AE, Monson WJ, Arata DA, Benton DA, Elvehjem CA (1953) Influence of various carbohydrates on utilization of low protein rations by white rat, comparison of several proteins and carbohydrates. J Nutr 51:523–537
84. Holt PR, Dominguez AA, Kwartier J (1979) Effect of sucrose feeding upon intestinal and hepatic lipid synthesis. Am J Clin Nutr 32:1792–1798
85. Macdonald I (1962) Some influence of dietary carbohydrates on liver and depot lipid. J Physiol (Lond) 162:334–344
86. Michaelis OE, Szepesi B (1973) Effects of various sugars on hepatic glucose-6-phosphate dehydrogenase, malic enzyme and total liver lipid of the rat. J Nutr 103:697–705
87. Baltzell JK, Berdanier CD (1985) Effect of the interaction of dietary carbohydrate and fat on the response of rats to starvation-refeeding. J Nutr 115:104–110
88. Carroll C, Williams K (1982) Choline deficiency in rats as influenced by dietary energy sources. Nutr Rep Int 25:773–782

Commentary

Würsch: The role of insulin in lipid metabolism has not been treated in this chapter. Many studies with diets high in dietary fibre have indicated that reducing the dietary stimulus to insulin secretion decreases serum cholesterol and triglyceride levels [1]. The use of Acarbose, an inhibitor of α-glucosidase, for example, produced a marked reduction within 2 weeks of fasting and day-long triglyceride, cholesterol and HDL cholesterol in type II diabetics [2]. Similar effects were also observed with gellifying polysaccharides in the diet. Comparative studies of pasta versus a pasta–guar diet [3], and cereal or potato versus bread–legumes [4,5] resulted in relative decreases in blood lipids in the slowly digestible carbohydrate diets compared with the control diet. In these studies the fibre–starch interaction has indeed a predominant effect on the digestion rate of the starch.

References

1. Wolever TMS, Jenkins DJA (1986) Effect of dietary fibre and foods on carbohydrate metabolism. In: Spiller (ed) CRC Handbook of dietary fibre in human nutrition. CRC Press, Boca Raton.
2. Baron AD, Eckel RS, Schmeisser L, Koltermann OG (1987) The effect of short term α-glycosidase inhibition on carbohydrate and lipid metabolism in type II diabetics. Metabolism 36:409–415
3. Gatti E, Catenazzo G, Camisasca E et al. (1984) Effects of guar-enriched pasta in the treatment of diabetes and hyperlipidemia. Ann Nutr Metab 28:1–10
4. Rivallese A, Ricardi G, Giacco A et al. (1980) Effect of dietary fibre on glucose control and serum lipoproteins in diabetic patients. Lancet II:447–450
5. Jenkins DJA, Wolever TMS et al. (1987) Low glycemic index diet in hyperlipidemia: use of traditional starch foods. Am J Clin Nutr 46:66–71

Booth: The studies of the "disaccharide effect" (Chapter 9, reference 52) did not exclude an oral difference as distinct from the postulated gut hormone mediation. The taste receptors stimulated by sucrose, and hence the cephalic visceral reflexes it evokes, have, I believe, yet to be shown to be identical to those stimulated by glucose or by fructose.

Chapter 10

The Effects on Energy Input and Output of Dietary Starches and Sugars

I. Macdonald

History

Rubner, in 1902 [1], published his observations on the influence of foodstuffs on metabolism. He had found that when 100 kcal (420 kJ) in the form of meat were ingested by a dog the heat production was increased by 30 kcal (126 kJ) over the resting state, and when 100 kcal (420 kJ) were given as cane sugar, the increase was only 5.8 kcal (24.4 kJ). When Rubner [2] gave an animal moderate quantities of food he found that the energy value of 100 g of fat was equivalent to 232 g of starch, 234 g of cane sugar or 256 g of glucose. Thus Rubner demonstrated not only that the energy output after a meal increased, depending on the type of food constituent, but also that not all dietary carbohydrates had the same energy equivalence when metabolised.

Further experiments on the short-term effects on metabolism of various dietary carbohydrates in man were carried out by Benedict and Carpenter [3] and reported in 1913. They found that the maximum increase in heat production after sucrose was materially greater than after fructose, though the total increment was similar. They also reported that the maximum increase in metabolic rate was similar after glucose and after fructose ingestion. It is, perhaps, of interest to note that the amount given of these two monosaccharides was low, due to the intestinal hurry caused by large doses of fructose.

Effects on Respiratory Quotient

The ratio of carbon dioxide produced to oxygen used gives an indication of the type of substrate being metabolised, and can also suggest which metabolic route is being taken by a dietary constituent. For these reasons it is useful to measure the respiratory quotient (RQ) after ingestion of various carbohydrates. Theoretically all dietary carbohydrates, after combining with oxygen, are broken down to carbon dioxide and water with the release of energy, the number of molecules used being equal to those produced as carbon dioxide. The theoretical RQ would therefore be 1.

In 1916 Higgins [4] reported that the RQ following the ingestion of fructose was greater than that after glucose, and from this he concluded that fructose showed a tendency or preference to change into fat in the body (and ultimately to carbon dioxide and water and energy). When various mono- and disaccharides were examined it was found that fructose, whether given by itself, with glucose, or as the disaccharide sucrose, always produced a significantly higher RQ than other sugars. This suggests that in the release of energy after fructose ingestion more of the carbohydrate is being broken down compared with the other sugars. The more rapid metabolism of fructose may be due to the fact that the hexokinase groups of enzymes are rate-limiting [5], whereas fructose metabolism follows a form typical of a first-order reaction, and its utilisation is proportional to concentration [6]. This suggestion is supported by the findings that fructose ingestion is followed by increases in plasma uric acid, lactate and pyruvate levels relative to those resulting from glucose ingestion [7], and that the physiological activity of keto-hexokinase is four times greater than that of hexokinase and glucokinase in the human liver [8].

Galactose is a monosaccharide that is not subject to rate-limitation in its metabolism in the liver and, like fructose, also results in a higher RQ than glucose after ingestion. The RQ value is not as high, though, as seen after ingestion of sucrose or glucose:fructose [9]. On the other hand lactose (glucose:galactose) does not have the high RQ seen after galactose ingestion. It is known that the handling of galactose by the body is greatly affected by whether or not it is accompanied by glucose, though the mechanism for this is not understood [10]. It is therefore not difficult to accept that more galactose seems to be used for energy in the short term when it is unaccompanied by glucose.

Variation in the rate of breakdown of the various sugars also receives support from a paper published in 1904 [11] in which it is stated that the rise in carbon dioxide output is greater after fructose than after glucose ingestion. The explanation that is given is that fructose is less readily retained in the liver and therefore "reaches the tissue in a larger stream" than does glucose under similar circumstances, and hence more completely replaces fat as a source of energy [12].

Thus glucose, the end product of starch digestion, does result in an RQ different from that of fructose or sucrose and, indeed, of galactose, indicating that in the hours after ingestion the contribution to the energy source of the body from starch and sugar is different (Fig. 10.1).

Effects on Dietary-Induced Thermogenesis

The cost of metabolising dietary protein is known to be high measured in terms of energy expenditure, while that for fat is lower than for dietary carbohydrate, presumably because dietary fat does not require much metabolic modification before being placed in the adipose tissue [13]. Studies in rats measuring the effect

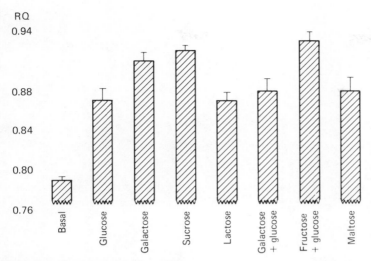

Fig. 10.1. The mean (± SE) respiratory quotient (RQ), in men, during 3 hours after ingestion of various sugars. Data from [9].

of various dietary carbohydrates on dietary-induced thermogenesis showed that the response varied depending on the carbohydrate consumed. Over a period of 3 hours after ingestion the increase in metabolic rate was greatest after sucrose and least with glucose and with fructose. However, after a glucose and fructose meal the increase in metabolic rate was greater than with isoenergetic amounts of either glucose or fructose alone. There seems no obvious explanation for this observation. When sucrose was compared with an equimolar mixture of glucose: fructose the increase in metabolic rate was greater after sucrose. This could be due either to the heat of hydrolysis of sucrose, or to the possibility that the absorption of fructose is more rapid after sucrose than after a comparable glucose:fructose mixture. The heat of hydrolysis of sucrose is very small [14], and in rats absorption after sucrose is such that it is unlikely to be different from that of a glucose:fructose meal [15].

   Similar studies have been carried out in man, where the resting metabolic rate was measured for 3 hours after the ingestion of various di- and monosaccharides and mixes of monosaccharides [9,16]. It was found that the increase in metabolic rate was significantly greater after sucrose than after an equivalent weight of glucose during the 3 hours of monitoring. In overweight men the difference in response between the two sugars was less marked.

   In studies using galactose, lactose, galactose:glucose, sucrose, glucose, fructose:glucose and maltose, the increase in metabolic rate over a 3-hour period was significantly greater after sucrose and after fructose:glucose than after any of the other sugars, in which the metabolic response was of a similar magnitude (Fig. 10.2). The mean percentage increase in metabolic rate following glucose ingestion in normal weight men over a 3-hour period was 10%, which was comparable to the 13% [17] and 9% [18] previously reported.

   When the thermogenesis induced by the various di- and monosaccharides was measured in healthy young women, sucrose again resulted in a greater response than the other sugars [9]. However, the response (to the same level of intake, i.e.

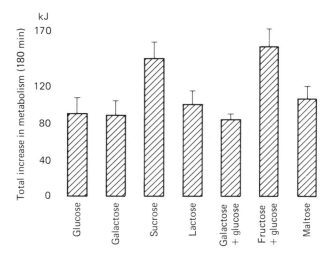

Fig. 10.2. The mean (± SE) total increase in metabolic rate following the ingestion of various sugars. Data from [9].

2 g sugar per kg body weight) was significantly less than in the men after galactose, sucrose and lactose. However, this difference disappears when the increase is expressed as a percentage of basal value.

Thus it seems that in the early thermogenic response (0–3 hours) in man to various simple sugars, fructose is outstanding, in that it engenders a greater response than any of the other carbohydrates, including glucose.

However, the metabolic response to the ingestion of a simple sugar may vary, depending on the previous diet of the individual [19]. Following a high fat, mixed or high carbohydrate intake for 3–6 days before the carbohydrate load, which was given in the post-absorptive state, it was found that the thermogenesis was greatest after a period on a high carbohydrate diet (8.6%) and least after a high fat diet (5.2%). It was suggested that after a high carbohydrate diet, though only a small proportion of the carbohydrate ingested was converted to fat, more fat was gained than after a high fat diet. The explanation given was that the carbohydrate load was replenishing the glycogen stores. This concept received support from the RQ, which was above 1.0 after the carbohydrate load on the carbohydrate and mixed diets, but tended to be less than 1.0 after the period on the fat diet.

The metabolic cost of dietary carbohydrate disposal can vary between 7% and 23% depending on whether the carbohydrate is being stored as glycogen or as fat [20].

Effects on Body Weight

It was for many years considered necessary in any weight-reducing plan to cut down on dietary starch, advice being given to reduce the intake of bread, potatoes, rice, etc. More recently these articles of the diet seem to be in vogue because they contain "complex" carbohydrates, and the spotlight has moved to sugars, the intake of which is considered to be synonymous with an increase in depot fat. To learn whether there is a difference between starches and sugars in this respect, it is necessary to consider the efficiency of the metabolic disposal of these substances. Classically the energy value of carbohydrate has been assumed to be that which can be obtained by calorimetry on the laboratory bench, and no cognisance has been taken off the efficiency with which this carbohydrate is handled by the body, or, in other words, how much of the energy content of a dietary constituent is available for the body's metabolism. The efficiency with which food energy is metabolically available does vary. For example, dietary fat can be stored as depot fat at a much lower cost than an isoenergetic amount of dietary carbohydrate (as determined in vitro) [13]. Are all dietary carbohydrates handled with equal efficiency by the body's metabolism? Rubner [1] implied that this was not so when discussing differences between starch and glucose, with the suggestion that some dietary carbohydrates will lead to greater body weight changes than others if consumed in isoenergetic amounts over long periods.

This hypothesis was tested in adult male rats given *ad libitum* diets containing 80% (by energy) carbohydrate in the form of glucose, fructose, partial hydrolysate of starch, or sucrose for 26 weeks [21]. It was found that the greatest body weight gain per 100 g food eaten was with sucrose, and the least with glucose. In a similar type of study using male baboons [22] it was shown that with similar levels of energy intake the animals consuming sucrose weighed more at the end of 26 weeks than those whose dietary carbohydrate was a partial hydrolysate of starch.

It would, perhaps, be unethical to attempt to increase body weight in normal weight human subjects, using diets high in carbohydrate, but the reverse situation is acceptable. To this end liquid formula diets were given by stomach tube for 12 days to healthy young adults. The diets contained 83% (by energy) sucrose or partial hydrolysate of starch, and the amount given was calculated to provide about 10% less energy than the energy output of the subjects [23]. It was found that the rate of weight loss in the men was significantly greater with glucose than with sucrose, a difference not apparent in the women. In a similar study in overweight men and women lasting 28 days, again the rate of weight loss was greater on the glucose diet [24]. As an extension to this experiment there was a cross-over of diets after the first 28 days, with the intake changing daily with the body weight loss, to 10% less than calculated needs, for a further 28 days. Contrary to expectation there was not a cross-over of rate of weight loss [24].

Similar experiments on large male rats were carried out in which sucrose or glucose was the principal constituent of the diet; again the energy intake was 10% less than that calculated to maintain energy balance [25]. The energy intake given to each animal was determined by bomb calorimetry, and the result was that, like the studies in man, the rate of weight loss was significantly greater on the glucose-rich compared with the sucrose-rich diet. Again, as in the human studies, an attempt to demonstrate a cross-over failed. An analysis of the amount of carcass fat of these animals revealed that more fat per unit of gross energy consumed was deposited while on the sucrose diet, again suggesting that sucrose is more efficiently handled metabolically, so that a smaller amount of sucrose than glucose is required to produce a unit of metabolisable energy.

Several levels of reduced energy intake varying from 9 to 15 kcal (38 to 62 kJ) per 100 g body weight were given to male rats, and at no level of intake was the body weight loss greater with sucrose than with glucose [26].

Studies with dietary starch as the principal source of dietary carbohydrate have not been carried out, because the high level of dietary carbohydrate as raw starch could give rise to increased bacterial fermentation in the colon, and cooked starch would be too bulky. However, use of a low dextrose equivalent maltodextrin (DE 40: see Chap. 1) would be evidence for the view that dietary starch might show results comparable with glucose.

Possible Explanations for the Differences Between the Effects of Dietary Carbohydrates on Body Weight Loss

There are several possible theoretical explanations for the differences between the effects of various carbohdrates on body weight loss, even including the possibility that dietary "fibre", which often accompanies dietary starch intake, may modify the absorption of the glucose end product, and hence lead to decreased post-prandial levels of plasma glucose and insulin [27]. Furthermore not all dietary starch is hydrolysed in the small intestine and it therefore may be converted to short chain fatty acids rather than glucose. This would lower the insulin response to the carbohydrate load, and thus contribute to differences in the thermogenic response.

It has been suggested that insulin release is associated with increased activity of the sympathetic nervous system and is the mechanism whereby sucrose and glucose stimulate the sympathetic nervous system in man and animals [28] (see Chap. 13). In man the level of noradrenaline in the plasma increases after a standard glucose tolerance test [29], with expected increases in heart rate and blood pressure using the hyperinsulinaemic clamp technique; this response seems to be dose-dependent [30].

If the stimulation of sympathetic activity is related to insulin output, and this increased sympathetic activity leads to increased thermogenesis, then it can be seen that dietary carbohydrates leading to high insulin responses would be more "inefficient" than those with lower insulin responses, because of the heat production. Thus, as ingestion of sucrose, compared with glucose, results in about half the plasma insulin response, weight for weight [7], sucrose would be more "efficiently" utilised by the body, and this could account for the greater weight loss in rats and man on diets containing glucose as compared with sucrose. Against this is the fact that in man β-adrenergic blockade does not seem to affect carbohydrate-induced thermogenesis after either acute [31] or chronic ingestion [32].

A second possible theoretical explanation could be in the operation of "futile" cycles. If a "futile" cycle affected glucose (e.g. the interconversion of fructose-6-phosphate and fructose-1,6-phosphate) but not fructose, then it can be seen that another possibility exists to account for the greater metabolic efficiency of fructose and sucrose.

The findings, both in rats and in man, that after a weight-reducing regime and a diet high in sucrose or glucose there is not a "cross-over" in the rate of weight loss when the two carbohydrates are reversed, is of some interest because it could lead to the speculation that in a weight-reducing regimen it may be advisable to have a high glucose (or glucose polymer such as starch?) diet before continuing on a more conventional diet. Of more scientific interest is a possible explanation of this "hang-over" effect. It cannot be due to enzyme induction, because the effect is too long-lasting (up to 28 days in man), and the possibility has been considered that it is in some way a slowly adapting humoral response; the endocrine gland studied has been the thyroid.

It is known that over- or under-feeding increases or decreases the peripheral conversion of T_4 to T_3 [33], and that after the acute ingestion of sucrose there is a significant fall in T_3 levels in the serum [34]. Recent observations in rats on *ad libitum* diets showed that their thyroid uptake of ^{123}I was significantly less than that of controls during 10 weeks on a high carbohydrate diet, and that animals on dietary fructose had significant falls in serum T_3, FT_3 and T_4 concentrations [35].

Effects of Carbohydrates on Energy Input

There is evidence to suggest that the type of sugar consumed may affect the appetite and food input several hours later. It has been shown that ingesting low concentrations of glucose increases the consumption of a nutritionally adequate diet in rats [36]. In man it has been reported that when fructose is given in the fasted state there was, at 2 hours, a significantly lower food intake compared with that after an isocaloric load of glucose [37]. Support for the hypothesis that the difference in insulin response to the two sugars influences subsequent food intake has recently been published [38]. However, this paper points out that when fructose ingestion is not associated with low post-prandial insulin levels in the serum, as found with a mixed meal, then there is no lower food intake due to the fructose ingestion.

Conclusions

Little work has been reported on the effect of dietary starch *per se* on energy input or output, as it has been assumed that starch is hydrolysed to glucose before it can be absorbed from the alimentary tract, and therefore a study of glucose would be synonymous with one of dietary starch. This is now known not always to be so. Nevertheless, there are differences in energy metabolism between various di- and monosaccharides, but whether dietary starch will exaggerate or minimise these changes remains to be discovered.

References

1. Rubner M (1902) Energiegessetze, p 410
2. Rubner M (1931) Sitzungsber d Preussischen Akad d Wissensch Physikal U Math Klasse 17:313

3. Benedict PG, Carpenter TH (1913) Food ingestion and energy transformation with special reference to the stimulating effects of nutrients. Carnegie Research Institute, Washington, Publication 261:47–250
4. Higgins HL (1916) The rapidity with which alcohol and some sugars may serve as nutriment. Am J Physiol 41:258–265
5. Heinz F (1973) The enzymes of carbohydrate degradation. In: Prog Biochem Pharmacol 8:1–56. S Karger, Basel
6. Smith LH, Ettinger RH, Seligson D (1953) A comparison of the metabolism of fructose and glucose in hepatic disease and diabetes mellitus. J Clin Invest 32:273–282
7. Macdonald I, Keyser A, Pacy D (1978) Some effects in man of varying the load of glucose, sucrose, fructose or sorbitol on various metabolites in blood. Am J Clin Nutr 31:1305–1311
8. Heinz F, Lamprecht W, Kirsch J (1968) Enzymes of fructose metabolism in human liver. J Clin Invest 47:1826–1832
9. Macdonald I (1984) Differences in dietary-induced thermogenesis following the ingestion of various carbohydrates. Ann Nutr Metab 28:226–230
10. Williams C, Phillips T, Macdonald I (1983) The influence of glucose on serum galactose levels in man. Metabolism 32:250–256
11. Johansson JE, Billstrom J, Heijl C (1904) Die kohlen saureabgabe bei zufuhr verschiedener zuckerarten. Skan Arch Physiol 16:263
12. Lusk G (1909) In: The elements of the science of nutrition. Saunders, Philadelphia, p 177
13. Wood JD, Reid JT (1975) The influence of dietary fat on fat metabolism and body fat composition in meat feeding and nibbling rats. Br J Nutr 34:15–24
14. Jackson RJ, Davis WB (1977) The energy values of carbohydrates: should bomb calorimeter data be modified? Proc Nutr Soc 36:90A
15. Dahlqvist A, Thomson DL (1963) The digestion and absorption of sucrose by the intact rat. J Physiol (Lond) 167:193–209
16. Sharief N, Macdonald I (1982) Differences in dietary-induced thermogenesis with various carbohydrates in normal and overweight man. Am J Clin Nutr 35:267–272
17. Pittet P, Chappius P, Acheson K, Techtermann FD, Jequier E (1976) The thermic effect of glucose in obese subjects studied by direct and indirect calorimetry. Br J Nutr 35:281–292
18. Golay Y, Schutz Y, Meyer HU et al. (1982) Glucose-induced thermogenesis in non-diabetic and diabetic obese subjects. Diabetes 31:1023–1028
19. Acheson KJ, Schutz Y, Bessard T, Ravussin E, Jequier E, Flatt JP (1984) Nutritional influences on lipogenesis and thermogenesis after a carbohydrate meal. Am J Physiol 246:E62–70
20. Flatt JP (1978) The biochemistry of energy expenditure: In: Obesity research II. Newman, London
21. Allen RJL, Leahy JS (1966) Some effects of dietary dextrose, fructose, liquid glucose and sucrose in the adult small rat. Br J Nutr 20:339–347
22. Brook M, Noel P (1969) Influence of dietary liquid glucose, sucrose and fructose on body fat formation. Nature 222:562–563
23. Macdonald I, Taylor J (1973) Differences in body weight loss on diets containing either sucrose or glucose syrup. Guy's Hosp, ref 122:155–159
24. Macdonald I (1986) Dietary carbohydrate and energy intake. Prog Biochem Pharmacol 21:181–191
25. Macdonald I, Grenby TH, Fisher MA, Williams CA (1981) Differences between sucrose and glucose diets in their effects on the rate of body weight change in rats. J Nutr 111:1543–1547
26. Macdonald I (1985) A non-linear relationship between energy intake and rate of weight loss in rats. J Physiol (Lond) 362:40P
27. Jenkins DJA, Leeds AR, Wolever TMS (1976) Unabsorbable carbohydrate and diabetes: decreased post-prandial hyperglycaemia. Lancet II:172–174
28. Landsberg L, Young JB (1983) The role of the sympathetic nervous system and catecholamines in the regulation of energy metabolism. Am J Clin Nutr 38:1018–1024
29. Welle S, Lilavivathana V, Campbell RG (1981) Themic effect of feeding in man: increased plasma norepinephyrine levels following glucose but not protein or fat consumption. Metabolism 30:953–958
30. Rowe JW, Young JB, Minaker KL, Stevens AL, Palotta J, Landsberg L (1981) Effect of insulin and glucose infusions on sympathetic nervous system activity in normal man. Diabetes 30:219–225
31. Zwillich C, Martin B, Hofeldt F, Charles A, Subryan U, Burman K (1981) Lack of effects of beta-sympathetic blockade on the metabolic and respiratory responses to carbohydrate feeding. Metabolism 30:451–456

32. Welle S, Campbell RG (1983) Stimulation of thermogenesis by carbohydrate overfeeding. J Clin Invest 71:916–925
33. Danforth E (1983) The role of thyroid hormones and insulin in the regulation of energy metabolism. Am J Clin Nutr 28:1006–1017
34. Sharief N, Marsden P (1981) Effects on serum tri-iodothyronine and reverse tri-iodothyronine concentrations of an acute intake of sucrose or glucose in man. Proc Nutr Sec 40:43A
35. Macdonald I (1989) Some effects of various dietary carbohydrates on thyroid activity in the rat. Ann Nutr Metab 26 (in press)
36. Merkel AD, Wayner MJ, Jolicoeur FB, Mintz RB (1979) Effects of glucose and saccharine solutions on subsequent food consumption. Physiol Behav 23:791–793
37. Spitzer L, Rodin J (1987) Differential effects of fructose and glucose on food intake. Appetite 8:135–145
38. Rodin J, Reed D, Jamner L (1988) Metabolic effects of fructose and glucose: implications for food intake. Am J Clin Nutr 47:683–689

Commentary

Würsch: Little work has been reported on the comparative effect of starchy foods on energy expenditure. Starch can be digested at various rates, resulting presumably in various rates of glucose absorption: which in turn can be reflected in the suprabasal energy expenditure rate and duration. The α-amylase inhibitor Trestatin decreased the suprabasal glucose oxidation rate after a starch load [1]. A slower rate was also observed with beans compared with potato, but the total carbohydrate oxidised over 6 hours was the same [2,3].

References

1. Tappy L, Würsch P, Randin JP, Felber JP, Jéquier E (1986) Metabolic effect of pre-cooked instant preparation of bean and potato in normal and in diabetic subjects. Am J Clin Nutr 43:30–35
2. Würsch P, Acheson K, Koellreutter B, Jéquier E (1988) Metabolic effects of instant bean and potato over six hours. Am J Clin Nutr 47:1470–1476
3. Tappy L, Buckert A, Griessen M, Golay A, Jéquier E, Felber J (1986) Effect of trestatin, a new inhibitor of pancreatic alpha-amylase, on starch metabolism in man. Int J Obesity 10:185–192

Booth: In early post-absorptive rats, Booth and Jarman [1] reported that intravenous fructose seemed acutely to be somewhat more satiating than glucose (and galactose less so). The satiating effects of gastric loads, hepatic portal infusion and vena caval infusion of glucose were not distinguishable from those of orally ingested maltodextrin or starch and a main caloric component of chow.

Reference

1. Booth DA, Jarman SP (1976) Inhibition of food intake in the rat following complete absorption of glucose delivered into the stomach, intestine or liver. J Physiol (Lond) 259:501–522

Marks: How reliable are the early observations of Rubner on the energy value of foods? Surely the energy values of foods vary according to the nutritional status of the animal, i.e. whether carbohydrate is being used as energy, converted into glycogen (energy cost possibly 7%) or into fat (energy cost possibly 23%)? I am also dubious about the significance of the figures relating to the energy value of 100 g fat equating with 232 g starch, 234 g cane sugar and 256 g glucose. Whilst this might be so, I doubt that Rubner's measurements were either precise enough, or sufficiently large in number, to make this statement anything more than a hint.

Southgate: The review of the historical evidence based on differences in RQ may be difficult to relate to recently ingested foods. However, recent studies of the biochemical pathways of energy transduction generally support the early ideas of Rubner on energy equivalents. Overall, however, I think the more conclusive evidence for differences in metabolic efficiency, such as that obtained from detailed energy balances, is more convincing. The level of precision required in these studies is very high and the meticulous control necessary is difficult to achieve in man. This is particularly true when either direct or indirect calorimetry is being used.

Booth: For the topic of energy balance, it must be noted that RQ is independent of oxygen consumption. Differences in fat metabolism and deposition cannot be extrapolated to differences in thermogenesis and net energy storage or body weight. Also, of course, as emphasised in Chap. 9, the effects of ingestion of a sugar in isolation, or in high doses, do not predict its effects in the normal diet. Only calorimetry over the full time course of differences between controlled normal diets can ascertain the effects of dietary carbohydrates on energy balance.

Marks: It is pointed out that the RQ is always higher when fructose is consumed than with other sugars. This, surely, very much depends on over how long a period of time following the ingestion of the sugar the carbon dioxide output and oxygen uptake are measured. In the final analysis the RQ for fructose and for glucose must always finish up as 1, unless there are excretory products, such as ketones, which occur with one sugar and not the other.

I would, moreover, question the whole validity of the RQ concept since it reflects changes occurring over only a very short period of vital activity rather than the prolonged or integrated period which is biologically meaningful. Moreover the relevance of RQ measurements made with pure compounds versus those obtained with real meals must be open to question.

Flatt: The statements that "insulin release is associated with increased activity of the sympathetic nervous system" and "If the stimulation of sympathetic activity is related to insulin output" are somewhat confusing. It is well known that increased activity of sympathetic nerves innervating the islets of Langerhans is associated with the inhibition of insulin release [1]. In fact, a component of the insulin secretory response to feeding is generally believed to be mediated by increased activity of local parasympathetic nerves. It is therefore important to distinguish between the various tissues in relation to changes in autonomic nervous tone.

Reference

1. Woods SC, Porte D (1974) Neural control of the endocrine pancreas. Physiol Rev 54:596–619

Marks: It is suggested that insulin release is associated with increased activity of the sympathetic nervous system, but the evidence for this is extremely flimsy. The only time that insulin activity has been shown to stimulate catecholamine release is when it is associated with hypoglycaemia. Hypoglycaemia sufficient to activate the sympathetic nervous system in people eating normal mixed diets occurs only rarely under the wholly unphysiological condition of a large oral glucose (or sucrose) load.

The idea that stimulation of sympathetic nervous activity is related to insulin output is, I believe, the reverse of the truth. Sympathetic nervous activity decreases insulin output. Increased activity of the parasympathetic nervous system on the other hand, and particularly stimulation of the vagus, does stimulate insulin secretion.

Booth: Do the contemporary hypotonic maltodextrin tolerance tests stimulate sympathetic activity? Glucose at 50 g/200 ml is five times isotonic, and even sucrose is twice isotonic, giving a rasping sensation in the throat and (at least in the rat) retention or sequestration of water in the upper gut [1] and conditioning aversions to even the sweetness of the drink [2]. Disaccharides would still have half the sympathetic effect of monosaccharides, but not mediated by half the insulin secretion.

References

1. Booth DA (1972) Satiety and behavioural caloric compensation following intragastric glucose loads in the rat. J Comp Physiol Psychol 78:412–432
2. Booth DA, Lovett D, McSherry GM (1972) Postingestive modulation of the sweetness preference gradient in the rat. J Comp Physiol Psychol 78:485–512

Riou: When indirect calorimetry is used to measure induced thermogenesis it must be remembered that glucose coming from the liver is still produced and utilised. In most studies the metabolic fate of exogenous versus endogenous glucose utilisation is not investigated. Moreover the final amount of glucose oxidised is dependent upon the degree of inhibition of lipolysis which is itself dependent upon the blood glucose and insulin. Thus, the more insulin present in the blood, the less fatty acids will be available for oxidation, and the more glucose will be oxidised. Therefore it is possibly not that the carbohydrate itself is utilised to a greater or lesser degree, but that its metabolic fate is dependent upon the degree of availability of endogenous fat as energy source.

Riou: Glucose is the major metabolic end product of fructose metabolism, but to my knowledge clear data on the rate of conversion of fructose to glucose during an oral fructose load are not available. In more general terms there are in most human studies no data on the appearance of the exogenous carbon in plasma glucose coming from the carbohydrate which is absorbed. New developments in mass spectrometry analysis, allowing one to measure tracer enrichment in ^{13}C in blood glucose, will help to clarify this important point.

Chapter 11

The Influence of Starch and Sugar Intake on Physical Performance

C. Williams

In this chapter the influence of starch and sugars on physical performance will be considered, especially the results of studies which have attempted to assess physical performance, and not only the biochemical responses to nutritional intervention. One of the confounding issues which prevents a clearer understanding of the link between sugar intake and exercise tolerance is the range of different exercise protocols used to investigate the topic. This makes it difficult to draw general conclusions from these particular studies about the benefits of carbohydrate intake. For this reason the exercise protocols involved have been described. In addition the description of the "fitness" of subjects is often inadequate, and so generalisations from, for example, studies which have employed trained individuals may not be applicable to the population at large. Therefore some of these issues have been addressed first before proceeding with an examination of the more informative studies on the influences of starch and sugar intake on physical performance.

Physical Performance

The metabolic response of an individual challenged by exercise of increasing intensity is an increase in oxygen consumption ($\dot{V}O_2$). A point is reached where, even though the exercise intensity can be increased still further, there is no accompanying increase in oxygen uptake (Fig. 11.1). This is called the maximum oxygen uptake ($\dot{V}O_2$max), and is related to size, age, sex, genetic potential and level of habitual activity of the individual [1]. When the oxygen cost of an activity is expressed as a percentage of $\dot{V}O_2$max it is called the "relative intensity". The relative exercise intensity reflects the physiological and psychological demands on an individual more accurately than, for example, the absolute values for work loads or running speeds. This is because the cardiovascular, thermoregulatory and metabolic responses to exercise occur with respect to the relative, rather than

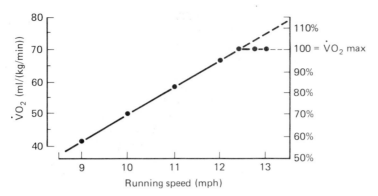

Fig. 11.1. Relationship between running speed (exercise intensity), oxygen consumption ($\dot{V}O_2$) and relative exercise intensity ($\%\dot{V}O_2$max).

the absolute exercise intensity [2]. Therefore it is not surprising that the exercise tolerance of an individual is also inversely proportional to the relative exercise intensity [3]. It is thus advisable to prescribe exercise of the same relative intensities for subjects taking part in studies designed to investigate the influence of nutritional intervention on exercise tolerance.

When considering studies on physical performance it is essential that the nature of the performance be defined. To this end it is helpful to have a clear understanding about the various terms which are commonly used in association with physical performance. For example, work capacity is most commonly assessed in the laboratory using cycle ergometry, which allows the external work performed to be monitored quite precisely [4]. Endurance capacity refers to the time taken to exercise to exhaustion, irrespective of the mode of exercise, whereas the term "endurance performance" should be restricted to reporting the time taken to perform a prescribed work load or distance. Performance or fitness tests have frequently employed a simple run over a given distance, with the "first past the post" being designated the fittest individual. This method is, of course, largely a reflection of the $\dot{V}O_2$max values of the individuals, and so the strong inverse relationship between performance times and maximum oxygen uptake has been used to predict $\dot{V}O_2$max values as an index of "fitness" [5].

Many studies have used $\dot{V}O_2$max values as a description of their subjects' fitness. This is not entirely satisfactory because it is, unfortunately, assumed that the size of an individual's $\dot{V}O_2$max value is a reflection of that individual's endurance fitness. This can be misleading, because while "nature" dictates the potential size of an individual's $\dot{V}O_2$max value, it is "nurture", through adequate nutrition and appropriate training, which leads to an improvement in "endurance fitness". Furthermore the endurance performance and capacity of already active individuals can improve significantly without accompanying improvements in their $\dot{V}O_2$max values [6,7]. Therefore simply reporting the $\dot{V}O_2$max value of an individual does not necessarily provide an insight into the endurance fitness or training status of that person. It is now well established that the endurance of an individual is determined by the capacity of the working muscles to cover their energy needs by aerobic metabolism of fatty acids and the limited carbohydrate stores of the body [8]. Thus the greater the "aerobic fitness" of an individual, the greater the contribution of fat metabolism to energy expenditure, which results in a more economical use of the limited muscle glycogen stores [9,10].

The improvement in the aerobic capacity of skeletal muscle as a result of training, is reflected by a delay in the accumulation of blood lactic acid during exercise. The point during submaximal exercise at which blood lactic acid concentrations increase has been described as the "lactic threshold" or "anaerobic threshold". The concept of an anaerobic threshold which reflects the training status of an individual is an attractive one. Wasserman and co-workers [11] proposed the term "anaerobic threshold", and provided evidence to suggest that it could be described by identifying the point at which there is a non-linear increase in ventilation during exercise of increasing intensity. There has, however, been some discussion on how closely the ventilatory response to exercise mirrors the changes in blood lactic acid concentration [12,13]. Nevertheless the exercise intensity at which there is a significant increase in blood lactic acid concentration may be regarded, for practical purposes, as the point at which the aerobic provision of energy is complemented by an increased contribution from the non-oxidative degradation of glycogen in the Embden–Meyerhof pathway. It has been suggested that lactate reference concentrations of 2 mmol/l or 4 mmol/l provide a useful alternative to the determination of "lactate thresholds" during exercise of increasing intensity [14]. The exercise intensity equivalent to 4 mmol/l, which has been referred to as the "onset of blood lactate accumulation" (OBLA) [15], has been shown to have a stronger correlation with endurance performance than $\dot{V}O_2max$ [16,15]. Endurance training increases the exercise intensity an individual can achieve for a given lactic acid concentration [17], which is the result of an increase in the aerobic capacity of the muscle cell [10]. The aerobic or endurance fitness of an individual is, therefore, best described in terms of the proportion of the $\dot{V}O_2max$ value which can be used before there is a significant increase in blood lactic acid concentration, irrespective of the absolute value of $\dot{V}O_2max$. Thus, while the absolute work capacity of an individual is directly related to the size of his or her maximum oxygen uptake, the aerobic or endurance fitness is reflected by the percentage $\dot{V}O_2max$ which can be achieved before there is a significant increase in blood lactate concentration.

Fatigue

Fatigue during prolonged submaximal exercise is associated with the depletion of the limited glycogen stores in active skeletal muscles [18,19]. However, it is important to recognise that skeletal muscle is composed of two main populations of motor units. The two populations of fibres are classified in terms of their speed of contraction and the metabolic source of their energy production. The Type I fibres are slow contracting, slow fatiguing oxidative fibres, which have a high mitochondrial density and capillary supply; the Type IIb fibres, on the other hand, also known as fast contracting, fast fatiguing, glycolytic fibres, have a low mitochondrial density and capacity for aerobic metabolism. The Type IIa fibres are fast contracting but have a significantly greater aerobic capacity than the Type IIb fibres, and so are more resistant to fatigue [20]. An alternative nomenclature for the different fibre populations is one which is frequently used in animal studies and describes Type I fibres as slow twitch oxidative (SO), Type IIa fibres as fast twitch oxidative and glycolytic (FOG) and Type IIb fibres as fast twitch glycolytic (FG).

The recruitment patterns of the different fibre populations have been assessed by following the decrease in glycogen concentration in each of the fibres, using histochemistry and light microscopy. For example, during cycle ergometer exercise, the post-exercise glycogen concentration is lowest in the Type I and Type IIa fibres, reflecting their recruitment during exercise [21–23]. Thus during prolonged submaximal exercise fatigue occurs when the glycogen concentration in the Type I and Type IIa fibres is reduced significantly. However, at the point of fatigue the glycogen concentration of the Type IIb fibres may still be high, and so it is important to be aware of the selective nature of muscle fibre recruitment when examining the link between muscle glycogen concentration and the onset of fatigue. Furthermore it is important to appreciate that when, during submaximal exercise, the exercise intensity is increased, then this change may require the recruitment of the fast contracting, fast fatiguing motor units, and so the ensuing fatigue may not simply be the result of inadequate substrate supply. Not surprisingly, individuals who have a high proportion of the Type IIb fibres tend to be better sprinters, but also produce more lactic acid during submaximal exercise and fatigue sooner than individuals with less Type IIb fibres [24].

Fatigue during brief periods of high-intensity exercise does not appear to be the result of a reduction in the limited carbohydrate stores in muscle. During these periods there is a rapid rate of glycogenolysis, yet fatigue occurs before the concentration of glycogen is reduced to low levels [25]. The high rate of glycogenolysis produces a parallel increase in lactic acid, and hence, in the concentration of hydrogen ions. The hydrogen ions have an inhibitory influence on the rate of glycolysis, as well as on the contractile process, via their competitive binding on the active sites of the regulatory protein troponin. More recently, however, there has been some support for the proposition that the more important influence of hydrogen ion activity is on the inhibition of ATP utilisation, rather than on the production of ATP [26]. Nevertheless there have been some studies which have investigated the possible benefits of the carbohydrate loading procedure on endurance capacity during high-intensity exercise. In one study carbohydrate loading was reported to improve exercise capacity by almost 2 minutes (4.87 ± 1.07 min vs. 6.65 ± 1.39 min) during cycling at an intensity equivalent to VO_2max [27]. However, a subsequent study, using a similar research design, was unable to confirm the dietary-induced improvement in performance, but suggested that the carbohydrate loading procedure may improve acid–base status of the blood, and so help improve performance during high-intensity exercise [28].

Dietary Carbohydrate Intake and Physical Performance

The significant improvement in endurance capacity which occurs following dietary manipulation of the body's carbohydrate stores is the most notable contribution of nutrition to improvements in physical performance. The early studies of Christensen and Hansen in 1939 were the first firmly to establish the link between a high carbohydrate diet and the improvement in endurance performance [29]. In these studies the endurance performance of a group of subjects was examined, on a cycle ergometer, after 3–4 days on either a normal mixed diet, a diet of fat and protein, or a diet rich in carbohydrate. After the high carbohydrate diet the endurance capacity of these subjects doubled in comparison with their endurance capacity following their normal mixed diet. In contrast the fat and protein diet reduced the exercise performance to almost half that achieved on a mixed diet [30].

The link between the changes in muscle glycogen concentration produced by diets high and low in carbohydrate was initially made possible by the classical study of Bergstrom and Hultman [31]. They used a modified form of the original Duchenne biopsy needle to obtain repeated samples of muscle, in order to follow the changes in glycogen concentrations after exercise. The experimental design was quite novel, in that it involved single-leg exercise with the contralateral limb acting as a resting control. The two authors contributed one exercise leg and one control leg, and exercised to exaustion, using a cycle ergometer placed between them. Samples of muscle from the vastus lateralis were obtained from the exercise and the control limbs before exercise, immediately after exercise, and at 24-hour intervals for 3 days. During the 3 days following the experiment the two subjects consumed a diet which was reported as "consisting almost exclusively of carbohydrate". The energy intake during the remainder of the experimental day and the following 2 days was reported as 2200 kcal (9.2 MJ) and 2600 kcal (10.9 MJ) for the two subjects respectively. Assuming that their diet did consist almost entirely of carbohydrate, as reported, then the two subjects would have consumed approximately 550 g and 650 g of carbohydrate a day respectively. The muscle glycogen concentration in the exercised legs was about 0.1 g/100 g muscle (wet weight) immediately after exercise, and increased to 2.0 g/100 g (ww) in the first 24 hours of recovery. This value of 2.0 g/100 g muscle (ww) was similar to the muscle glycogen concentration of the control leg, which had not changed during the experiment. On the third day after the exercise to exhaustion the muscle glycogen concentrations in the exercised legs of the two subjects had increased to approximately 4.0 g/100 g muscle (ww), which was twice as great as the glycogen concentration in the control legs. This "local phenomenon" of an increase in muscle glycogen concentration which occurs when a high carbohydrate diet is consumed after exercise to exhaustion has been called "glycogen supercompensation".

Bergstrom and Hultman [32] also examined the influence of different nutritional states on the resynthesis of glycogen during recovery from prolonged exhaustive exercise. In their study they found that fasting and/or a fat and protein diet for 2–3 days produced a delayed muscle glycogen resynthesis, whereas a high carbohydrate diet, for the same period of time, produced rapid resynthesis, leading to glycogen supercompensation. They also went on to show the link between the endurance performance, during cycle ergometry, and pre-exercise muscle glycogen concentration [33]. The correlation between initial muscle glycogen concentrations and endurance times for their nine subjects was 0.68, whereas the correlation between glycogen concentration, corrected for body mass, and total work done during the experiment was a more impressive 0.87.

The optimum conditions for the glycogen supercompensation were also explored by Ahlborg et al. [34], who determined the muscle glycogen concent-rations of young healthy military conscripts after a combination of exercise and three dietary conditions. The experimental design was such that all subjects were required to exercise to exhaustion on a cycle ergometer following a period on a normal mixed diet. After the exercise test to exhaustion, the first dietary condition required the subjects to consume a high carbohydrate diet (3000 kcal (12.6 MJ); less than 5% fat and less than 5% protein) for several days, which was followed by another exercise test to exhaustion. The second dietary condition involved the consumption of a fat and protein diet (3000 kcal (12.6 MJ); 30%–40% fat and 60%–70% protein) for 1 day after exercise to exhaustion, which was followed by another period of exercise to exhaustion, and then 3–4 days on the high carbohydrate diet. The third dietary condition consisted of 3 days on the fat/protein diet, followed by another bout of exercise to exhaustion, and then 3 days on a high carbohydrate diet. Muscle glycogen concentrations increased by 64%, 82% and 92% following the three exercise and dietary manipulations respecti-vely. The exercise times to exhaustion after the carbohydrate phase of the dietary experiment were all significantly greater than the performance times obtained at the beginning of the experiment when the subjects were on their normal mixed diets. The authors reported improvements of 157% in endurance times when their subjects exercised to exhaustion after 3 days on a high carbohydrate diet, and of 156% following 3 days on the fat/protein diet and then 3 days on the high carbohydrate diet. However, a closer examination of their results shows that these changes were not as large as suggested, because the authors took the performance time for the initial exercise test as 100%. Therefore the actual improvements following these two methods of increasing skeletal muscle glyco-gen were 56%–57%.

These dietary conditions were subsequently used to examine the relationship between pre-exercise muscle glycogen concentration and the endurance perfor-mance of a group of active subjects. This involved an endurance performance test on a cycle ergometer, followed by 3 days on a fat/protein diet (protein, 1500 kcal (6.3 MJ); fat 1300 kcal (5.5 MJ), and then exercise to exhaustion; this was followed by 3 days on a high carbohydrate diet (carbohydrate, 2300 kcal (9.7 MJ); protein 500 kcal (2.1 MJ) and then a final exercise test to exhaustion as a measure of endurance capacity [35]. For all six subjects strong correlation was obtained ($r = 0.92$) between the pre-exercise glycogen concentration and the exercise time to exhaustion, when all three conditions were examined collectively.

It is important to recognise, however, the the glycogen supercompensation phenomenon occurs only in previously active muscle [31]. Further exploration of this interesting phenomenon has shown an inverse relationship between the activity of the active form of glycogen synthetase (EC 2.4.1.11) and the concentration of glycogen in muscle [36,37]. The specificity of this relationship has been highlighted in a study which examined the activity of this enzyme, both biochemically and histochemically, in samples of the different populations of fibre types obtained from subjects who had exercised to exhaustion. The activity of the active form of glycogen synthetase was greatest in the Type I fibres, i.e. the slow contracting oxidative fibres [38]. From more recent work it can be concluded that this population of fibres is most involved during submaximal cycling to exhaustion. While the carbohydrate loading procedure for increasing muscle glycogen stores is now generally well known, it is often overlooked that the increased carbohydrate intake also produces a similar response in the storage of liver glycogen [39]. Glucose, derived from liver glycogen, does contribute to carbohydrate metabolism in working muscles, especially towards the onset of fatigue [40–42]. However, the proportional contribution of liver glycogen to improved endurance performance is difficult to separate from the influence of an increased muscle glycogen store.

The applicability of the carbohydrate loading procedure for improved physical performance during athletic competition was demonstrated by Karlsson and Saltin [43]. They addressed themselves to the question of whether or not an increased pre-exercise glycogen concentration improves running speed as well as endurance capacity. The study was conducted under race conditions, with their subjects divided into two groups. One group underwent the carbohydrate loading procedure before a race over a distance of 30 km (19 miles), while the other group remained on their normal mixed diet. In the second part of the study the race conditions were re-created some 3 weeks later, when the dietary preparation of each group was reversed. They found that the time to complete the 30-km race was improved by 8 minutes (5.6%) when the subjects increased their pre-race muscle glycogen concentrations through the carbohydrate loading procedure. The running speed of each subject was not increased during the early part of the race. However, as a result of carbohydrate loading they were able to sustain their optimum pace during the latter part of the race. A closer examination of the results of this study shows that the subjects were a mixture of experienced and recreational runners. The experienced runners had higher pre-race muscle glycogen concentrations than the recreational runners before the start of the experiment. It is therefore not surprising that the recreational runners had the largest increase in their pre-race glycogen concentrations, and recorded the greatest improvements in running times (12 min vs. 5 min). Trained individuals, irrespective of their sport, appear to have higher resting muscle glycogen concentrations than untrained individuals [44,45]. More specifically, it has been shown in studies in which only one limb has been trained that it is the trained limb which has the higher glycogen concentration [38].

Although the now traditional method of carbohydrate loading has been clearly shown to increase muscle glycogen concentrations and improve physical performance during endurance activities, the low carbohydrate period is not a dietary condition which is well tolerated by athletes. More recently evidence has been provided which suggests that the low phase of the traditional carbohydrate loading procedure need not be included. A diet rich in carbohydrate consumed for 3 days before competition, along with a decreased training intensity, results in an increased muscle glycogen concentration which is no different from that achieved as a result of the traditional carbohydrate loading procedure [45]. It appears that the rest period provides time for the glycogen stores of the active individuals to be restored to their normal high levels.

Starch and Sugar Intake and Performance

Almost without exception, the carbohydrate loading studies reported in the literature at best report only the total amount of carbohydrate consumed during the high carbohydrate phase of the dietary manipulation. There appears to have been no attempt to provide information about the amounts of sugar and starches consumed during the normal diet, nor during the high carbohydrate phase of the carbohydrate loading procedure. In our own studies on the normal diets of male and female runners, using 7-day weighed food intake analyses, we found that the amounts of sugars and starches consumed were similar (Table 11.1)

Table 11.1. Energy and macro-nutrient intakes of male ($n = 29$) and female ($n = 22$) runners

Group	Energy (MJ)	Energy (kcal)	Protein (g)	Protein (%)	Fat (g)	Fat (%)	CHO (g)	CHO (%)	Sugar (g)	Starches (g)
Males	13	3174	105	14	122	34	414	49	184	196
± SD	3	760	29	2	32	5	128	6	55	112
Females	8	1958	70	16	79	35	244	48	117	114
± SD	2	441	22	3	24	7	59	6	36	42

CHO, carbohydrate.

The question of whether or not there is a difference in the physical performance of individuals consuming either mainly sugars or mainly starches has received little attention. However, there is a limited amount of information on the influences of sugars and starches on glycogen repletion. Costill et al. [46] followed the glycogen resynthesis in the gastrocnemius muscles of a group of runners for 48 hours after prolonged heavy exercise. The energy intake of the subjects during the first 24 hours was reported as 3700 kcal (15.5 MJ), and consisted of 70% carbohydrate, 20% fat and 10% protein; the total carbohydrate intake for both groups, namely those consuming simple carbohydrates and those consuming complex carbohydrates, was 648 g. On the second day the energy intake was 2383 kcal (9.96 MJ), with apparently the same ratio of carbohydrate, fat and protein, but with a total carbohydrate intake of only 415 g. The authors provide no information about the nature of the carbohydrates consumed during the 48 hours, other than that the simple carbohydrate group consumed "glucose" and the complex carbohydrate group consumed "starch". After the first 24 hours of recovery there was no difference between the muscle glycogen concentrations of the two groups. The muscle glycogen concentration of the simple carbohydrate group immediately after exercise was 56.1 ± 7.1 mmol/kg wet weight (241.2 ± 30.5 mmol/kg dry weight) whereas the value for the complex carbohydrate group was 53.4 ± 7.5 mmol/kg wet weight (229.6 ± 32.3 mmol/kg dry weight). After the first 24 hours of recovery there was no difference in the glycogen concentrations between the two groups, which was approximately 135 mmol/kg wet weight (580.5 mmol/kg dry weight). During the second 24 hours of recovery the muscle glycogen concentration of the simple carbohydrate group increased by only 7.8 mmol/kg wet weight (33.5 mmol/kg dry weight), whereas the glycogen concent-ration of the complex carbohydrate group increased by a significantly greater ($p<0.05$) amount, namely 22.1 mmol/kg wet weight (95.0 mmol/kg dry weight). It is worth noting that the muscle glycogen concentration after 48 hours of recovery for the subjects consuming starch was 165 mmol/kg wet weight (710 mmol/kg dry weight) and for the group consuming the simple carbohydrates 143 mmol/kg wet weight (615 mmol/kg dry weight). Both these values are very high, in absolute terms, for muscle glycogen concentrations.

Costill et al. [46] also investigated the influence of size and frequency of carbohydrate intake on the muscle glycogen concentration of the gastrocnemius 24 hours after prolonged exercise. They found that the more important of these two variables was the amount of carbohydrate consumed. Both studies showed that muscle glycogen concentration could be replaced over a 24-hour recovery period when the carbohydrate consumed was of the order of 500–650 g. In this study the focus of attention was the resynthesis of muscle glycogen, and no attempt was made to examine the effects of the two types of carbohydrate consumed on exercise tolerance. The rationale is that once the glycogen has been replaced, irrespective of the nature of the carbohydrates consumed, then there is unlikely to be any difference in the rate at which it is used during a subsequent period of exercise.

A more recent study compared the amount of glycogen resynthesised following the consumption of either simple or complex carbohydrates in four groups of runners [47]. Two groups consumed a high carbohydrate diet, of which 85% was made up of complex carbohydrates. One of these groups undertook prolonged exercise and a low carbohydrate diet (<15% energy intake from carbohydrate) for 3 days before consuming the diet high in complex carbohydrate for 3 days. The other group did not undergo the exercise and low carbohydrate phase, which was designed to reduce muscle glycogen concentration, and maintained a mixed diet consisting of 50% carbohydrate for 3 days before consuming a diet high in complex carbohydrate for 3 days. The other two groups followed a similar pattern, with the exception that they consumed a carbohydrate diet of which 85% was composed of simple rather than complex carbohydrates. Muscle samples were obtained on the day before exercise, and again after the 3 days on the high carbohydrate diet. The results reported show that the dietary and recovery condition which produced the most significant increase in glycogen concentration was that involving the consumption of a diet high in simple carbohydrate, and without the preceding exercise and low carbohydrate intake. It is important to note that in this study there was a significant increase in the glycogen concentrations of all four groups of subjects after 3 days on a high carbohydrate diet, irrespective of the nature of the carbohydrates consumed. All the distance runners who took part in this study maintained some training, albeit reduced, during the experimental period in an attempt to maintain an activity pattern which was close to their normal lifestyles. The results of this study are consistent with those of a previous study which suggested that glycogen supercompensation can be achieved simply by reducing training and increasing the carbohydrate content of the diet for 3–4 days before competition [45]. Thus the period on a low carbohydrate diet, traditionally recommended as a necessary prelude to the 3 days on a high carbohydrate diet [30], is unnecessary for glycogen supercompensation.

One of the advantages of introducing simple carbohydrate drinks into the recovery diets of active individuals is that they are easily assimilated. The most rapid repletion of muscle glycogen occurs immediately after exercise, and it is during this period that substrate availability is most important [48–50]. Providing easy-to-consume carbohydrate immediately after exercise has recently been shown to produce a greater increase in glycogen concentration than when carbohydrate consumption is delayed for as little as 2 hours after exercise [51]. Furthermore, carbohydrates in the form of sugars are more readily available to active individuals immediately after exercise than are complex carbohydrates, and they are usually more palatable, especially in liquid form. It appears that sucrose is as effective as glucose solutions in producing a rapid rate of glycogen resynthesis [52].

Recognising the importance of the timing of carbohydrate intake for maximum glycogen repletion, and the accessibility and acceptability of confectionery products, we undertook a series of studies to investigate the influence of carbohydrate loading, using either complex or simple carbohydrates, on physical performance [53]. In this study 30 recreational runners were required to run to exhaustion at 70% $\dot{V}O_2$max on a treadmill. Thereafter they were divided into three dietary groups for 3 days, before running to exhaustion in a second trial. The normal diets of the 15 male and 15 female runners were determined from 7-day weighed food intake analyses 2 weeks before the start of the study, and again during the 3 days between the first trial and the second trial. The three dietary groups were: (a) normal diet supplemented with complex carbohydrates, namely additional bread, potatoes and pasta; (b) normal diet supplemented with simple carbohydrates in the form of confectionery products, most of which were chocolate bars; and (c) normal diet with additional fat and protein in order to match the energy intakes of the two carbohydrate groups. The daily carbohydrate intake increased from approximately 300 g to almost 500 g, and the average energy intake increased by 36% from 2350 kcal (9.8 MJ) to 3150 kcal (13.3 MJ); as might be expected, there was a small increase in body weight. After 3 days on the high carbohydrate diets there was a significant improvement in exercise time to exhaustion of 26% and 23% for the complex and simple carbohydrate groups respectively, whereas there was no significant increase in the endurance time of the control group. Thus the results of this study showed that the supplementation of a normal diet with simple carbohydrates improves endurance capacity to the same extent as supplementation with complex carbohydrates. The direct effect of the carbohydrate supplementation on muscle glycogen concentration was not assessed in this study, but it is reasonable to assume that the improvements in endurance capacity of the two groups were the result of an increased availability of stored carbohydrate.

In a subsequent study we investigated the influence of a similar carbohydrate supplementation on endurance performance (Brewer et al. unpublished). In this study 18 recreational runners completed a distance of 30 km (18.7 miles) on a laboratory treadmill in as fast a time as possible. The runners were then divided into two groups. One group supplemented their normal diet with simple carbohydrates, in the form of confectionery, while the other consumed additional protein and fat in order to achieve a similar energy intake. Seven days after the first treadmill trial the subjects again ran the same distance, and attempted to improve on their previous running times. There was no significant improvement in the overall running times of the subjects on the high carbohydrate diet, but their speeds over the last 10 km were significantly faster than the values recorded during the control trial. The carbohydrate group maintained their blood glucose concentrations throughout the time trial, whereas the control group had a significant reduction in their blood glucose concentrations during the latter part of the run. The daily carbohydrate intake during the first 3 days following the first trial was increased from a pre-exercise intake of 334 ± 54 g to 566 ± 88 g, representing an increase of 63%. After 3 days the amount of carbohydrate prescribed was reduced to 452 ± 79 g and maintained until the day before the second trial. This particular dietary preparation for the 30 km treadmill time trial produced some improvement in performance, and also prevented the fall in blood glucose concentration which occurs when only water is consumed. Maintaining blood glucose concentrations during a similar 30 km treadmill time trial, by providing either a glucose polymer solution or a similar solution which also contained fructose did not, however, improve the performances of a group of recreational runners (Figs. 11.2 and 11.3) (Williams et al. unpublished). The results from these two dietary studies show that simple carbohydrates can contribute effectively to carbohydrate loading, and as such produce an improvement in both endurance capacity and endurance performance during running. While blood glucose homeostasis during exercise is maintained following dietary

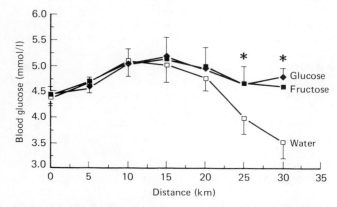

Fig. 11.2. Blood glucose concentrations during 30-km treadmill time trials. Carbohydrate solutions containing a maltodextrin plus glucose or a maltodextrin plus fructose were consumed before and during the time trials; water was consumed during the control trial. (An asterisk denotes significant difference; $p < 0.05$.)

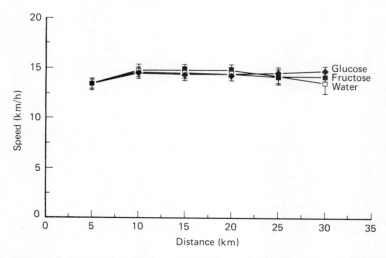

Fig. 11.3. Running speeds over 30 km during treadmill time trials following the consumption of a maltodextrin plus glucose or a maltodextrin plus fructose solution before and during the time trials; water was consumed during the control trial.

carbohydrate loading, simply maintaining blood glucose concentrations during exercise does not appear to improve endurance performance under simulated race conditions.

Pre-exercise Carbohydrate Intake

If during prolonged exercise the reduction of skeletal muscle glycogen stores is the main contributor to the onset of fatigue, then attempts to supplement the limited carbohydrate stores are indeed logical. Attempting to complement existing carbohydrate stores by ingesting glucose solutions in the hour before exercise appears however, to have the potential to produce an earlier onset of fatigue. When a glucose solution is ingested before exercise, the concomitant increase in insulin concentration depresses the mobilisation of fatty acids, and so deprives the working muscles of this substrate. As a result of this interaction there is, of necessity, an increased contribution of glycogen to muscle metabolism. This somewhat "paradoxical" increase in glycogenolysis has the potential for causing an earlier onset of fatigue. Costill et al. [54] showed that when a group of runners exercised for 30 minutes on a treadmill, after ingesting 75 g of glucose in 300 ml of water 45 minutes before exercise, they developed persistent hypoglycaemia, and their muscle glycogen concentrations were lower than the values recorded when water alone was consumed. In a subsequent study, involving cycling rather than running, they found that the endurance capacity of a group of subjects was reduced by 19% after they had ingested the same concentration of glucose solution as in the previous study [55]. The results confirmed the predictions of the earlier study that the ingestion of glucose solutions within an hour of the start of exercise decreased, rather than increased, endurance capacity.

In an attempt to provide additional carbohydrate shortly before the start of exercise, without producing an earlier onset of fatigue, fructose has been proposed as an alternative to glucose. The insulin response to fructose ingestion is less pronounced than it is to glucose, and so theoretically it would not be expected to depress fatty acid mobilisation to the same extent as a glucose solution. Furthermore the hypoglycaemia which occurs during exercise following ingestion of a glucose solution is avoided when fructose rather than glucose is the pre-exercise carbohydrate [56–58]. However, some [59] but not all [60] authors have reported a slower rate of metabolism of fructose than of glucose during submaximal exercise. Nevertheless there is some evidence of an improved endurance capacity following the consumption of fructose before the start of submaximal exercise [61,62]. In a recent study, a group of subjects underwent endurance training for 6 weeks before taking part in two exercise trials to examine the influence of ingesting fructose on endurance capacity [62]. The trained subjects consumed a prescribed lunch 4 hours before ingesting either a fructose solution (65 g or 85 g in 500 ml of water) or the same volume of placebo of sweetened water. An hour after ingesting the fructose or placebo solutions, the subjects exercised to exhaustion on a cycle ergometer at an initial intensity of 62% $\dot{V}O_2$max, which was increased to 72% $\dot{V}O_2$max after an hour, and then to 81% $\dot{V}O_2$max after 2 hours of exercise. The fructose ingestion produced a significantly greater endurance time than did the placebo (145 ± 4 min vs. 132 ± 3 min). The hypoglycaemia reported during exercise following the ingestion of glucose solutions before exercise, did not occur in this study, neither was there any evidence of a decreased rate of fatty acid metabolism. This study is the only one to have clearly shown a significant improvement in endurance capacity. Some studies [61,63,64], but by no means all [57,65–68] have reported a "glycogen sparing" influence of fructose ingestion before exercise.

Notwithstanding this recent report on the improved exercise tolerance resulting from the pre-exercise ingestion of fructose, its contribution to glycogen resynthesis following exercise appears to be significantly less than that after the infusion of glucose [69]. Post-exercise fructose infusion produced a muscle glycogen concentration which was only 56% of the value achieved when glucose was used. An explanation for this reduced contribution to muscle glycogen resynthesis is that fructose makes a significantly greater contribution to the liver's store of glycogen than does glucose [70]. Interestingly, sucrose solutions consumed immediately after exercise appear to make a greater contribution to glycogen resynthesis than does the ingestion of fructose. Blom et al. [52] reported that the glycogen resynthesis rate following fructose ingestion was only 68% of that achieved when sucrose was ingested.

The provision of additional carbohydrate before exercise has also been explored, using solid rather than liquid supplementation, in an attempt to improve exercise tolerance. Hargreaves et al. [63] reported an improvement in work capacity at the end of 4 hours of intermittent cycling exercise when the subjects had ingested 43 g of solid carbohydrate (chocolate bar; 43 g sucrose, 9 g fat, 3 g protein plus 400 ml of water) immediately before exercise and at hourly intervals for the first 3 hours of cycling. Thus these subjects consumed 172 g of solid carbohydrate in 1.2 litres of water over the 4 hours of exercise. The subjects performed exercise, on each trial, which consisted of 20 minutes of cycling at 50% $\dot{V}O_2$max, followed by 10 minutes of intermittent exercise of high intensity. During the high-intensity exercise periods the subjects performed 30 seconds at a work load equivalent to their $\dot{V}O_2$max, followed by 2 minutes of rest. This procedure was followed for each half-hour period, with the exception that at the end of the 4 hours the high-intensity exercise was maintained until exhaustion. Exercise time to exhaustion during the final all-out sprint was improved by 40 seconds, and the amount of muscle glycogen used was less when compared with values obtained in the control experiment, in which water was ingested at similar time intervals [63].

In an extension of this latter study, and using the same exercise protocol, the benefits of frequency of feeding solid carbohydrate were investigated [66]. In one trial the subjects consumed 10.75 g carbohydrate (25% of a chocolate bar plus 200 ml of water) every 30 minutes during the 4-hour exercise test, whereas in the other trial 21.5 g (50% of a chocolate bar plus 400 ml of water) was consumed every hour. The more frequent feeding of the solid carbohydrate (86 g of chocolate and 1.6 litres of water) prevented the fall in blood glucose concentration which occurred in the control trial with water alone, and also resulted in a greater endurance capacity. In this particular study the subjects were required to exercise to exhaustion, at an intensity equivalent to their maximum oxygen uptake, at the end of the 4 hours of exercise. The improvement in performance time was again 40 seconds, but there was no evidence that the more frequent feeding produced a glycogen sparing effect.

A more recent study by Costill et al. [71] compared the influence on work capacity of liquid (45 g) and solid carbohydrate (45 g, chocolate bar) supplementation with that of water alone. In these experiments the carbohydrate supplementation was taken 5 minutes before the start of exercise rather than the 30–45 minutes as was the case in earlier studies [54]. The exercise test employed in these experiments required the subjects to perform 45 minutes of cycling at an exercise intensity of 77% $\dot{V}O_2$max, followed by a final 15 minutes of exercise during which the subjects were required to perform as much work as possible. Both the liquid and the solid carbohydrate trials resulted in a greater amount of work being achieved than when water alone was ingested. In an additional trial the subjects consumed a carbohydrate meal (200 g) 4 hours before exercise, and then ate the solid carbohydrate immediately before the exercise test. Interestingly, the work load achieved during the last 15 minutes of the standard cycling test, following this latter dietary trial, was significantly greater than those achieved during trials using the solid and the liquid carbohydrate supplementation alone. There was no evidence, however, of a glycogen sparing effect similar to that reported by Hargreaves et al. [63,64].

In contrast to the positive benefits reported for solid carbohydrate feeding in this latter study, no such improvements were found when a group of untrained subjects were exposed to a similar pre-exercise treatment. When carbohydrate, in the form of a snack bar (260 kcal; 1.1 MJ), was ingested pre-exercise, exercise capacity was not improved, neither was muscle glycogen spared [72]. The exercise intensity was 70% $\dot{V}O_2$max, and the duration was less than 60 minutes. Under these conditions glycogen depletion was probably not the limiting factor, but rather the high lactic acid concentrations, reflecting even higher muscle lactate concentrations. The blood lactate concentrations after 15 minutes of exercise were 7.6 ± 0.74 mmol/l for the control trial, before which sweetened water was consumed, and 8.02 ± 0.95 mmol/l for the carbohydrate trial. Both values are high for submaximal exercise of this intensity, and this supports the proposition that the limiting factor was probably not insufficient carbohydrate. Furthermore this study also highlights the fact that in almost all the studies in which carbohydrate intake before and during exercise has been shown to have a positive influence on physical performance, the subjects have been either trained or physically active. A trained individual is able to perform submaximal exercise long enough to challenge the limited glycogen stores, whereas this is not the case for untrained and sedentary individuals. It is also important to note that the timing of the pre-exercise carbohydrate intake is crucial. For example, ingesting carbohydrate solutions immediately before exercise does not appear to provoke the same degree of hyperinsulinaemia, nor such large changes in blood glucose concentrations, as occurs when carbohydrate is consumed within the hour before exercise. The explanation appears to be that at the onset of exercise the increase in circulating catecholamines suppresses the insulin response to carbohydrate ingestion [73].

Carbohydrate Intake During Exercise

The problem of fluid replacement during exercise is perceived as one in which gastric emptying is the limiting factor to adequate replacement. Gastric emptying rate is influenced by a number of factors, such as volume of fluid ingested, energy content of the solution, osmolality of the solution, temperature of the solution and of the environment, as well as exercise intensity. The rate of gastric emptying of water is, within limits, proportional to the volume ingested. The initial clearance rate is quite rapid, and it then falls off in an exponential way. Electrolytes lost in sweat can normally be replaced by consuming a normal balanced diet, and there is no good evidence to suggest that electrolyte drinks need be consumed for this purpose alone.

The general view is that the addition of electrolytes and/or carbohydrate to water will result in a decreased gastric emptying, and so impede the process of fluid replacement. Therefore, as a result of earlier experiments on the clearance rates of glucose solutions of varying concentrations, it was concluded that concentrated glucose solutions empty more slowly from the stomach than water, and thus only dilute solutions should be recommended for fluid and carbohydrate replacement during exercise [74]. One of the overriding concerns when considering the composition of fluid replacement solutions has been its osmolality. Considerable effort has been expended in attempting to produce fluid replacement solutions, in order to achieve low osmolality while still offering a reasonable amount of carbohydrate. When glucose polymers have been used to increase the carbohydrate content of these solutions, there is some evidence to suggest that their clearance rates are at least as rapid as those of solutions containing glucose alone [75,76]. In a comparison of the gastric clearance rates of solutions of glucose and glucose polymers during rest and during running, the clearance rates were found to be greater during exercise than they were at rest [77]. The authors suggested that the improvement in the clearance rates of these carbohydrate solutions during exercise may be explained by the greater mechanical movement of fluid in the stomach during running, which is not the case during cycling. More recently evidence from the same laboratory has shown that when glucose polymer solutions are ingested before and during intermittent cycling exercise, the gastric emptying was no different for the concentrated solutions (e.g. 6% and 7.5% carbohydrate) than for water alone. In addition each of the carbohydrate solutions ingested (1336 ml) resulted in the performance of a greater amount of work than was achieved during the water control trial [78]. The intermittent exercise involved seven 12-minute periods of cycling at an intensity of 70% $\dot{V}O_2$max. During the last period of exercise the cyclists attempted to complete as much work as possible.

Although a considerable amount of attention has been paid to the formulation of fluid replacement solutions to achieve optimum osmolality, there is evidence to suggest that the more important influence on gastric emptying rates is the energy content rather than osmolality per se [79]. While a number of the studies attempted to determine the optimum composition of carbohydrate replacement solutions, others have simply explored the influences of quite concentrated carbohydrate solutions on exercise tolerance. In one such study, the subjects walked on a treadmill to the point of fatigue, ingesting either a glucose polymer solution or water alone. The concentration of the glucose polymer solution was 20%, and the total amount consumed was equivalent to approximately 120 g of carbohydrate, which was administered during the first of $3\frac{1}{2}$ hours of exercise. Endurance time was increased by 11.5% (299.0 ± 9.8 min vs. 268 ± 11.8 min) during the glucose polymer trial compared with the water trial, but following the prolonged walk there were no differences in either running time to exhaustion or performance during a series of psychomotor ability tests [80]. A similar improvement in work capacity was reported for a group of trained cyclists who exercised for 3 hours at an intensity equivalent to 74% $\dot{V}O_2$max, while consuming a glucose polymer solution. An analysis of the results showed that a significant improvement in work capacity occurred only in the subjects for whom a fall in blood glucose concentration had been recorded during the control experiment [81]. In this experiment the subjects consumed approximately 140 ml of a 50% glucose polymer solution after the first 20 minutes of exercise. Thereafter they consumed 300 ml of a 6% solution of the glucose polymer every 30 minutes.

Murray et al. [82] have also examined the influence of carbohydrate supplementation on work capacity, and found that during 90 minutes of intermittent cycling their subjects were able to complete a set amount of work (a prescribed number of revolutions) faster during the carbohydrate trials than during the control (water) trial. They examined three carbohydrate solutions, consisting of 5% glucose polymer, 6% glucose/sucrose and 7% glucose/fructose polymer. In all three carbohydrate trials the time taken to perform the prescribed amount of work was faster than that achieved during the water trial. The most dramatic improvement in endurance capacity was, however, achieved during an experiment using trained cyclists who ingested a glucose polymer solution during exercise to fatigue at an intensity equivalent to 71% $\dot{V}O_2$. In this experiment the subjects ingested approximately 300 ml of a 50% carbohydrate solution (which provided 2 g/kg body weight) after the first 20 minutes of exercise, and approximately 300 ml of a 10% solution (to provide 0.4 g/kg body weight) every 20 minutes for the remainder of the exercise period. Fatigue occurred after 3 hours of exercise during the control experiment, and after 4 hours during the carbohydrate trial. Interestingly the rate of muscle glycogen metabolism was the same for the first 3 hours of exercise during both trials. The authors concluded that the trained cyclists were able to utilise the carbohydrate supplied during the extra hour of exercise, and so the improvement in endurance was not due to glycogen sparing *per se* [42]. Therefore, while the results of this particular study clearly demonstrate a beneficial effect from carbohydrate supplementation during prolonged exercise, it is worth noting that the improvement in performance was obtained during cycling rather than running. Thus the quantities of solid and liquid consumed in order to produce these improvements in performance may only be effective in activities in which the body weight is supported. The gastrointestinal discomfort which would inevitably result during weight-bearing activities, such as running, would probably make the use of this approach to carbohydrate supplementation unacceptable.

Summary

An increase in dietary carbohydrate intake following prolonged exercise, or after a period of training, will increase muscle and liver glycogen concentrations. This dietary procedure can employ additional starches and/or additional sugars to supplement the normal carbohydrate intake with almost equally good results. Physical performance during prolonged submaximal exercise following dietary carbohydrate loading is improved, and this is a widely accepted practice by athletes preparing for endurance competitions. Supplementation of the endogenous carbohydrate stores with solutions of sugars before and during exercise can provide additional substrate for muscle metabolism [40,41]. However, this contribution to muscle metabolism occurs when the muscle glycogen concentration has been reduced as a result of prolonged exercise. When the carbohydrate stores have been increased as a result of carbohydrate loading before exercise, then further carbohydrate intake during exercise in the form of sugar solutions, appears not to be effective [83,41]. Ingesting glucose solutions within the hour before the start of exercise produces a paradoxical reduction in blood glucose concentrations during the early part of exercise, and an apparent increased rate of glycogen degradation [54]. Furthermore the reported decreased endurance capacity following the pre-exercise ingestion of glucose solutions [55] prompted investigations into the possible benefits of using fructose rather than glucose [61,63,64,68] as an exogenous source of carbohydrate. While fructose solutions do not provoke such a significant insulinogenic response as do glucose solutions [56, 57, 67], only one study has shown an improvement in exercise endurance following their ingestion [62]. A recent re-examination of the apparent detrimental influence on endurance performance of ingesting glucose solutions in the hour before exercise [54,55] has shown that there are no differences between the endurance performances following the ingestion of glucose or fructose solutions [65]. The beneficial influence on physical performance of sugar solutions taken immediately before and also during exercise has been shown mainly with trained, or at least recreationally active subjects. Furthermore, with one exception [42], most of the studies which have reported an improvement in physical performance have used exercise protocols which require the subjects to perform periods of submaximal exercise followed by periods of high intensity exercise. Each of these studies has used different exercise protocols, and so it is imprudent to make wide generalisations about the benefits of this form of carbohydrate supplementation on performance. Further studies are required to understand more fully the influence of starch and sugar intake on carbohydrate metabolism in general and on physical performance in particular.

References

1. Astrand PO, Rodahl K (1970) Textbook of work physiology. McGraw-Hill Book Company, New York, pp 305–315
2. Rowell LB (1974) Human cardiovascular adjustments to exercise and thermal stress. Physiol Rev 54:75–159
3. Davies CTM, Thompson MW (1979) Aerobic performance of female marathon and male ultramarathon athletes. Eur J Appl Physiol 41:233–245
4. Lakomy HKA (1986) Measurement of work and power using friction loaded cycle ergometers. Ergonomics 29:509–514
5. Cooper KH (1968) A means of assessing maximum oxygen uptake JAMA 203:135–138
6. Daniels J, Yarborough RA, Foster C (1978) Changes in VO$_2$max and running performance with training. Eur J Appl Physiol 39:249–254
7. Williams C, Nute MG (1986) Training-induced changes in endurance capacity of female games players. In: Watkins J et al. (eds) Sports science. E & F Spon, London, pp 11–17
8. Davies KJA, Packer L, Brooks GA (1981) Biochemical adaptations of mitochondria, muscle and whole animal respiration to endurance training. Arch Biochem Biophys 209:538–553

9. Rennie MJ, Winder WW, Holloszy JO (1976) A sparing effect of increased plasma fatty acids on muscle and liver glycogen content in exercising rat. Biochem J 156:647–655
10. Gollnick PD, Saltin B (1982) Significance of skeletal muscle oxidative enzyme enhancement with endurance training. Clin Physiol 2:1–12
11. Wasserman K, Whipp BJ, Koyal SN, Beaver ML (1973) Anaerobic threshold and respiratory gas exchange during exercise. J Appl Physiol 35:236–243
12. Davis JA (1985) Anaerobic threshold: review of the concept and directions for future research. Med Sci Sports Exer 17:5–18
13. Brooks GA (1985) Anaerobic threshold: review of the concept and direction for future research. Med Sci Sports Exer 17:22–31
14. Kindermann W, Simon G, Keul J (1979) The significance of the aerobic–anaerobic transition for the determination of work load intensities during endurance training. Eur J Appl Physiol 42:25–34
15. Jacobs J (1986) Blood lactate: implications for training and research. Sports Med 3:10–25
16. Williams C, Nute MG (1983) Some physiological demands of a half-marathon race on recreational runners. Br J Sports Med 17:152–161
17. Hurley BF, Hagberg JM, Allen WK et al. (1984) Effect of training on blood lactate level during submaximal exercise J Appl Physiol 56:1260–1264
18. Hermansen L, Hultman E, Saltin B (1967) Muscle glycogen during prolonged severe exercise. Acta Physiol Scand 71:120–139
19. Gollnick PD, Ianuzzo CD, Williams C, Hill R (1969) Effect of prolonged severe exercise on the ultrastructure of human skeletal muscle. Z Angew Physiol 27:257–265
20. Essen B, Jansson E, Henrrikson J, Taylor AW, Saltin B (1975) Metabolic characteristics of fibre types in human skeletal muscle. Acta Physiol Scand 95:153–165
21. Gollnick PD, Piehl K, Armstrong RB, Saubert CW, Saltin B (1972) Diet, exercise and glycogen changes in human muscle fibres. J Appl Physiol 33:421–425
22. Gollnick PD, Armstrong RB, Saubert CW, Sembrowich WL, Shepherd RE, Saltin B (1973) Glycogen depletion patterns in human skeletal muscle fibres during prolonged work. Pflugers Arch 344:1–12
23. Vollestad NK, Vaage O, Hermansen L (1984) Muscle glycogen depletion patterns in type I and subgroups of type II fibres during prolonged severe exercise in man. Acta Physiol Scand 122:433–441
24. Boobis LH (1987) Metabolic aspects of fatigue during sprinting. In: Macleod D et al. (eds) Exercise: benefits, limitations and adaptations. E & F Spon, London, pp 116–140
25. Hardman AE, Kabat MGL, Williams C (1981) Blood lactic acid concentrations in male and female subjects during treadmill running. J Physiol 325:53–54P
26. Hultman E, Spriet LL, Sodelund K (1987) Energy metabolism and fatigue in working muscle. In: Macleod D et al. (eds) Exercise: benefits, limitations and adaptations. E & F Spon, London, pp 63–80
27. Maughan RJ, Poole DC (1981) The effects of a glycogen-loading regimen on the capacity to perform anaerobic exercise. Eur J Appl Physiol 46:211–219
28. Greenhaff PL, Gleeson M, Maughan RJ (1987) The effects of dietary manipulation on blood acid–base status and performance of high intensity exercise. Eur J Appl Physiol 56:331–337
29. Christensen EH, Hansen O (1939) Arbeitsfahigkeit und Ehrnahrung. Skand Arch Physiol 81:160–175.
30. Astrand PO (1967) Diet and athletic performance. Fed Proc 26:1772–1777
31. Bergstrom J, Hultman E (1966) Muscle glycogen synthesis after exercise: an enhancing factor localized to the muscle cells in man. Nature 210:309–310
32. Bergstrom J, Hultman E (1967) Muscle glycogen synthesis in relation to diet studied in normal subjects. Acta Med Scand 182:109–117
33. Ahlborg B, Bergstrom J, Ekelund L-G, Hultman E (1967) Muscle glycogen and muscle electrolytes during prolonged physical exercise. Acta Physiol Scand 70:129–142
34. Ahlborg B, Bergstrom JB, Brohult J, Ekelung L-G, Hultman E, Maschio G (1967) Human muscle glycogen content and capacity for prolonged exercise after different diets. Forsvarmedicin 3:85–99
35. Bergstrom J, Hermansen L, Hultman E, Saltin B (1967) Diet, muscle glycogen and physical performance. Acta Physiol Scand 71:140–150
36. Hultman E, Bergstrom J, Roch-Norland AE (1971) Glycogen storage in human muscle. In: Pernow B, Saltin B (eds) Muscle metabolism during exercise. Plenum Press, New York, pp 273–288
37. Adolphsson S (1973) Effects of contractions in vivi on glycogen content and glycogen synthetase activity in muscle. Acta Physiol Scand 88:189–197

38. Piehl K, Adolfsson S, Nazar K (1974) Glycogen storage and glycogen synthetase activity in trained and untrained muscle of man. Acta Physiol Scand 90:779–788
39. Nilsson LH:Son, Hultman E (1973) Liver glycogen in man: the effect of total starvation or a carbohydrate-poor diet followed by carbohydrate refeeding. Scand J Clin Lab Invest 32:325–330
40. Wahren J (1973) Substrate utilization by exercising muscle in man. Prog Cardiol 2:255–280
41. Bonen A, Malcolm SA, Kilgour RD, MacIntyre KP, Belcastro AN (1981) Glucose ingestion before and during intense exercise. J Appl Physiol 50:766–771
42. Coyle ED, Coggan AR, Hemmert ME, Ivy JJ (1986) Muscle glycogen utilization during prolonged strenuous exercise when fed carbohydrate. J Appl Physiol 61:165–172
43. Karlsson J, Saltin B (1971) Diet, muscle glycogen and endurance performance. J Appl Physiol 31:203–206
44. Gollnick PD, Armstrong RB, Saltin B, et al. (1973) Effect of training on enzyme activity and fiber composition of human skeletal muscle. J Appl Physiol 34:107–111
45. Sherman WM, Costill DL, Fink W, Miller J (1981) Effect of exercise–diet manipulation on muscle glycogen and its subsequent utilization during performance. Int J Sports Med 2:114–118
46. Costill DL, Sherman WM, Fink WJ, Maresh C, Witten M, Miller JM (1981) The role of dietary carbohydrates in glycogen synthesis after strenuous running. Am J Clin Nutr 34:1831–1836
47. Roberts KM, Noble EG, Hayden DB, Taylor AW (1987) Simple and complex carbohydrate-rich diets and muscle glycogen content of marathon runners. Eur J Appl Physiol 57:70–74
48. Piehl K (1974) Time course for refilling glycogen stores after exercise induced glycogen depletion. Acta Physiol Scand 90:297–302
49. MacDougall JD, Ward GR, Sale DG, Sutton JR (1977) Muscle glycogen repletion after high intensity exercise. J Appl Physiol 42:129–132
50. Maehlum S, Hermansen L (1978) Synthesis of muscle glycogen during recovery after prolonged severe exercise in fasting subjects. Scand J Lab Invest 38:557–560
51. Ivy JL, Katz AL, Cutler CL, Sherman WM, Coyle EF (1988) Muscle glycogen synthesis after exercise: effect of time of carbohydrate ingestion. J Appl Physiol 64:1480–1485
52. Blom P, Vollestad NK, Hermansen L (1984) Diet and recovery process. S Karger, Basel, pp 148–160 (Medicine and Sports Science 17)
53. Brewer J, Williams C, Patton A (1988) The influence of high carbohydrate diets on endurance running performance. Eur J Appl Physiol 57:698–706
54. Costill DL, Coyle ED, Dalsky G, Evans W, Fink W, Hoopes D (1977) Effects of elevated plasma FFA and insulin on muscle glycogen usage during exercise. J Appl Physiol 43:695–699
55. Foster C, Costill DL, Fink WJ (1979) Effect of pre-exercise feeding on endurance performance. Med Sci Sports 11:1–5
56. Koivisto VA, Karonen S-L, Nikkila EA (1981) Carbohydrate ingestion before exercise: comparison of glucose, fructose and sweet placebo. J Appl Physiol 51:783–787
57. Koivisto VA, Harkonen M, Karonen S-L (1985) Glycogen depletion during prolonged exercise: influence of glucose, fructose, or placebo. J Appl Physiol 58:731–737
58. Hasson SM, Barnes WS (1987) Blood glucose levels during rest and exercise: influence of fructose and glucose ingestion. J Sports Med 27:326–332
59. Massicotte D, Peronnet F, Allah C, Hillaire-Marcel C, Ledoux M, Brisson G (1986) Metabolic response to (^{13}C) glucose and (^{13}C) fructose ingestion during exercise. J Appl Physiol 61:1180–1184
60. Decombaz J, Sartori DJ, Arnaud MJ, Theliu AL, Schürch P, Howald H (1985) Oxidation and metabolic effects of fructose or glucose ingested before exercise. Int J Sports Med 6:282–286
61. Levine L, Evans WJ, Cadarett BS, Fisher EC, Bullen BA (1983) Fructose and glucose ingestion and muscle glycogen use during submaximal exercise. J Appl Physiol 55:1767–1771
62. Okano G, Takeda H, Morita I, Katoh M, Mu Z, Miyake S (1988) Effect of pre-exercise fructose ingestion on endurance performance in fed men. Med Sci Sports Exerc 20:105–109
63. Hargreaves M, Costill DL, Coggan A, Fink WJ, Nishibbata I (1984) Effect of carbohydrate feedings on muscle glycogen utilization and exercise performance. Med Sci Sports Exerc 16:219–222
64. Hargreaves M, Costill DL, Katz A, Fink WJ (1985) Effect of fructose ingestion on muscle glycogen usage during exercise. Med Sci Sports Exerc 17:360–363
65. Hargreaves M, Costill DL, Fink WJ, King DS, Fielding RA (1987) Effect of pre-exercise carbohydrate feedings on endurance cycling performance. Med Sci Sports Exerc 19:33–36
66. Fielding RA, Costill DL, Fink WJ, King DA, Hargreaves M, Kovaleski JE (1985) Effect of carbohydrate feeding frequencies and dosage on muscle glycogen use during exercise. Med Sci Sports Exerc 17:472–476
67. Fielding RA, Costill DL, Fink WJ, King DA, Kovaleski JE, John P (1987) Effects of pre-exercise

carbohydrate feedings on muscle glycogen use during exercise in well trained runners. Eur J Appl Physiol 56:225–239

68. Bjorkman O, Sahlin K, Hagenfeldt L, Wahren J (1984) Influence of glucose and fructose ingestion on the capacity for long-term exercise in well trained men. Clin Physiol 4:483–494

69. Bergstrom J, Hultman E (1967) Synthesis of muscle glycogen in man after glucose and fructose infusion. Acta Med Scand 182:93–107

70. Nilsson LH:Son, Hultman E (1974) Liver and muscle glycogen in man after glucose and fructose infusion. Scand J Clin Lab Invest 33:5–10

71. Neufer DP, Costill DL, Flynn MG, Kirwan JP, Mitchell JB, Houmard J (1987) Improvements in exercise performance: effects of carbohydrate feedings and diet. J Appl Physiol 62:983–988

72. Devlin JT, Calles-Escandon J, Horton ES (1986) Effects of pre-exercise snack feeding on endurance cycle exercise. J Appl Physiol 60:980–985

73. Galbo H (1983) Hormonal and metabolic adaptations to exercise. George Thieme Verlag, New York

74. American College of Sports Medicine (1984) Position statement: the prevention of thermal injuries during distance racing. Med Sci Sports Exerc 16:ix

75. Foster C, Costill DL, Fink WJ (1980) Gastric emptying characteristics of glucose and glucose polymers. Res Q Sports Exerc 51:299–305

76. Wheeler KB, Banwell LB (1986) Intestinal water and electrolyte flux of glucose-polymer electrolyte solutions. Med Sci Sports Exerc 18:436–439

77. Neufer DP, Costill DL, Fink WJ, Kirwan JP, Fielding RA, Flynn MG (1986) Effect of exercise and carbohydrate composition on gastric emptying. Med Sci Sports Exerc 18:658–662

78. Mitchell JB, Costill DL, Houmard JA, Flynn MG, Fink WJ, Beltz D (1988) Effects of carbohydrate ingestion on gastric emptying and exercise performance. Med Sci Sports Exerc 20:110–115

79. Murray R (1987) The effects of consuming carbohydrate-electrolyte beverages on gastric emptying and fluid absorption during and following exercise. Sports Med 4:322–351

80. Ivy JL, Miller W, Dover V et al. (1983) Endurance improved by ingestion of a glucose polymer supplement. Med Sci Sports Exerc 15:466–471

81. Coyle ED, Hagberg SA, Hurley WH, Eshani JM, Hollszy JO (1983) Carbohydrate feeding during prolonged strenuous exercise can delay fatigue J Appl Physiol 55:230–235

82. Murray R, Eddy DE, Murray TW, Seifert JG, Paul GL, Halaby GA (1987) The effect of fluid and carbohydrate feedings during intermittent cycling exercise. Med Sci Sports Exerc 19:597–604

83. Flynn MG, Costill DL, Hawley JA, (1987) Influence of selected carbohydrate drinks on cycling performance and glycogen use. Med Sci Sports Exerc 19:37–40

Commentary

Southgate: The use of relatively compex mixtures such as chocolate confectionery to increase simple carbohydrate intakes does seem to raise questions regarding the effects of the other components. It is unfortunate that in many of the studies reported the carbohydrate intakes were relatively poorly characterised; however, in view of the small differences observed this is probably of little significance. I think that the concept of "solid" carbohydrate needs expansion; these presumably are solid in contrast to solutions. However, the conclusion that in a dietary context there is little or no difference between simple and complex carbohydrates is reasonably well substantiated.

Chapter 12

Insulin, Glucagon and Catecholamine Responses to the Ingestion of Various Carbohydrates

M. Laville, S. Picard, S. Normand and J.P. Riou

The hormonal and metabolic events occurring during an oral glucose load have been extensively characterised in man. The use of tracers, the splanchnic hepatic balance technique and indirect calorimetry [1] have permitted the measurement of most, if not all the kinetic parameters of glucose metabolism. Despite this extensive work the mechanisms involved in the regulation of glycaemia are still the subject of debate and research. Most of the sophisticated methodology used still needs to be applied to various carbohydrate nutrients which are usually transformed inside the body into glucose before being oxidised, stored or recycled. The aim of this short review is not to provide extensive data on the glycaemic index of various carbohydrates, but to focus on the respective roles of hormones and glucose *per se*, and to underline some remaining methodological problems in this kind of study.

Regulation of Glycaemia During an Oral Glucose Load

Need for Kinetic Measurements of Glucose Metabolism

The rapid increase in arterial blood glucose which follows the absorption of 75 g of glucose is due to an increased rate of appearance of glucose which transiently overcomes the capacity of the body to utilise it. The increased blood glucose stimulates insulin secretion and decreases glucagon secretion by the pancreatic islets of Langerhans. Early elevation of plasma insulin inhibits endogenous hepatic glucose production, thereby decreasing the total rate of appearance of glucose in the blood [2–6]. Both increased glycaemia *per se* [7] and increased plasma insulin [8] are involved in this suppression of hepatic glucose production. Nevertheless it should be emphasised that although hepatic glucose production is completely suppressed in healthy subjects by an insulin level of 30 μu/ml during a euglycaemic hyperinsulinaemic glucose clamp [8], this is never the case during an oral glucose load [3–6]. Under this condition hyperinsulinaemia as high as 100 µu/ ml failed to suppress hepatic glucose production completely.

The decrease in blood glucose occurring 60–90 minutes after the load is related to a lowered rate of appearance of exogenous glucose and an increase in glucose utilisation by the body. At least three physiological events are involved in this phenomenon. Firstly, the increase in blood glucose alone stimulates glucose transport in both insulin-dependent and insulin-independent tissues. The physiological apparent K_m for glucose transport in vivo is in the range 8–12 mM [9]. Therefore glucose uptake is far from being saturated at physiological plasma glucose levels and any increase in plasma glucose will stimulate glucose utilisation, thereby decreasing plasma glucose concentration. Secondly, the rise in plasma insulin to over 30 µu/ml stimulates glucose utilisation [8] in insulin-dependent tissues, mainly the muscles. In these tissues insulin stimulates not only glucose transport but also glucose storage and glucose oxidation, this pathway being rapidly saturated [10]. Thirdly, the increased glucose utilisation produced by the elevated blood glucose and plasma insulin is facilitated by the fall in circulating free fatty acids and ketone bodies occurring during the oral glucose load. The role of plasma fatty acids in regulating glucose utilisation by insulin-dependent tissues was demonstrated more than 20 years ago in animals [11]. In humans this important physiological event has been clearly demonstrated in the last 4 years [12–13].

Most clinical studies on glucose and insulin response to various carbohydrates are unfortunately performed for only 3 hours. It is usual to consider that all the metabolic parameters have returned to basal levels by 150–180 minutes after the oral glucose load. This is a wrong assumption. Three hours after the load, although blood glucose has returned to basal levels, glucose oxidation is still going on at an increasing rate, and lipid oxidation is still low. Glucose carbon of exogenous origin recirculates in the whole body for up to 6 hours [3,6,14]. These events must be taken into account when the metabolic responses to various carbohydrates are studied.

This brief summary has emphasised the fact that the glycaemic index is a poor indication of the metabolic fate of ingested carbohydrates, and also that kinetic measurements of glucose metabolism are necessary in order to ascertain which parameters are affected. If one needs to know the metabolic fate of sugars in man, there is an urgent need for quantitative data on exogenous glucose appearance rate, endogenous glucose production, glucose utilisation, glucose storage and glucose oxidation during the absorption of various carbohydrates.

Need for Kinetic Measurements of Insulin Secretion

The secretory activity of pancreatic beta cells in response to an oral glucose load is usually evaluated by measurements of serum insulin concentration. This approach has often yielded conflicting results [15]. In fact insulin secretion rates cannot be directly calculated from peripheral insulin concentrations, because of the large and variable hepatic extraction, which is dependent on the size of the glucose load [15–19]. Insulin secretion rates may be determined either by invasive techniques using sampling catheters in the portal and hepatic veins (impossible in man) or by an indirect approach based on mathematical analysis of peripheral C peptide concentrations. Insulin and C peptide are co-secreted by the beta cell in equimolar concentrations [17]. The hepatic extraction of C peptide is negligible and its metabolic clearance is constant over a wide range of physiological concentrations. Therefore secretion rates of insulin can be derived from C peptide concentrations, and the C peptide:insulin ratio provides an accurate measurement of the hepatic extraction of insulin [20]. These considerations should be kept in mind when the acute insulin response to various carbohydrates is to be determined. This phenomenon could, at least in part, be one explanation of the classical finding that orally administered glucose evokes a greater insulin response than does intravenously administered glucose [21,22]. It has been suggested that, when glucose is administered orally, an "incretin" factor is secreted which stimulates insulin secretion [23] (see Chap. 6).

However, study of the C peptide:insulin ratio shows that the higher insulin response during an oral glucose load could be explained by a lower hepatic insulin extraction [24].

Intra-islet Regulation

When the hormonal response to carbohydrates is studied in vivo a simplified scheme is used in which glucose stimulates insulin secretion by the beta cells and decreases the glucagon secretion by the A cell, thus creating an optimal endocrine response for the rapid return of blood glucose levels to those before stimulation. This description, although true, hides the important finding of the "social organisation of the beta cell" [25]. In this model the ability of glucose to stimulate insulin release depends not only on the direct effect of glucose *per se*, but also on its interaction with signals which originate locally and at a distance from other cell types. For example it now seems clear that glucagon acts by a direct effect on the insulin secretory activity of pure homogeneous isolated beta cells. On the other hand insulin secretion by the beta cells directly reduces glucagon secretion by the A cells. Therefore, when studying insulin and glucagon responses to carbohydrate feeding, it is important to keep in mind that the supposedly carbohydrate-related effect could in fact be due to a secondary hormonally mediated effect, and not only to a direct effect of the hexose on the A and/or beta cells of the islets.

Finally, this concept should be extended to the whole pancreatic gland, since it has been shown that basal insulin secretion has a direct effect on pancreatic enzyme output. This suggests that an insulo-acinar axis may play an important role in the regulation of acinar cell function [26]. If this finding is confirmed in man it could be of importance for the handling of glucose and starchy food.

Acute Insulin Response to Carbohydrates

Data obtained in vitro have permitted clarification of the direct effect of carbohydrates on insulin secretion. Insulin secretion in response to glucose is not a simple linear function of the glucose concentration: a biphasic response is found in the perfused rat pancreas [27]. The response consists of a rapid early peak of insulin, followed by a second, more slowly rising output. Data from in vitro systems indicate that the early phase represents release of pre-formed insulin, the later phase depending in part on its *de novo* synthesis. The effect of hexoses on insulin secretion has been studied extensively by perfusing the pancreas in rodents and in isolated islets. It has been shown that D-glucose, D-mannose and to a lesser extent D-fructose are oxidised by pancreatic islets, whereas D-galactose is not [28]. Galactose has been shown not to stimulate insulin secretion [29]. D-fructose has been found to have no direct stimulating effect on insulin secretion, but potentiates insulin release in response to D-glucose [30] or D-mannose [31]. Limited studies which have been performed with isolated human islets support these observations [32].

In vivo in man, after a 75 g oral glucose load, insulin is released and the peak elevation of plasma insulin is observed between 30 and 60 minutes later; the return to basal levels usually occurs within 180–240 minutes. However, it has been shown that the insulin response may be delayed in non-insulin-dependent diabetes, and that a return to basal level may not be reached even after 5 hours [33].

Furthermore the insulin response after oral glucose varies widely among normal individuals without any variation in the blood glucose response. The insulin response is correlated with insulin sensitivity measured during a hyperglycaemic hyperinsulinaemic clamp [33]. It is also well known that insulin release is dependent on the diet in the days preceding the load, a reduction in carbohydrate leading to alterations in insulin secretion. The size of the load has little influence on the magnitude of the blood glucose response, while in contrast the insulin response appears to be proportional to the amount of glucose given [35–37].

Many studies have demonstrated that different simple or complex sugars induce different glucose and insulin responses. Glucose and other hexoses can stimulate insulin secretion directly or indirectly through the release of insulogenic gastrointestinal hormones, such as gastro intestinal peptide (GIP) and GLP-1 7–36 [38]. Ingestion of these sugars may also activate parasympathetic nerves innervating the islets, with the stimulation of insulin release [39]. Galactose given orally induced a moderate insulin release associated with GIP secretion. The insulin response to oral galactose is greatly enhanced by hyperglycaemia induced by intravenous glucose infusion [38]. Rapid intravenous infusion of galactose (0.5 g/kg in 3 minutes) results in a significant increase in plasma insulin, the peak approaching that after glucose infusion. Ingestion of 50 g of fructose induced a small rise in plasma insulin but no secretion of GIP. Ingestion of 50 g of mannose induces a rise in neither insulin nor GIP [40]. The insulin response to intravenous glucose was only moderately raised after oral fructose or mannose compared with that after intravenous glucose alone. Insulin secretion is greater when subjects are given sucrose rather than the monosaccharide equivalent. It is the disaccharide effect which is not explained by differences in GIP response [41].

The insulin response to complex carbohydrates is closely related to the digestion and/or absorption of these sugars. Consumption of potato starch yields insulin levels similar to those after ingestion of a similar "glucose equivalent" amount of glucose, whereas rice grains give a relatively flat insulin response [42]. A mixed meal results in a much larger insulin response than that observed with a similar amount of carbohydrate given as glucose alone [42]. This greater insulin response could be partly explained by the protein content of the meal, which potentiates the insulin secretion. Insulin stimulation is highest for oatmeal, is large and nearly similar for bread, rice and low-amylose muffins, while it is low for lentils, kidney beans and high-amylose muffins [43]. All these findings are difficult to interpret, as most of the studies were performed without kinetic measurements of glucose metabolism, and usually for too short a length of time. We have recently studied the appearance of ^{13}C in blood glucose from naturally ^{13}C-labelled starches prepared as biscuits (extruded), pasta or corn semolina. Although some differences were observed in blood glucose (lower with pasta), no obvious differences were observed in the appearance in the blood of ^{13}C-labelled glucose of exogenous origin. The main finding was that labelling of the glucose pool with carbon of exogenous origin lasted for more than 6 hours, even though all the metabolic parameters had returned to normal [44]. A similar methodological approach was used to show that synthetic glucose polymers branched in the $\alpha1-4$ or $\alpha1-6$ position were metabolised in the same way by the human body (J.P. Riou, unpublished observations).

Acute Glucagon Response to Carbohydrates

As beta cells and alpha cells from the endocrine pancreas have a close functional relationship [25] it must first be emphasised that the direct action of carbohydrates on glucagon secretion is often difficult to separate from an indirect effect mediated by changes in insulin secretion. Indeed insulin secretion has a direct effect on the glucagon secretion of the A cells. When plasma glucose is increased to about twice the normal basal level, glucagon secretion is inhibited by the concurrent changes in beta cell activity, rather than by a direct effect of glucose or circulating insulin on the alpha cells [45].

Nevertheless, glucose seems to have a direct effect on glucagon secretion, independent of its ability to stimulate insulin release. Inhibition of glucagon secretion occurs at a lower glucose concentration than that at which insulin secretion occurs, and glucose effects can be demonstrated in the absence of beta cells [47]. Increased plasma glucagon in response to low blood sugar levels is also mediated by an adrenergic mechanism in the pancreas itself, rather than through the central nervous system, as has been shown in isolated rat pancreas [48].

In Vitro Data

Secretion of glucagon is inhibited by glucose, and evidence of an inverse relationship between glucagon output and glucose concentration has been obtained in the perfused dog pancreas and confirmed in isolated pancreatic islets [46–48]. Suppression can be observed with glucose concentrations as little as 2.8 mmol/l. A half-maximal effect is obtained at 5.6 mmol/l, and a maximal effect at 11.2 mmol/l [48]. The time-dependent effects of glucose on glucagon and insulin secretion seem to be similar [49]. Effects of sugars other than glucose on glucagon secretion in vitro have also been investigated using pancreatic islets. When concentrations of 1 g/l are used, glucose, galactose, fructose, xylose, xylitol and ribose have the same effects on glucagon release, and mannose seems to stimulate glucagon secretion weakly but significantly. When inhibitors of glucose metabolism (such as 2-deoxy-D-glucose or mannoheptulose) are added to glucose there is no significantly different effect on glucagon secretion [48].

In Vivo Data

Carbohydrates have an inhibitory action on glucagon secretion as early as the first week of life [50]. Study of the acute response of glucagon to carbohydrates in vivo is difficult, primarily because of the presence of circulating peptides of various molecular weights that cross-react immunologically with antibodies to pancreatic glucagon [46]. This could be the reason why the decrease in the apparent plasma concentration of glucagon is only of about 50% during a 75 g oral glucose tolerance test [2]. During this test, the lowest level of plasma glucagon concentration is reached at about 90 minutes, and reduced values are still observed at 150 minutes. Moreover glucagon, like insulin, is taken up by liver and rapidly removed. Hepatic retention of glucagon is found to be about 80% [2].

Lastly, studies have to take into account factors other than carbohydrates which could influence glucagon secretion. Other substrates are amino acids, fatty acids, and also hormones, particularly insulin and catecholamines. Thus changes in glucose-induced glucagon secretion are much greater in the presence of arginine, which stimulates both glucagon and insulin secretion [48].

All these facts result in difficulties in studying the glucagon response to carbohydrates in vivo, and in comparing results with in vitro studies. Thus, with oral galactose but with neither intravenous galactose nor fructose, there is a significant increase in total serum glucagon in man [46,48], and mannose lowers plasma glucagon level [48].

Acute Catecholamine Response to Carbohydrates

The involvement of catecholamines in the acute response to an oral load of various carbohydrates in man has recently been clearly demonstrated. Oral or intravenous glucose raises noradrenaline levels without early significant change in adrenaline. The increase in noradrenaline is specific for carbohydrate absorption, but is probably not related to insulin secretion, as it also occurs after xylose or mannitol absorption, sugars which are devoid of any insulin secretory effect [51–53]. The increased plasma noradrenaline has been implicated in the thermic effect of glucose, in which a β-adrenergically mediated effect seems to be responsible for the excess rise in energy expenditure [54]. Adrenaline has been implicated in the late hypoglycaemic response to an oral glucose load [55]. In this case the stimulus is in fact not dependent upon the carbohydrate but only on the intensity of the late hypoglycaemia phase. The involvement of catecholamine in the acute response to carbohydrate feeding is probably of minor importance, but the sympathetic nervous system probably plays a crucial role in the long-term effect of carbohydrate feeding on the insulin–glucagon response to the ingestion of glucose and starches.

Long-term Effect of Carbohydrate Feeding on the Metabolic and Hormonal Response to a Glucose Load

The lack of reproducibility of most of the routine evaluation of glucose tolerance in humans is probably related to uncontrolled dietary factors, and mainly to the percentage of energy arising from carbohydrate-containing food in the days preceding the oral glucose tolerance test. A high carbohydrate diet increases, and a low carbohydrate diet decreases, the insulin sensitivity of the whole body glucose metabolism. The increase in sensitivity is in fact almost a short-term effect of glucose feeding. In recent work [56] it has been shown in normal humans that glucose given 4 hours before euglycaemic insulin clamp increases in a dose-dependent manner the insulin sensitivity of glucose metabolism. This finding strongly supports the old Straub–Traugott effect, suggesting that insulin was more active in diabetics "pre-fed" with glucose. When the amount of carbohydrate in the total energy absorbed by normal subjects was increased for 3–17 days from 40% to 70%, the general metabolic effects were improvement in carbohydrate tolerance, and stimulation of carbohydrate oxidation and lipogenesis [57,58]. In all these studies plasma insulin levels are unchanged for most of the time, but insulin secretion rate was not measured. One possible explanation is that by increasing basal insulin secretion, and thereby decreasing lipolysis, which is highly sensitive to insulin in man, high carbohydrate intake increases insulin sensitivity.

It is also possible that the sympathetic nervous system is implicated in this long-term metabolic effect of carbohydrates. It has been clearly shown that the catecholamine turnover rate is dependent upon the nutritional status of both animals and man [59,60]. High carbohydrate feeding, by increasing plasma noradrenaline, may be responsible for the improved glucose oxidation, even though acute administration of both adrenaline and noradrenaline results in large glucose intolerance.

Measurement of the involvement of catecholamine in man is not easy, since catecholamine turnover studies are difficult to perform, and plasma measurements are a poor reflection of the metabolism of catecholamines which mainly occurs at the nerve endings. In spite of these reservations the data in animal models and man [60] strongly support a major role of the sympathetic nervous system in the regulation of both insulin secretion and insulin sensitivity. Obviously this field requires additional studies in humans.

Future Research and Overview

Even though thousands of studies have been performed to determine the hormonal and metabolic responses to carbohydrate feeding, it is still a large field of investigation which obviously needs new investigative tools. The biochemistry of glucose metabolism in the body has been completely changed by the discovery that fructose-2,6-bisphosphate regulates the glycolytic pathway in the liver, and by the still open question of the "glucose paradox" [61].

Data on glucose metabolism in man have been taken much further by the use of such new tools as insulin clamp, stable isotopes, nuclear magnetic resonance [1] etc., and by the better use of old tools such as indirect calorimetry [1]. The possibility of using naturally ^{13}C-labelled carbohydrates or starch to determine the rate of exogenous glucose appearance, and/or the interconversion of one carbohydrate to another, in combination with the kinetic measurement of glucose metabolism, should open up new fields in our understanding of carbohydrate metabolism in humans.

References

1. Alberti KGM, Home PD, Taylor R (1987) Technique for metabolic investigation in man, Clin Endocrinol Metab 1:773–1071
2. Waldhausl WK, Gasic S, Bratusch-Marrain P, Nowotny P (1983) The 75 g oral glucose tolerance test: effect on splanchnic metabolism of substrates and pancreatic hormone release in healthy man. Diabetologia 25:484–495
3. Ferrannini E, Bjorkman O, Reichard GA et al. (1985) The disposal of an oral glucose load in healthy subjects. A quantitative study. Diabetes 34:580–588
4. Abumrad NJ, Cherrington AD, Williams PE, Lacy WW, Rabin D (1982) Absorption and disposition of a glucose load in the conscious dog. Am J Physiol 242:E398–E406
5. Kelly D, Mitrakou A, March H et al. (1988) Skeletal muscle glycolysis, oxidation and storage of an oral glucose load. J Clin Invest 81:1563–1571
6. Jackson RA, Rosmansa RD, Hawa MI, Sim BM, Disilvio I (1986) Impact of glucose ingestion on hepatic and peripheral glucose metabolism in man: An analysis based on simultaneous use of forearm and double isotope techniques. J Clin Endocrinol Metab 63:541–549
7. Adkins BA, Myers SR, Hendrick GL, Stevenson RW, Williams PE, Cherrington AD (1987) Importance of the route of intravenous glucose delivery to hepatic glucose balance in the conscious dog. J Clin Invest 79:557–565
8. Rizza RA, Mandarino LJ, Gerich JE (1981) Dose response characteristics for effects of insulin on production and utilisation of glucose in man. Am J Physiol 240:E630–E639
9. Gottesman I, Mandarino L, Gerich J (1984) Use of glucose uptake and glucose clearance for the evolution of insulin action in vivo. Diabetes 33:184–191
10. Jacot E, Defronzo A, Jequier E, Maeder E, Felber JP (1982) The effect of hyperglycemia, hyperinsulinemia and route of glucose administration on glucose oxidation and glucose storage. Metabolism 31:922–990
11. Randle PJ, Newsholme EA, Garland PB (1964) Regulation of glucose uptake by muscle; effects of fatty acids, ketone bodies and pyruvate and of alloxan diabetes and starvation on the uptake and metabolic fats of glucose in rat heart and diaphragm muscles. Biochem J 93:657–685
12. Ferrannini E, Garett EJ, Beuilacqua J, Defronzo RA (1983) Effect of fatty acids on glucose production and utilization in man. J Clin Invest 72:1737–1747
13. Jequier E, Felber JP (1987) Role of fats in obesity and type II diabetes mellitus. In: Horisberger M, Bracco U (eds) Lipids in human nutrition. Raven Press, pp 205–211
14. Ebiner JR, Acheson KJ, Doerner A (1979) Carbohydrate utilization in man using indirect calorimetry and mass spectometry after an oral load of 100 g naturally labelled ^{13}C glucose. Br J Nutr 41:419–429
15. Dix D, Cohen P (1980) Interpretation of the glucose tolerance test. Diabetologia 19:488–494
16. Kaden M, Harding P, Field JB (1973) Effect of intraduodenal glucose administration on hepatic extraction of insulin in anesthetized dog. J Clin Invest 52:2016–2028
17. Rubinstein AH, Clark JL, Melani F, Steiner D (1969) Secretion of proinsulin and C peptide by pancreatic beta cells and its circulation in blood. Nature 224:697–699
18. Polonsky K, Jaspan J, Pugh W et al. (1983) Metabolism of C-peptide in the dog: in vivo demonstration of the absence of hepatic extraction. J Clin Invest 72:1114–1123
19. Eaton RP, Allen RC, Schade DS (1983) Hepatic removal of insulin in normal man: dose response to endogenous insulin secretion. J Clin Endocrinol Metab 56:1294–1301

20. Polonsky K, Licino-Paixao J, Given BD et al. (1986) Use of biosynthetic human C-peptide in the measurement of insulin secretion rates in normal volunteers and type I diabetic patients. J Clin Invest 77:98–105

21. Faber OK, Madsbad S, Kehlet M, Binder C (1978) Pancreatic beta cell secretion during oral and intravenous glucose administration. Acta Med Scand [Suppl] 624:61–64

22. Elrick M, Stimmler L, Mcao C, Arai Y (1964) Plasma insulin response to oral and intravenous glucose administration. J Clin Endocrinol Metab 24:1076–1082

23. McIntyre N, Holdsworth C, Turner P (1965) Intestinal factors in the context of insulin secretion. J Clin Endocrinol 25:1317–1324

24. Madsbad O, Kehlet M, Hilsted J, Tronier B (1983) Discrepancy between plasma C-peptide and insulin response to oral and intravenous glucose. Diabetes 32:436–438

25. Pipeleers D (1987) The biosociology of pancreatic B cells. Diabetologia 30:277–291

26. Trimble ER, Bruzzone R, Gjsnovci A, Renold AE (1985) Activity of the insulo-acinar axis in the isolated perfused pancreas. Endocrinology 117:1246–1252

27. Curry DL, Bennett LL, Grodsky GM (1968) Dynamics of insulin secretion by the perfused pancreas. Endocrinology 83:572–584

28. Jarrett RJ, Keen H (1968) Oxidation of sugars, other than glucose, by isolated mammalian islets of Langerhans Metabolism 17:155–159

29. MacDonald MJ, Ball DH, Patel TN, Lauris V, Steinke J (1975) Studies of insulin release and rat pancreatic islet metabolism with diastereoisomers of D-glucose. Biochim Biophys Acta 385:188

30. Ashcroft SJ, Basset JM, Randle PJ (1972) Insulin secretion mechanisms and glucose metabolism in isolated islets. Diabetes [Suppl 2] 21:538–545

31. Curry DL (1974) Fructose potentiation of mannose-induced insulin secretion. Am J Physiol 226:1073–1078

32. Grant AM, Christie MR, Ashcroft SJH (1980) Insulin release from human pancreatic islets in vitro. Diabetologia 19:114–117

33. Gannon MC, Nuttall FQ (1987) Factors affecting interpretation of postprandial glucose and insulin areas. Diabetes Care 10:759–763

34. Hollenbeck C, Reaven G (1987) Variations in insulin-stimulated glucose uptake in healthy individuals with normal glucose tolerance. J Clin Endocrinol Metab 64:1169–1173

35. DeNobel E, Van 'T Laar A (1978) The size of the loading dose as an important result of the oral glucose tolerance test. Diabetes 27:42–48

36. Mosora F, Lefebvre B, Lacroix M (1981) Glucose oxidation in relation to the size of oral glucose loading. Metabolism 30:1143–1149

37. Bratusch-Marrain PR, Waldhaus WK, Gasic S, Korn A, Nowotny P (1980) Oral glucose tolerance test: effect of different glucose load on splanchnic carbohydrate and substrate metabolism in healthy man. Metabolism 29:289–295

38. Morgan LM, Wright JW, Marks V (1979) The effect of oral galactose on GIP and insulin secretion in man. Diabetologia 16:235–239

39. Woods SC, Porte D (1974) Neural control of the endocrine pancreas. Physiol Rev 54:596–619

40. Ganda OP, Soeldner SJ, Gleason R, Cleator IGM, Reynolds C (1979) Metabolic effects of glucose, mannose, galactose and fructose in man. J Clin Endocrinol Metab 49:616–622

41. Ellwood K, Michaelis O, Hallfrischt J, O'Dorisio T, Cataland S (1983) Blood insulin, glucose, fructose and gastric inhibitory peptide levels in carbohydrate-sensitive and normal men given a sucrose or invert sugar tolerance test. J Nutr 113:1732–1736

42. Crapo PA, Reaven GM, Olefski JM (1977) Post-prandial glucose and insulin responses to different complex carbohydrates. Diabetes 26:1178–1183

43. Krzowski PA, Nuttfall FQ, Gannon M, Billington C, Parker S (1987) Insulin and glucose response to various starch-containing foods in type II diabetic subjects. Diabetes Care 10:205–212

44. Riou JP, Normand S, Kahlfallah Y (1988) Measurement of 13 glucose in the blood arising from starch food naturally labelled, with gas chromatograph isotope ratio mass spectrometer. Starch Workshop INRA, Nantes, November 1988

45. Asplin CM, Hollander PM, Palmer JP (1984) How does glucose regulate the human pancreatic A cell in vivo? Diabetologia 26:203–207

46. Porte Jr D, Halter JB (1981) The endocrine pancreas and diabetes mellitus. In: Williams RH (ed) Textbook of endocrinology. 6th ed. Saunders, Philadelphia, pp 716–843

47. Hisatoni A, Maruyama H, Orci L, Vasko M, Unger RH (1985) Andrenergically mediated intrapancreatic control of the glucagon response to glucopenia in the isolated rat pancreas. J Clin Invest 75:420–426

48. Fao PP The secretion of glucagon (1972) In: Geiger SR (ed) Handbook of physiology, section 7: endocrinology, vol 1: Endocrine Pancreas. Williams and Wilkins, Baltimore pp 261–277
49. Grill V, Adamson U, Rundfeldt M, Anderson S (1979) Glucose memory of pancreatic B and A cells. J Clin Invest 64:700–707
50. Molinari D, Angeletti G, Santeusanio F, Falorni A (1982) Blood glucose plasma insulin and glucagon response to intravenous administration of glucose in premature infants during the first week of life. J Endocrinol Invest 5:169–171
51. Young JB, Landsberg L (1977) Stimulation of the sympathetic nervous system during glucose feeding. Nature 269:615–617
52. Welle S, Lilavivat U, Campbell RG (1981) Thermic effect of feeding in man: increased plasma norepinephrine levels following glucose but not protein on fat consumption. Metab Clin Exp 30:353–358
53. Tse TF, Clutter WE, Shah JD, Miller JP, Cryer PE (1983) Neuro-endocrine responses to glucose ingestion in man; specifically temporal relationships and quantitative aspects. J Clin Invest 72:270–277
54. Defronzo RA, Thorin D, Felber JP et al. (1984) Effect of Beta and Alphadrenergic blockade on glucose induced thermogenesis in man. J Clin Invest 73:633–639
55. Tse TF, Clutter WE, Shah JD, Cryer PE (1983) Mechanisms of post-prandial glucose counter regulation in man, physiologic roles of glucagon and epinephrine vis-à-vis insulin in the prevention of hypoglycemia late after glucose injection. J Clin Invest 72:278–286
56. Kingston WJ, Livingston JN, Moxley-Ill RT (1986) Enhancement of insulin action after oral glucose injection. J Clin Invest 77:1153–1162
57. Welle SL, Campbell RG (1983) Improved carbohydrate tolerance and stimulation of carbohydrate oxidation and lipogenesis during short-term carbohydrate overfeeding. Metabolism 32:889–893
58. Acheson KJ, Schutz T, Bessard T, Ravussin E, Jéquier E, Flatt JP (1984) Nutritional influences on lipogenesis and thermogenesis after a carbohydrate meal. Am J Physiol 246:E62–70
59. Landsberg L, Young JB (1980) The role of the sympatho-adrenal system in modulating energy expenditure. Clin Endocrinol Metab 13:475–499
60. Bray GA (1984) Integration of energy intake and expenditure in animals and man: the autonomic and adrenal hypothesis. Clin Endocrinol Metab 13:521–546
61. Katz J, McGarry JD (1984) The glucose paradox: is glucose a substrate for liver metabolism? J Clin Invest 76:1901–1909

Commentary

Marks: Intra-islet regulation of hormonal secretion is much more complicated than was originally thought [1]. Not only do the A cells, through the glucagon they secrete, stimulate the B cells to secrete insulin and vice versa through a paracrine relationship, but there is also an intra-islet endocrine control exerted through the intra-islet capillaries. Blood flow within the islet is mainly from the B cell core to the D cell mantle, passing through the A cell layer. The insulin-secreting B cells exercise a tonic suppressive effect upon the glucagon-secreting A cells, which themselves exercise a tonic stimulatory effect upon the D cells. These in turn are responsible for the production of somatostatin [2].

GIP, which has a direct stimulatory effect upon the B cells, also stimulates the A cells to produce glucagon. So too do the sulphonylurea drugs. This probably explains why, in insulin-dependent diabetes mellitus patients, in whom the suppressive effect of insulin in glucagon secretion by A cells is lacking, both oral glucose and sulphonylurea drugs (the former by liberating GIP from the gut) stimulate, rather than inhibit, pancreatic glucagon release.

References

1. Samols E, Tyler JM, Marks V (1972) Glucagon–insulin interrelationships. In: Lefebvre PJ, Unger RH (eds) Glucagon: molecular physiology, clinical and therapeutic implications. Pergamon Press, Oxford
2. Samols E, Stagner JI, Ewart RBL, Marks V (1988) The order of islet microvascular perfusion is B–A–D in the perfused rat pancreas. J Clin Invest 82:350–353

Peeters: The "incretin" concept is, in my view, a more important factor in determining changes in plasma levels of insulin than hepatic extraction.

Würsch: It is of interest to note that the plasma glucose level after a starchy meal already decreases when only one quarter of the food has left the stomach [1]. Would this suggest that the decrease is not caused by a decrease in glucose absorption rate, but from insulin starting to act on glucose utilisation in muscle?

Reference

1. Torsdottir I, Alpsten M, Andersson H et al. (1989) Effect of different starchy foods in composite meals on gastric emptying rate and glucose metabolism. Hum Nutr Clin Nutr (in press)

Marks: The work of Masbad et al. and of Gibby and Hales [1] on the relative insulin stimulatory effects of hyperglycaemia resulting from oral versus intravenous glucose, are flawed, because the comparisons were made using matched venous, rather than arterial blood. If matched "arterialised" blood comparisons are made [2] there is clear evidence of alimentary enhancement of insulin secretion as well as of increased fractional extraction of insulin following ingestion of an oral glucose load. In other words, increased insulin secretion is a more important cause of alimentary augmentation of insulin release, as first suggested by McIntyre et al. [3], than reduction in the fraction extraction of insulin by the liver.

References

1. Gibby OM, Hales CN (1983) Oral glucose decreases hepatic extraction of insulin. Br Med J 286:921–923
2. Hampton SM, Morgan LM, Tredger JA, Cramb R, Marks V (1986) Insulin and C-peptide levels after oral and intravenous glucose. Contribution of the entero-insular axis to insulin secretion. Diabetes 35:612–616
3. McIntyre N, Holdsworth CD, Turner DS (1964) New interpretation of oral glucose tolerance. Lancet II:20–21

Chapter 13

The Effect of Dietary Starches and Sugars on Satiety and on Mental State and Performance

D.A. Booth

This chapter reviews the scientific information available on ways in which sugars and starches in ingested foods and beverages affect satiety. Their effects on behaviour unrelated to eating, such as expressed mood and ability to concentrate, are also considered briefly. It also deals with the practical questions arising: whether either simple or complex carbohydrates can effect total energy intake, and hence obesity, or can ameliorate or exacerbate problems with behaviour.

Satiety Effects of Carbohydrates

Foods affect psychological states and processes by many different mechanisms. There are to date no absolute measurements nor satisfactory comparisons of this diversity of satiating or other behavioural effects of sugars and starches in the human diet. Observations of the overall effects of particular foods provide no sound basis for interpretation. We thus have to adduce adequately controlled findings in laboratory animals, as well as background considerations for humans.

This dearth of good data, despite long-standing interest in the issue, is due to the technical problems of designing such experiments. Analysis of the behavioural effects of dietary starches and sugars involves food technology and human biology as well as psychology.

Multiple Actions of Food Sugars and Starches

Sugars are very different from grain, legume or tuber starches in their impact on the senses and in their actions as secretogogues and metabolic substrates. Thus, there are often no unambiguous control conditions available for experiments on particular satiating actions of, for example, sucrose or maize starch, whether in isolated form or in their dietary context.

Furthermore sugars and starches have mostly quite different roles in food technology and in dietary patterns. As a consequence, one may wonder whether it would ever be possible to compare the behavioural effects of common sugars and whole starches, because of the confounding effects of other components of the necessarily diverse test foods.

We also need to take into account much non-behavioural research on differences in physiological effects that are relevant to the known mechanisms of satiety. For example, the differences in time taken to digest and absorb glucose from maltodextrin and from various cooked forms of flour, rice, beans, potato, etc., can be predicted to affect the timing of the post-ingestional satiating effects of different polysaccharides compared with the di- and monosaccharides.

Moreover, such controlled investigation of the compositional, sensory and physiological variables has to be combined with a fully scientific approach to objective measurement of the processes going on in people's minds, whether during experiments or as they go about their daily activities [1]. This means that it is insufficient to be collecting numbers from people. The research design must identify specifically the psychological process of interest, be it satiety or some aspect of mental performance.

Measurement of Satiety and its Mechanisms

Satiety is the motivational state in which the interest in eating, i.e. the appetite for food or hunger, has been suppressed as a result of the ingestion of food or drink. There have, however, been long-standing and persisting attempts to "define" satiety and appetite, hunger, thirst and palatability as other than different aspects and strengths of the observable disposition to eat and drink, whatever may be causing the desire. These verbal moves reflect a poor understanding of the nature of the scientific problems, which has weakened the experimental analysis of the different mechanisms of satiation.

The Disposition and the Sensations Allied with It

The behavioural disposition of satiety has long been widely confused with the bodily sensations that no doubt sometimes contribute to it in some people. Some profess not to experience particular sensations when they want to eat or drink or to stop. Whatever the empirical implications of such claims, they are perfectly logical. Therefore it is fallacious to define satiety, hunger or thirst as sensations. Furthermore, asking people about sensations will not be the best way of measuring the tendency to eat. There are other influences on the desire to eat, such as differences between foods on offer, and the approach of an habitual mealtime. The experience of an epigastric pang is usually clearly distinguishable from the experience of abdominal fulness, but it is naïve to ignore the power of suggestion and the direction of attention by circumstances on the subjective impression.

The strength of a sensation is much harder to measure than is usually presumed. A notion of direct estimation of subjective magnitudes has dominated psychophysics and the use of ratings generally. This is based on a complete misunderstanding of the language we use to express our experiences without committing ourselves about what is actually happening [2]. Despite their grammatical form, descriptions and intensity values of fulness are not reports and estimates from an infallible inspection of causally efficacious mental phenomena, let alone consequences of visceral events [1,2]. In particular, nobody has yet shown how well, and under what conditions, scores for the strengths of sensations of satiety, hunger or thirst are distinguished from ratings of the overall disposition to eat. The ratings of bodily sensations have to be shown to be independent of the many other possible influences on appetite for food. Such identification of mental processes requires causal analysis at the individual level [2]. One must determine the relationships between the sensation ratings and measures of the overall disposition to eat and of other influences on it; and preferably also of as many as possible of the sources of those influences, such as sensed food constituents, the gastrointestinal or parenteral afferent stimulation, or motor reafference [3], that give rise to the sensations, and the habitual circumstances under which eating occurs.

Such causal analysis can be done without the sensation ratings [2]. Thus, these experiments on the influence of sensed bodily signals on satiety are not confined to human subjects, as often claimed, and this chapter will refer to animal data where human data are inadequate or non-existent.

The Disposition and Intake Parameters

A more recent confusion has identified satiety with aspects of food intake, such as slowing of eating towards the end of a bout of eating, reduction in meal or snack size, and time interval to the next meal. These parameters are the results of variations in the disposition to eat, and so they can be used in appropriately controlled and calculated comparisons to measure the strength of some influence on that disposition. Yet, by the same token, the slowing of eating, meal termination, or time interval to the next meal, cannot be defined as satiety. Even the proportional effect on meal size is not a measure of satiety or palatability: the strength of an influence is measured by the effect as a proportion of normal variation or, better still, as the discriminative sensitivity of meal size to the influence [2,4], in the way that engineers measure the sensitivity of a function of a system.

Measuring the Mechanisms of Satiety

The disposition to eat or not to eat is thus most directly measured as the behaviour of the moment, emitted by an individual faced with an array of foods that is usual to the occasion.

At the risk of reduced realism, but avoiding the effects of actual eating on the appetite that one wishes to measure, people can be asked what, and how much, they would eat from a normal choice, and their answers used to calculate a wished-for intake [4–6].

It turns out to be just as validly predictive of eating to ask people how pleasing it would be to eat particular tasted, presented or even just named foods. Indeed, the average of the ratings of the momentary pleasantnesses of eating a mouthful of several staple foods is the most sensitive measure of a food-general inhibition of eating, whether the satiety arises from post-ingestional effects of maltodextrin [4], from immediately prior sensory experience [7] or from supposed caloric content of a pre-load [4].

The mechanisms of satiety can thus be elucidated by jointly measuring one or more discrete influences on satiety, and one of the indexes of the overall disposition to eat [1,2,4]. The scientific question is how an observed input affects the observed output. Valid evidence as to what is actually going on cannot be obtained merely by asking people how much they think it is, for example, their body, the foods or the time on the clock that disposes them to accept or to refuse food [2].

Satiety Effects of Sugar Solutions

Much early animal research on satiety, and some still continuing, used solutions of D-glucose as the source of nutrient intended to satisfy appetite. This strategy appears to have been based in part on the notion that scientific experiments require physically pure and conceptually clean manipulations that produce big effects. In fact, laboratory studies taking such an approach may never make contact with the natural phenomena they purport to be addressing. The effects of drinking unflavoured solutions of glucose, sucrose or saccharin may bear very little relation to the consumption of sweetened beverages, let alone of solid foods out of which sugars have to be dissolved by saliva, by gastric secretions and by accompanying drinks.

Pleasantness Rating of Sweetness

Over the last 20 years, many experiments have been done on pleasantness ratings of plain sweet solutions, and occasionally of food aromas.

The original assumption was that such ratings measure the intensity of sensual pleasures provided by the tastes [8]. However, the normal meaning of being pleased with a drink or food is "being disposed to ingest (more of) it", and it is highly implausible to suppose that plain saccharin or sugar solutions can generate gastronomic thrills. Rather, the pleasantness of sweet water is a measure of appetite for any drinks or indeed foods of which it may be reminiscent. Other workers more recently have regarded the pleasantness ratings of ordinary foods as "subjective appetite", i.e. a verbal expression of the disposition to ingest.

Sweet water provides a very insensitive test for normal degrees of suppression of appetite for foodstuffs, because of its unreality and its beverage character [9]. Indeed, in some people it may need an influence as powerful as nausea from a large hypertonic load to reduce these ratings appreciably. Also, much-repeated tasting of such a simple stimulus is likely to induce boredom, i.e. a satiety specifically for sweet water or perhaps for sweet things more generally [10]. This sweet-taste satiety has indeed been demonstrated to play a major part in the effect of a drink of a sugar solution on pleasantness ratings of plain sugar- or saccharin-sweetened water [11]. Nevertheless, a statistically non-significant maximum reduction of pleasantness was seen about 20 minutes after the drink [11]. This corresponds to the time-delay at which intake of even an unpalatable test food was suppressed by a flavoured drink of 50% glucose, compared with an energy-free drink of the same flavour made even sweeter with saccharin and cyclamate [12,13]. Thus, sugar solutions satiate post-ingestionally as well as sensorially.

"Negative Alliesthesia"

The pleasantness scores for sweet water are often reduced by drinking concentrated solutions of glucose or other sugars or by receiving them by gavage. This effect has been called "negative alliesthesia" (literally, downward change in sensation, although it is the reaction to sweetness that is changed, not the sensation), defined as a presumed post-ingestionally induced reduction in ratings of the pleasantness of sweet-tasting fluid [8].

On the basis of such results it was proposed that the alleged pleasures, and their reduction by ingested nutrient, serve a biological function in regulating energy balance [8]. Of course, it has never been doubted that the inhibition of eating by eating is adaptive. Nevertheless, the superficial novelty of the proposal and the pseudo-objectivity of the technique led to many experiments looking in obese people for higher sweetness pleasantness rating means, slopes or peaks than in normal weight people, or, more logically, for a relatively reduced suppression of the ratings by the concentrated glucose load in the obese. Even psychologists working on obesity, or on food preferences, took this very strange route to testing whether sweeteners made people eat more and get fat, and/or whether the obese were less sensitive to boredom with saccharin tastes or to nausea in old-style glucose tolerance tests. It is no less strange for behavioural scientists to rely on body weight to be associated with satiety deficits. It should be more productive to relate them to failure to reduce weight that has been identified as arising from difficulties in keeping energy intake down [3].

It took more than a decade of negative results to kill off this fashion. In fact, the approach is still not quite dead, because the disillusionment derives from confused results, rather than from appreciating the inappropriateness of the methodology to the theoretical or practical questions.

The issues that laboratory tests could resolve are the mechanisms [13] by which the drinking of sweet beverages induces a reduced or indeed increased inclination to drink them, or to drink or eat other comestibles. More relevant to the present point, how does satiety arise from the sugars in the drinks and foods commonly consumed? The real-life issues are whether obesity is caused to any extent by effects of the taste of sugar on total energy intake or expenditure, and/or by some defect in the post-ingestional satiating effects of carbohydrate or indeed of other energy sources.

Unfortunately, these issues have been addressed in the laboratory largely by "models" of obesity and of pure sugar diets. Instead, we need controlled analyses of the mechanisms by which sugars and starches, as normally consumed, can affect the control of intake and the regulation of energy balance. Studies of the body weight of lean and obese strains of rats or monkeys presented with glucose, sucrose, maltodextrin, dextrin, bread or whole wheat grains, in or with the diet, give a great variety of divergent results. There is little or no gain in understanding, let alone prospect of reducing or preventing obesity, until investigators begin to analyse both the physiological and the psychological processes by which the diets are affecting intake and/or energy expenditure.

Intriguing mechanisms are then sometimes hypothesised, such as an oral receptor for oligosaccharides in the rat which is not that for sweetness or for viscosity [14], and the countering of the ability of maintenance diets to keep rats slim by wetting the diet [15]. Experimental analysis of the influences on eating in the rat would not only advance our understanding of that species; it could also elucidate mechanisms that the rat is likely to share with our own species [13,16].

This justification does not apply to the study of satiety mechanisms in hypothalamic obesity. When the specifically post-ingestional and largely metabolic [17] effect of carbohydrate on the subsequent intake of chow is tested in rats made obese with lesions in the ventromedial hypothalamic area (VMH), the satiety effect is no less than in intact rats and may even be stronger [18]. Indeed, similar results were obtained with sugars 20 years ago, and refuted then the still-persisting myth that the VMH is one of the regions in the brain critical to the inhibition of eating behaviour, or is even the satiety centre [19].

Mechanistic analysis in the rat showed that the post-ingestional action(s) of hypertonic sugar solutions, that further suppress consumption of the solutions or of maintenance diet, arose from their colligative properties [20,21]. As might therefore be expected, osmotic effects in the duodenum appear to be the major post-ingestional contributor to "negative alliesthesia" from sugar solutions that are much more concentrated than normal beverages [22].

Satiety Effects of Disguised Maltodextrin

Because concentrated sugar solutions have abnormal effects, maltodextrins with low dextrose equivalent have been used in most of the interpretable experiments on human (and animal) satiety carried out since the appearance of these compounds in the late 1960s.

Multiple replications of dose-ranges of flavoured maltodextrin solution were used in the first studies, which have so far been only briefly reported [23]. High concentrations were shown to produce a slightly delayed and transient suppression of intake of sandwich pieces in the test meal. At 65 g/100 ml, even a mere 10 g of maltodextrin gave a reliable suppression of intake after 30 minutes. The energy-intake suppression was substantially greater than the energy in this low dose of maltodextrin. At higher doses, however, the satiety effect achieved approximate constancy in total energy intake, i.e. caloric "compensation", as seen in the rat [21] and monkey [24].

This pattern of results supports an interpretation, based on animal data, that rapid absorption of glucose generates a major satiety signal based on hepatic intermediary metabolism [17,21,25]. The observed short delay is necessary for initial passage of the maltodextrin load to the duodenum and for its digestion and absorption to get under way. The supracaloric satiety at a very low dose may be a "threshold" effect of the initial dumping of a liquid meal [26]: at the same time as the products of digestion bring gastric emptying under control by the carbo-hydrate concentration passing from the stomach [27], they generate a rapid initial rate of glucose absorption, no less than will a higher dose. In addition, the higher the dose, the longer is the rapid absorption sustained. The satiety effect may last for 10–20 minutes after the period of rapid absorption, thus continuing during the test meal after larger pre-loads. As a result, there is a dose-response relationship at a particular delay from pre-load to test, such as 30 minutes, and a time course after which the satiety effect ceases to be reliably detectable.

Subsequent experiments have used lower concentrations, and similar or smaller amounts of maltodextrin in pre-loads that are normal foodstuffs such as soups and various desserts, and have also used non-caloric bulking agents to match the mouth-feel of maltodextrin. Rating or intake tests that are sensitive to general satiety (i.e. tests on staple foods) have detected satiating effects of the disguised maltodextrin within 10–40 minutes in adults [4,23,28]. The effects are seen at 1 hour in children [29], but they may absorb loads more slowly relative to dose per body weight.

Results with maltodextrin and with glucose, therefore, indicate that ingested sugars or starches will have a satiating effect during and shortly after any rapid absorption of glucose they give rise to. From the animal data, rapid absorption of fructose is also likely to satiate, perhaps more strongly [17], because it is oxidised more rapidly by the liver. In consequence, sucrose could produce a satiety effect which is stronger than that of even a starch that is as rapidly and completely digested and absorbed. The animal data also open the possibility that lactose, on the other hand, is more weakly satiating than starch over the whole time course of assimilation, if galactose is not used for energy by the human liver [17].

Satiety Effects of Sweetness and of Energy Metabolism

Thus there is clear evidence for two of the several mechanisms by which dietary sugars and starches are likely to induce satiety: (1) sweetness in foods and drinks immediately reduces the appetite for sweet things for a short while [10]; (2) while energy production continues as a result of rapid absorption of carbohydrate digestion products, appetite in general for ordinary foods is reduced [4,12,23,28,29].

The "Sweetness by Calories" 2 × 2 Design

The results bearing on such mechanisms were obtained when people's reactions to real foods were assessed following the consumption of foods or drinks comprising at least two of four controlled conditions. The satiating foods were very similar to the senses but contained: (a) sucrose, glucose or gluco-oligosac-charides (i.e. low glucose maltodextrin) plus intense sweetener(s); (b) gluco-oligosaccharides that are not sweet and without addition of intense sweetener; (c) various intense sweeteners, providing a taste similar to that of the sugars but without the utilisable carbohydrate; and (d) no sweetener nor appreciable amounts of carbohydrate or other energy. Both (c) and (d) need non-caloric bulking agent to match the "body" of substantial concentrations of maltodextrin or of higher concentrations of sugars. Some non-caloric osmotic agent may be needed in (b) if (a) contains hypertonic sugars.

Some information on effects of sweetness is provided by comparison of (a) with (b) and (c) with (d).

Similarly, the whole range of post-ingestional effects of small molecular weight carbohydrates can be monitored by comparing (a) with (c) and (b) with (d). It must be emphasised that, although low glucose and low maltose maltodextrin preparations are more similar to starch than to some sugars in lack of sweetness and of osmotic effects, the insoluble dietary starches, as cooked and eaten, have to undergo many additional mechanical and enzymatic digestion processes before thereby more slowly, and perhaps incompletely, yielding maltose and glucose to absorption [30]. Maltodextrin is not a model for starch. It is an experimental tool for studying the actions of glucose in and beyond the wall of the duodenum.

Sensory and Cognitive Interaction Artefacts

There are, however, severe limitations to this 2×2 design, of high or low sweetness crossed with high or low carbohydrate calories. Few foods or beverages can be prepared in all four forms and still be acceptable to most people, let alone seem normal to them.

Even when problems of sensory quality are overcome, cognitive interactions between sensory and caloric effects can destroy the uncoupling of sweetness and calories that superficially appears to be offered by the design. Such complications are evidently occurring when the differences between pairs of conditions are not the same: the effects of sweetness may differ between the presence and absence of calories, for example because the post-ingestional calorific satiety supports the expectations generated by strongly sweet taste. Such cognitive coupling can be dealt with only by individual causal analysis [2,31]. This requires the measurement of graded interactions of sweetness and calories in ranges near the individual's norm for the food and eating occasion being tested.

Thus, this four-condition design cannot be regarded as "balanced" [32]. The same mistake was made in research on effects of alcohol, where the four combinations of alcohol presence/absence and alcohol instructed/uninstructed were once thought to achieve a "balanced placebo design". Experienced drinkers, however, simply do not add up instructions and effects of alcohol; either disparity between them produces its own surprise. Similarly with foods, each person is used to particular strengths of sweetness and texture, and may have learned that ingestion of certain amounts of the comestible having those sensory characteristics is liable to assuage hunger to some degree for a certain time.

Expectations from Sweetness

In other words, sweetness is like an instruction as to the presence of energy. The closeness of the sweetness to its habitual level contributes to palatability [33,34] but the sweetness level also creates expectations, and these are liable to have effects over and above any behavioural effect of autonomic reflexes triggered by certain tastants [35].

As a result, the experimental subject could feel that odd things are happening as time passes after a very sweet drink not known to be low in calories, or after an unsweetened variant of a dessert nevertheless high in calories. Such interactions could stimulate eating, instead of the sweetness or energy producing part of the satiety that they normally produce together. For example, unattributed disturbance and distress can by itself cause overeating in some people [36]. Surprises can also create suspicious attitudes to the experiment and affect the behaviour observed.

Moreover, effects of experimentally false expectations, for example of prolonged satiety from a very sweet but non-calorific drink, are not mechanisms relevant to everyday life. Any invalid expectation would be corrected by experience. Indeed, marketing would not usually be allowed to induce false hopes. Goods must be labelled accurately.

Thus experiments on the satiating effects of sugars and other sweeteners must monitor the effects of sweetness on beliefs about effects of eating the food. No report to date has included such essential information.

Yet even measuring the expectancy is not enough: the effects of the beliefs are not predictable without further information about the eater's habits and attitudes. The supposition that a food has a high calorie content can in itself have a substantial satiating effect [4]. On the other hand, those who are worried about their weight, and trying to restrain their eating, can be triggered into giving up their diet or even into an emotional binge [37], by believing that they have consumed more calories than they should have done, even if the belief is false [38]. Thus, if some of the experimental subjects are restraining eating to some extent, false beliefs about energy content, created by experimental inclusion of a low-calorie sweetener, could produce a so-called stimulation of appetite by the sweetener, which is an artefact of a breakdown of dietary restraint.

Thus, an essential part of any experiment on satiety is the analysis of the potentially relevant judgements that each subject makes about the pre-loaded foods, relative to each subject's characteristics of the sort that might predict the effect of the expectation [39]. This is still almost never done, despite increasingly widespread acknowledgement that "cognitive satiety" and interpretations of the laboratory test situation have effects on food intake and appetite ratings.

We do not scientifically understand the effect of sweetness on eating until we have distinguished between the various possible actions of the sweetener and shown that their time courses account for the overall effect observed.

An Overview of Carbohydrate Actions in Satiety

It is clear, therefore, that much more and better research is needed before we can hope to reach any sound conclusions concerning the actions of the different dietary carbohydrates on appetite. Nonetheless, it is possible to construct some working hypotheses from background considerations and the limited data on effects of food ingestion on people's eating behaviour.

The potential sources of difference in the satiating effects of dietary starches and sugars in themselves might, in my view, be summarised as follows.

Metabolic "Memory"

Metabolically or hormonally satiating effects could in theory arise from the mobilisation of differential deposition of the glucose from starches and sugars and of the fructose or the galactose peculiar to the sucrose and lactose abundant in the diet.

For example, the different profiles of insulin secretion might affect lipid deposition. "Background satiety" effects resulting from adipose mass, however, are in theory [40,41] very small, albeit powerfully persistent; and the relative impact on total energy store of a difference between the sugar and starch components of intake will be less still. Differential sequestration of carbon in liver glycogen or muscle protein has been suggested [25] as a potential source of differences in the time course of satiety, but the size and even existence of such mechanisms remain to be demonstrated.

Another possible mechanism by which assimilation of an experimental pre-load could produce after-effects is by indirect action on cerebral processing. A great deal of attention has been paid to possible differences between low and high protein loads, in their effects on brain levels or release of the transmitter serotonin (5HT). Insulin tends to sequester branched-chain amino acids into muscle and so reduce competition for transport of tryptophan into the brain, where supply is more nearly rate-limiting to 5HT synthesis than is precursor supply for catecholamine transmitters. The protein-free carbohydrate loads sometimes used in satiety experiments have not been studied in this connection.

After an overnight fast, 50 g of fructose was found to decrease intake at a buffet as long as 135 minutes later [42]. This is consistent with the idea that fructose metabolism may be strongly satiating [17]. However, an increase in intake was induced by 50 g glucose, with or without the addition of aspartame in an amount to equate sweetness to the fructose. The glucose would have stimulated insulin secretion far more effectively than the fructose during the period of its absorption – perhaps about an hour. The directions of these intake effects were the same in men and women, but only one dose was used, one pre-load test interval, and the sizes of effect varied substantially. If the effect is robust and occurs after insulin has returned to baseline, neurohumoral after-effects are worth considering.

Energy Supply

The major and basic physiological control of appetite may well be hepatic energy production from glucogenic and ketogenic substrates, neurally or humorally signalled to the brain [17,25,43–45]. Possibly signals from the intestine wall serve "anticipatory" functions [46,47].

Intestinal or parenteral differences in satiety signals, or in their timing, could arise from slower and less complete digestion, diffusion and absorption of glucose from starches than of monosaccharides from sugars. The short chain fatty acids, absorbed in the large intestine from enzymatically resistant starch, might contribute to delaying the rise of appetite.

Glucose-Specific Gut Wall Receptors

Glucose-specific receptors have been seen in the duodenal wall and gastric antrum in the rabbit [47]. There is as yet no behavioural evidence that afferents from these receptors inhibit central mechanisms of eating. Nevertheless, if such receptors lie behind the brush border disaccharidases in human beings, and do signal satiety, then sucrose and lactose would generate less satiety than maltose, glucose and readily digested starch at this point.

Duodenal Wall Stretch and Osmoreception

Very high concentrations of sugar solution, drunk on an empty stomach, may be dumped before the dilution that normally occurs. They might then stimulate osmoreceptors that trigger nausea and emesis [21,22]. Osmotic retention or diffusion of water may transiently increase the volume in the duodenal lumen. Resulting tension in the wall may be a cause of appetite-suppressant pain, nausea or strong sensations of fulness. These syrups also give aversive tearing sensations in the oropharynx.

Stomach Volume and Retention

It is not clear that, as consumed, starches contribute more than sugars to gastric satiety. Sugars and starches in isolated form are not retained differently by the stomach [48]. As to accompanying non-nutritive bulk, large amounts of water are more often consumed with sugars than with starches. Although, when eaten fairly dry, starches may be accompanied by substantial amounts of dietary fibre more often than are sugars, a major use of sugars is to add them to the foods highest in fibre content in order to make them edible (breakfast cereals and baked goods with low-extraction flour).

What may produce more protracted gastric retention of starchy food is the time necessary to comminute indissoluble particles down to the size that the pylorus will pass. Whole grain, and hard-baked, low-sugar forms of starch may therefore maintain gastric volume for longer.

Nevertheless, only very high levels of gastric pressure are recognised as satiety sensations [49]. However, more normal levels of gastric distension may serve as part of an appetite-suppressive stimulus complex, after learning from readily assimilated starches and sugars in the early parts of meals ("conditioned satiety") [4,9,28].

Textures and Sweet Tastes

Animals show cephalic reflexes to tastes like that of glucose; rats in particular show an insulin secretory reflex to the taste of saccharin, and the metabolic sequelae of that appear to stimulate food intake, directly or via learning [35]. Cephalic reflexes might conceivably be elicited by the firm or coarse textures of starches in foods, as well as by the sweetness of sugars. However, the cephalic insulin reflex to foods has proved hard to elicit reliably in human subjects, and the salivary reflex is not very robust, other than to sour tastes. Such a physiological mechanism might be peculiar to stimulation by saccharin, for that sweetener has been reported more than once to increase food intake by some comparisons, whereas acesulfame and aspartame had less or no effect on intake [32].

Boredom with food of particular sensory characteristics (food-specific satiety) should be similar for sugary and starchy foods, but this remains to be tested.

Both strong sweetness and the texture and aroma of baked goods are frequently associated with post-ingestional moderation of appetite for periods related to the amount of such stimulation. The supposition that there is a lot of sugar or starch in a food, confused with dietary fibre and fat, may, as a result, divert interest from eating for a substantial period. The "rich" aroma and mouth-feel of finely ground cereal triggers strong expectations of satiation (S. Rawle and D.A. Booth, unpublished). The effect of drinking aspartame solution on appetite ratings after 40–60 minutes might, it has been suggested, be a "disappointment" reaction to lack of the expected protracted suppression of hunger sensations [50]. There is no reason to expect these cognitive reactions to be learned any differently for sweetness and for textures characteristic of starches. Thus, sweetness, "body" and chewiness of foods may trigger similar attitudes to subsequent offerings.

A genuinely held attitude can have a powerful effect. A mere guess, randomly correct or incorrect, as to which of two loads was higher in calories, produced, at 20 minutes after loading, as strong a reduction in intake-predictive appetite ratings of several sorts as did post-ingestional action of a 50-g difference in maltodextrin loaded [4]. The "low fat" label makes people rate the food as less filling [1]. It must not be assumed, however, that merely labelling the load with a caloric content, even correctly, creates an effective expectancy: just telling people a food is high in calories had no effect on intake in our hands or in others' [51].

Sugary Versus Starchy Foods

The likely effects of such attitudes to familiar foods unfortunately vitiate the only tests to date that purport to compare the satiating effects of sugars and starches directly [6]. The high-starch pre-load was 390 g of pasta shells in tomato sauce. The equicaloric high sucrose pre-load was also 390 g, but volume was equated by drinking water with 94 g of lemon-flavoured Turkish Delight. The carbohydrate doses were approximately the same. There were three other sorts of food pre-load, but no low-calorie control.

The results showed that the sucrose-rich pre-load was much less satiating than the wheat-starch-rich pre-load. However, the immediate and successive appetite ratings, and the energy intake from the test menu presented at 2 hours, all showed the same relative sizes of satiety effect among the five pre-loads. Furthermore, the relative sizes of the effects of the snacks were what would be expected from their substantiveness as it is likely to be commonly perceived, and as indeed could have been measured in these subjects. Some such confounding of nutrition with habit is clearly demonstrated by the immediate effect on rated satiety. Thus this result provides no evidence whatsoever that sucrose in the diet is less satiating than starch.

A gel confection or dessert is indeed one of the few types of food in which it might be practicable to test equicaloric pre-loads of sugar and whole starch and a lower-energy pre-load, all three of which were hard to distinguish between occasions in volume, mouth-feel, sweetness, flavouring or visual appearance (E.L. Gibson, C.J. Wainwright and D.A. Booth, work in progress, 1988).

Satiety for Carbohydrate

It might be best to mention at this point that the experiment on the satiating effects of different macronutrients [6] also provided no evidence that carbo- hydrate intake selectively satiated appetite for the carbohydrate content of the test foods.

The notion of nutrient-specific dietary selection has been even more direly misconceived than that of general satiety. As a result, all such attempts to look for nutrient-specific satieties, as well as for "carbohydrate cravings" and the like, have been confounded by subjects' reactions to differences in the sensory characteristics of the test foods differing in nutrient composition [52].

Preferences between textural differences can affect the consumption even of equally palatable diets. Subjects' categorisations of food as entrées and desserts, and as nutritious and "junk" foods, will affect what they choose to eat after having eaten one category. Thus, the analysis of data for nutrient and energy intakes should be extended to analysis for much more likely determinants of eating, namely, the perceptions, attitudes and intentions that are informing the food choices and intakes. These can be no less objectively analysed from behavioural performance [2].

Practical Implications of Satiety Effects

The main practical interest is whether, in our present dietary culture, or at any other specified time and place, sugars as used stimulate an individual's total energy intake on average more than do starches as consumed. This is a descriptive, historical question. It can only be answered in a scientifically sound manner by elucidating the mechanisms by which starches and sugars satiate, and then measuring the prevalence of the influence of those mechanisms in current eating habits [13,50,53].

In fact, the relationships between experiments on satiety, and the practical issues on food intake and obesity, have generally not been well conceived. No amount of extrapolation from experimental results or compiling of population statistics from survey data, nor even their traditional "combination", can provide reliable answers. Unobtrusively collected field data must be designed so that it can be subjected to theory-testing multivariate analysis of the timing and nature of the energy intake that habitually follows the consumption of foods and drinks of known sugar and starch content [13].

Satiety and Obesity

None of the currently considered sources of a difference in the satiating effects of dietary starches and sugars in themselves seems a plausible candidate for a substantial effect on total intake.

For example, there is no evidence for an insatiable appetite for sugar, in which it differs from starch. That is to say, the results of behavioural experiments do not support current imputations on such grounds that sugar especially causes obesity, any more than they support the imputations of the not so distant past, still widely held, that starchy foods are particularly fattening.

There is scientific reason, however, for believing that technical and cultural constraints on the uses of sugars or starches tend to bypass the adaptive function of satiety, and so are contributing to obesity [13,33,50]. That is, neither sugars nor starches are particularly fattening but the consumption of any energy at the wrong time can be. This view is based on the experimental work on post-ingestional satiety mechanisms, and is now also increasingly supported by mechanistically interpretable field data [13,53].

Both sugars and starches, together with fats and alcohol, and indeed protein, are consumed after meals and between meals in amounts that are modest enough to have long been assimilated by the next time an occasion arises for eating a mixture of foodstuffs, whether snack or meal. Hence the satiating effect of that drink and/or "nibble" will have passed before food intake recurs. As a result the energy will have slipped in past the main mechanism for compensating caloric intake for prior consumption and hence for generating some moderation of total energy consumed. It could therefore be to the advantage of consumers liable to weight gain to have the option of attractive energy-free fluids and solids at drink breaks [50,53].

The most obvious possibility of a relevant difference between the uses of sugars and starches is a greater contribution of sugars to the energy in drinks and their accompaniments between and after meals.

This is not a foregone conclusion. Potato, maize and wheat starches are frequently part of the solid accompaniment of drinks, as well as sucrose. On the other hand, only sucrose, glucose syrups and fruit sugars can be included in the drinks themselves. They are frequently drunk in substantial amounts.

Hence, given the sensory characteristics that are traditional for what we drink and eat between meals in Westernised cultures, partly for the preservative effects of reducing water activity, sugars as used probably contribute more than starches as used to a satiety that does not affect subsequent intake. Thus, such eating habits abuse sugar by permitting a larger total intake, so making it liable to contribute to obesity. There is, however, only limited point in providing energy-free drinks without also providing energy-free "nibbles", if the habit is for drinks to be consumed with an energy-rich confectionery or snack food item.

It is therefore necessary to distinguish between sugars and/or starches themselves and the different ways in which they are used, i.e. at meals and between meals. This distinction between total intakes and habitual patterns has not been adequately made in professional or public discussions of applied nutrition. The distinction has major implications for food marketing and for health education in support of healthy eating [53]. For example, the emphasis in dietary guidelines and dietetic advice on weight reduction should shift from the total amounts of sugar that the nation or a person consumes [54], and from details of the composition of the manufactured convenience products or snackfoods as such [55]. Instead, nutrition education and new product development should be guided primarily by the patterns of timing and sequence of ingestion of energy-containing beverages and food items of all sorts, and by the cultural and economic factors supporting the demand for such beverages and foods from such lifestyles.

Improved Analysis of Satiety Mechanisms

From the above discussion, various developments in methodology seem to be desirable, both for fundamental elucidation of mechanisms and to enable more informative results to be obtained from epidemiological and market studies.

The Pre-load Paradigm

The very concept of satiation implies the experimental design of observing the effects on eating of prior consumption of food. This requires a "pre-load" and a "test meal". The situation is no different if the experimenter is attempting to measure the effects of eating on eating at the same meal: the design and interpretation still have to distinguish between causal "pre-load" factors and "test-meal" consequences; otherwise no scientific information is obtained.

Thus, from the earliest days of research on mechanisms by which eating inhibits eating, nutrient loads of known volume and composition were infused intra-venously (for parenteral effects), intubated gastrically (for gut and parentery) or, running the whole gamut of satiety effects, voluntarily ingested, and then subsequent food intake was measured [56].

Double-Dissociated Dose-Responses

Statistical comparisons between experimental and control means are but the minimum criterion for mechanistic analysis. Cause–effect relationships are far more effectively identified by sensitivity measurements: how much causal difference is necessary to make a reliable difference in response [1,2,31,34]. Sensitivity (causal strength) can be calculated from two-condition designs, or from dose-response designs over the system's dynamic range. However, even just the mean values of a dose-response relationship provide information on the range of causal operation that is missed by the experimental/control comparison.

The acme of mechanistic analysis, evidence for specificity both of input and of output, is the four-way contrast of double dissociation, i.e. two uncorrelated inputs differentially relating to two uncorrelated outputs. The dissociation may be obtained in sensitivities, in dose-responses or even just in experimental/ control comparisons, so long as the interactions are analysed and "balance" is not assumed.

The earliest pre-loading experiments recorded immediate food intake after graded sizes and nutrient concentrations of loads delivered by gastric fistula; the results implicated gastric or gastrointestinal "bulk" in dogs [56]. Gavage of sugar solutions to rats was much used subsequently to seek nutritive influences on satiety. Dose-controlled comparison of glucose and salt loads on both food and water intake demonstrated a gastrointestinal "osmotic" satiety mechanism, however [20]. Hence, finally, rats were gavaged with different concentrations of glucose, food withheld to allow most of it to be absorbed and bulk and osmotic effects to decline, and then the dose/time-course of parenteral effects inhibiting the tendency to eat was measured as reductions in cumulative food intake, delays in the initiation of meals and terminations of meals after eating less [21].

Satiety Values

Unfortunately, this "pre-load" paradigm has been traduced by the confusions between the objectives of scientific experiments and of real-life investigations, coupled with misunderstandings of the nature of behaviour, and hence of phenomena such as satiation and palatability. An example that is highly relevant here is the notion that a food, rather than a mechanism, could be assigned a satiety value. The more common idea, that a level of palatability is inherent in a food, is incoherent for the same reasons.

Satiety, palatability and other aspects of appetite or hunger are not in the foods. They are aspects of the organisation of behaviour towards food, i.e. causal relationships between foodstuffs and eating behaviour. They are neither a quasi-physical property of a foodstuff, nor a physical aspect or result of eating behaviour, such as ingestive movements [57], microbehavioural eating parameters or meal patterns. These physical responses, and indeed verbal responses that predict them, can all be used as measures of the output side of appetite, satiety or palatability. Nevertheless, they only become measurements of behavioural processes when they are related to measures of the input side. To implicate an input causally, control of ingestion must be specifically associated with its measure, be it the physicochemical nature of the foods that might be eaten in the test of appetite, how such potential comestibles would be sensed, and, in the case of satiety, how previously ingested food is acting at the time of test.

A quantitative satiety value can thus only be the extent to which a particular satiating mechanism contributes to a suppression of the disposition to eat. The basic datum is the degree of suppression of eating of specified foods, in an individual at a particular point in time, that is attributable to current or past action of a controlled and measured cause of that suppression.

Most of the data reported thus far on human satiety are group mean suppressions of intake from one particular test menu at one particular time, or of one highly intercorrelated set of ratings of appetite, induced by one level of a pre-load relative to some more or less controlled comparison. Much more is needed before any conclusion can be drawn about the strength of the satiating effect of a particular amount or concentration of a sugar or starch in food. We must have full control of particular mechanisms, dose relationships and time courses, and allow as necessary for the individuality of eaters too.

The Metric for Satiety Mechanisms

In order to apply the experimental results to field observations, a food-general measure of satiety-effect values must be found. The one that would be most relevant to the practical issue of overeating is suppression of energy intake. Most intake data from satiety experiments are indeed reported in energy units (kcal or MJ, rather than g or ml). Megajoule suppression, however, has yet to be proved to work as a consistent measure across pre-load and test foods, across times relative to habitual eating and drinking occasions and, of course, across people in a given population. Thus the caloric effects of the operation of a particular mechanism at a pre-load–test delay will have to be replicated in different subject samples, at different times of day and with other pre-loading procedures and test foods.

In addition, two further dimensions have to be plotted before a satiety mechanism has been measured. Pre-load doses in the normally ingested range have to be sampled and the timing of their major consequences assessed.

Time-course of Action of a Satiety Mechanism

Information on the time-courses of action of different "doses" of a satiating influence is essential to the adequate design of investigations of the role of satiety in free-living energy intakes.

Such information can also be highly illuminating as to mechanism. An example of instructive time-course data is appetite ratings that reach their lowest point immediately the pre-load has been administered and remain in similar relative positions as appetite rises, i.e. as satiety declines. Such satiety effects are most likely to be almost entirely "cognitive" [4,58]. The initial effects of equal-volume pre-loads that have not had time for differential emptying from the stomach have to be based on expectation, because evidence for gastric chemoreceptor satiety is lacking (B.J. Baker and D.A. Booth, in preparation). If the subject's expectations are veridical, then the later post-ingestionally generated satiety effects will be correlated with them, but the experimental design will have confounded cognitive and physiological satiety effects.

It must be noted that this is not a weakness of using ratings to assess satiety. The weakness is poor control of the pre-load, so that differential expectations can arise. Expectancy could equally affect intake tests.

It must be noted, though, that an expectation is not necessarily present and evident in satiety immediately after the pre-load: someone might expect appetite to be postponed longer without any immediately greater effect. Nevertheless, the procedure of rating at the end of pre-loading might get the subject thinking as s/he would not otherwise have done and therefore possibly committed to some judgements of the pre-load that might not otherwise have been made. Even random guesses after the event account for a considerable part of rated satiety [4]. Also, expectations can act as part of an autosuggestive process: that is, they can "amplify" real effects (e.g. a small post-ingestionally induced satiation) to a size they would not have reached simply by addition of the expectations to the sensed effects [3].

Dose-time Functions of Satiety Mechanisms

Thus, the basic scientific objective for a programme of experiments on satiety should be to seek a dose-dependent time-course of suppression of the tendency to eat foods in general, that results from the action of a particular aspect of ingestion of foods, such as the energy metabolism of sugars or the expected time-course of appetite suppression following ingestion of a beverage of a certain sweetness.

This function could be extracted from the data only if there was qualitative, and even roughly quantitative, additivity with people's food choices. Any exact satiety function depends on each person being self-consistent in relative palatability of foods. This is only approximately true at best, to the extent that palatability has not been contextualised, especially with satiety itself [1,4,7,9,10]. Moreover, the availability of a single function also depends on the suppression of intake being similar for highly preferred and less preferred foods, which is certainly not the case in preschool children [29].

If the interactions with differences in palatability and eating habits among foods and among people are too great, then not even a rough answer to the practical question is available. In that case, the question does not relate to the actual phenomena and is, in a sense, misconceived. If, however, a set of such satiety functions, or differences in satiety function, for major mechanisms of action of dietary starches and sugars can be measured, they can then be applied in the causal analysis of observations of real-life eating patterns. We could hope to assess whether the effects of sugars and starches differ appreciably and are large or small relative to normal intake and its variation.

Qualitative versions of this operation have already been attempted in the theoretical argument for and retrospective survey test of the "zero-calorie drink-break option" for reduction of obesity [13,50,53]. Systematic quantitative research has yet to begin, because the requirements of such work have yet to be clearly recognised in academic science, let alone in the applied areas that need such a database.

Effects on Mental Performance and Mood

Slightly calorific drinks are considered to be "refreshing" by some consumers. The everyday concept of liquid refreshment is, however, very broad, extending from fresh-tasting and thirst-slaking to removal of fatigue and restoration of mental and/or physical psychological energy. It does not distinguish actual effects of the drink from what might be expected to occur on a conventional occasion for taking a rest and indulging in enjoyment.

Glucose drinks in particular have been marketed for use in convalescence or, more recently, during sports activity. The question whether dietary carbohydrate affects physical performance is considered by Williams (Chap. 11).

Glucose tablets alone, and not just caffeine drinks that may also contain sugar, are used by some in intellectually demanding tasks, e.g. students during examinations and in other study crises. The question therefore arises whether small amounts of readily assimilated carbohydrate really improve any aspects of mental performance or at least foster a subjective sense of alertness, ability to concentrate, or reduced mental fatigue.

Scientific Evidence

Improvements in performance at a vigilance task, but not other mental challenges, have recently been confirmed for caffeine, although there was a puzzling lack of dose-response relationship [59]. Such effects have been reported also for glucose at doses of 15–50 g, but the data are not so fully documented [60,61].

As mentioned earlier, drinking a hypertonic solution of free hexose can be mildly stressful, rasping the throat and inducing nausea. This might arouse some experimental subjects to attend more to the mental task set. Lower in the dose range, however, this may be evidence that a small amount of rapidly assimilated carbohydrate, consumed by itself, acts physiologically to produce a transient improvement in ability to concentrate under certain circumstances. Such effects conceivably have their consequences for mood. A performance effect and, even more, any mood effect would, however, be highly susceptible to augmentation (or reduction) by ways of thinking the user has acquired from personal experience or social convention. Such "placebo" effects can arise from sheer expectation or from an autosuggestive working up of real but less specific effects (side-effects or sensory cues).

Large, meal-approximating doses of carbohydrate, on the other hand, especially when ingested with little or no protein, generally produce sedative effects, if any effect other than perhaps satiation [62,63]. There is no scientific basis for supposing that sucrose or any other carbohydrate will stimulate or aggravate unruly behaviour. Overt manipulation of the diet by authority might have effects, however, because it is interpreted as a disciplinary measure.

Theoretical Mechanisms

It is hard to believe that hepatic supply of glucose to the brain in normally eating people would become limiting on neuronal function, even in a highly active region of the brain [64]. Some effect on brain function secondary to oral or gastrointestinal stimulation, or the absorption of the glucose, should be considered. As mentioned above, large amounts of carbohydrate absorbed with very little protein (4% or less [65]) have been claimed to affect brain serotonin activity, and so a protein-free glucose drink after the last meal has been absorbed may perhaps have a neural effect.

Practical Implications

At present, the evidence for mental concentration-improving effects of glucose drinks is insufficiently documented for such a claim to be justifiable in marketing such products. Nevertheless, there may be little harm in attempting to meet any such existing demand, so long as such marketing does not have the effect of encouraging potentially cariogenic or fattening uses of such drinks. This caveat may be difficult to meet, given the prevalence of sucrose drinks, with or without caffeine, which in modest amounts may have similar effects on mental performance but as great a risk of cariogenicity and greater risk of contributing to obesity.

It may therefore be in producers' interests to encourage adequately designed experimental work on this phenomenon and to monitor major patterns of use of such drinks in their current positioning for sports people or convalescents. If it is confirmed that more substantial doses of glucose as often as not have sedative effects, as well as putting the teeth and the waistline at greater risk, then careful control may have to be kept on influences that increase the frequency or the amount taken on an occasion in glucose drinks.

References

1. Booth DA (1987) Cognitive experimental psychology of appetite. In: Boakes RA et al. (eds) Eating habits. Wiley, Chichester, pp 175–209
2. Booth DA (1987) Objective measurement of determinants of food acceptance: sensory, physiological and psychosocial. In: Solms J et al. (eds) Food Acceptance and Nutrition. Academic Press, London, pp 1–27
3. Booth DA (1980) Acquired behavior controlling energy intake and output. In: Stunkard AJ (ed) Obesity. Saunders, Philadelphia, pp 101–143
4. Booth DA, Mather P, Fuller J (1982) Starch content of ordinary foods associatively conditions human appetite and satiation, indexed by intake and eating pleasantness of starch-paired flavours. Appetite 3:163–184
5. Hill AJ, Blundell JE (1986) Model system for investigating the actions of anorectic drugs. In: Advances in the Biosciences. Pergamon, Oxford, pp 377–389
6. Rolls BJ, Hetherington M, Burley VJ (1988) The specificity of satiety: the influence of foods of different macronutrient content on the development of satiety. Physiol Behav 43:145–154
7. Rolls BJ, Rolls ET, Rowe EA, Sweeney K (1981) Sensory specific satiety in man. Physiol Behav 27:137–142
8. Cabanac M (1971) Physiological role of pleasure. Science 173:1103–1107
9. Booth DA (1977) Appetite and satiety as metabolic expectancies. In: Katsuki et al. (eds) Food Intake and Chemical Senses. University of Tokyo Press, Tokyo, pp 317–330
10. Rolls BJ (1984) Pleasantness changes and food intake in a varied four-course meal. Appetite 5:337–348
11. Wooley OW, Wooley SC, Dunham RB (1972) Calories and sweet taste: effects on sucrose preference in the obese and non-obese. Physiol Behav 9:765–768
12. Booth DA, Campbell AT, Chase A (1970) Temporal bounds of post-ingestive glucose-induced satiety in man. Nature 228:1104–1105
13. Booth DA (1988) Mechanisms from models – actual effects from real life: the zero-calorie drink-break option. Appetite 11 [Suppl 1]:94–102
14. Sclafani A (1987) Carbohydrate taste, appetite, and obesity: an overview. Neurosci Biobehav Rev 11:131–153
15. Ramirez I (1987) When does sucrose increase appetite and adiposity? Appetite 9:1–19
16. Simopoulos AP (1988) Obesity and the control of food intake. Appetite 11 [Suppl]: 1–4
17. Booth DA, Jarman SP (1976) Inhibition of food intake in the rat following complete absorption of glucose delivered into the stomach, intestine or liver. J Physiol (Lond) 259:501–522
18. Duggan JP, Booth DA (1986) Obesity, overeating and rapid gastric emptying in rats with ventromedial hypothalamic lesions. Science 231:609–611
19. Rabin BM (1972) Ventromedial hypothalamic control of food intake and satiety: a reappraisal. Brain Res 43:317–325

20. Jacobs HL (1964) Evaluation of the osmotic effects of glucose loads on food satiation. J Comp Physiol Psychol 57:309–310
21. Booth DA (1972) Satiety and behavioural caloric compensation following intragastric glucose loads in the rat. J Comp Physiol Psychol 78:412–432
22. Cabanac M, Fantino M (1977) Origin of olfacto-gustatory allesthesia: intestinal sensitivity to carbohydrate concentration? Physiol Behav 18:1039–1045
23. Booth DA (1981) The physiology of appetite. Br Med Bull 37:135–140
24. McHugh PR, Moran TH, Barton CN (1975) Satiety: a graded behavioral phenomenon regulating caloric intake. Science 190:167–169
25. Booth DA (1972) Post-absorptively induced suppression of appetite and the energostatic control of feeding. Physiol Behav 9:199–202
26. Booth DA, Davis JD (1973) Gastrointestinal factors in the acquisition of oral sensory control of satiation. Physiol Behav 11:23–29
27. Hunt JN, Stubbs DF (1975) The volume and content of meals as determinants of gastric emptying. J Physiol (Lond) 245:209–225
28. Booth DA, Lee M, McAleavey C (1976) Acquired sensory control of satiation in man. Br J Psychol 67:137–147
29. Birch LL, McPhee L, Sullivan S (1989) Children's food intake following drinks sweetened with sucrose or aspartame: time course effects. Physiol Behav (in press)
30. Southgate DAT, Johnson IT (1987) New thoughts on carbohydrate digestion. Contemp Nutr 12(10):1–2
31. Booth DA, Blair AJ (1988) Objective factors in the appeal of a brand during use by the individual consumer. In: Thomson DMH (ed) Food acceptability. Elsevier Applied Science, London, pp 329–346
32. Blundell JE, Rogers PJ, Hill AJ (1988) Uncoupling sweetness and calories: methodological aspects of laboratory studies on appetite control. Appetite 11 [Suppl]:54–61
33. Booth DA, Conner MT, Marie S (1987) Sweetness and food selection: measurement of sweeteners' effects on acceptance. In: Dobbing J (ed) Sweetness. Springer, Berlin Heidelberg New York, pp 143–160
34. Conner MT, Haddon AV, Pickering ES, Booth DA (1988) Sweet tooth demonstrated: individual differences in preferences for both sweet foods and foods highly sweetened. J Appl Psychol 73:275–280
35. Tordoff MG (1988) How do non-nutritive sweeteners increase food intake? Appetite 11 [Suppl]:5–11
36. Slochower J, Kaplan SP (1980) Anxiety, perceived control, and eating in obese and normal weight persons. Appetite 1:75–83
37. Herman CP, Polivy J (1980) Restrained eating. In: Stunkard AJ (ed) Obesity. Saunders, Philadelphia, pp 208–225
38. Polivy J (1976) Perception of calories and regulation of intake in restrained and unrestrained subjects. Addict Behav 1:237–243
39. Booth DA, Toase AM (1983) Conditioning of hunger/satiety signals as well as flavour cues in dieters. Appetite 4:235–236
40. Booth DA, Mather P (1978) Prototype model of human feeding, growth and obesity. In: Booth DA (ed) Hunger models. Academic Press, London, pp 279–322
41. Booth DA (1988) A simulation model of psychobiosocial theory of human food-intake controls. Int J Vit Nutr Res 58:55–69
42. Spitzer L, Rodin J (1987) Effects of fructose and glucose preloads on subsequent food intake. Appetite 8:135–145
43. Booth DA, Toates FM, Platt SV (1976) Control system for hunger and its implications in animals and man. In: Novin D et al. (eds) Hunger. Raven Press, New York, pp 127–142
44. Friedman MI, Stricker EM (1976) The physiological psychology of hunger: a physiological perspective. Psychol Rev 83:409–431
45. Russek M (1981) Current status of the hepatostatic theory of food intake control. Appetite 2:137–143
46. Smith GP, Gibbs J (1979) Postprandial satiety. Prog Psychobiol Physiol Psychol 10:179–242
47. Mei N (1987) Physiological effects of nutrients on internal sensors. In: Solms J et al. (ed) Food acceptance and nutrition. Academic Press, London, pp 221–228
48. Hunt JN (1960) The site of receptors slowing gastric emptying response to starch in test meals. J Physiol (Lond) 154:270–276
49. Boring EG (1915) The sensations of the alimentary canal. Am J Psychol 26:1–57

50. Booth DA (1987) Evaluation of the usefulness of low-calorie sweeteners in weight control. In: Grenby TH (ed) Developments in sweeteners: 3. Elsevier Applied Science, London, pp 287–316
51. Wardle J (1987) Hunger and satiety: a multidimensional assessment of responses to caloric loads. Physiol Behav 40:577–582
52. Booth DA (1987) Central dietary "feedback onto nutrient selection": not even a scientific hypothesis. Appetite 8:195–201
53. Booth DA (1988) Relationships of diet to health: the behavioral research gaps. In: Manley CH, Morse RE (eds) Healthy eating – a scientific perspective. Wheaton, Exeter, pp 39–76
54. National Advisory Committee on Nutrition Education (1983) Proposals for nutritional guidelines for health education in Britain. Health Education Council, London
55. BNF Task Force on Sugars and Syrups (1986) Report. British Nutrition Foundation, London
56. Janowitz HD, Grossman MI (1949) Effect of variations in nutritive density on food intake in the dog. Am J Physiol 184–193
57. Grill H, Berridge K (1986) Ingestive motor patterns and palatability in the rat. Prog Psychobiol Physiol Psychol 11:36–58
58. Wooley SC (1972) Physiologic versus cognitive factors in short-term regulation in the obese and non-obese. Psychosom Med 34:62–68
59. Lieberman HR et al. (1987) The effects of low doses of caffeine on human performance and mood. Psychopharmacology 92:308–312
60. Keul J, Huber G, Lehmann M, Berg A, Jakob E-F (1982) Einfluss von Dextrose auf Fahlleistung, Konzentrationsfaehigkeit, Kreislauf und Stoffwechsel im Kraftfahrzeug-Simulator. Akt Ehnaer 7:7–14
61. Azari NP (1988) Effects of glucose on human memory. Soc Neurosci Abst 14:861
62. Spring B, Chiodo J, Bowen DJ (1987) Carbohydrates, tryptophan, and behavior: a methodological review. Psychol Bull 102:234–256
63. Rosen LA, Booth SR, Bender ME, McGrath ML, Sorrell S, Drabman RS (1988) Effects of sugar (sucrose) on children's behavior. J Consult Clin Psychol 56:583–589
64. Siebert G, Gessner B, Klasser M (1986) Energy supply of the central nervous system. Bibl Nutr Dieta 38:1–26
65. Teff KL, Young SN (1988) Effect of protein and carbohydrate breakfasts on plasma tryptophan ratio in males. Soc Neurosci Abst 14:530

Commentary

Levin: Because some people profess "not to experience particular sensations when they want to eat or stop eating" can hardly be the hardest piece of evidence for saying that they are not "sensations", because firstly you have not *defined* what a sensation is. Perhaps they are verbally inadequate or simply do not recognise them as sensations. In the sexual area of sensations some people do not recognise that they are experiencing specific sensations because they have not been taught to recognise them. Perhaps because I am unfamiliar with the type of language used by psychologists, I found this article very difficult to comprehend.

Author's reply: Some physicians would find the chemical chapters very hard going and a chemical engineer may have difficulty grasping the full implications of the physiological biochemistry, even though they eat and assimilate sugary and starchy foods every day of their lives. The fact that we all have experiences and perform thinking and actions continually does not necessarily make it any easier to get on top of the scientific issues about the actual psychological mechanisms involved in our behaviour.

With help from the editor and commentators, the exposition of the psychological issues about effects of sugars and starches should be about as clear as the technical facts permit. I hope that the material will be used by those who have to carry out or evaluate research on such issues, because the chapter illustrates how badly the sensory, physiological and marketing sides of food research have suffered from incomprehension of the disciplines necessary in experimental work on psychological phenomena.

Behavioural nutrition has been neglected by psychologists, unfortunately, as well as by food scientists and nutritional research. Yet everybody concerned with research on food as food needs to understand the psychology which is central to any realistic scientific approach.

Flatt: In considering the mode of action of carbohydrates in inducing satiety, the involvement of gastrointestinal hormones and neuropeptides in the regulation of appetite and in mediating the effects of dietary starches and sugars on satiety, cannot be overlooked. This subject has been covered in a recent review which cites 497 references [1]. For example, there is evidence that insulin, glucagon, CCK and gastrointestinal hormones released in response to feeding act as humoral satiety agents.

Reference

1. Morley JE (1987) Neuropeptide regulation of appetite and weight. Endocr Rev 8:256–287

Author's reply: The roles of insulin and cholecystokinin in satiety or hunger have been grossly overplayed in the last decade or so. Measurements merely of correlates of ingestion and assimilation have been overinterpreted, not establishing any causal role for the hormonal changes in eating. The ambiguities of effects of exogenous administration have been neglected.

As it happens, I was one of the first to study systematically the interactions of exogenous insulin and glucose on hunger [1,2,3] and on satiety [4]. Subsequent work has not solved the mystery of the mechanism by which insulin can augment the satiating effect of absorbed glucose, and human experiments have not consistently shown such an effect.

Also, gastric emptying rate was identified as the main modulator of post-prandial metabolic satiety in the rat [5,6], complementing the presumed role of gastric wall stretch afferents in the satiety that ends a meal. CCK at physiological levels was therefore likely to be a modulatory rather than fundamental satiating signal by slowing of gastric emptying, and this has finally been demonstrated to be a major part of the CCK effect in rats and monkeys. Perhaps the other part of its effect is on gastric tone or its signalling, and hence also modulatory during meals normally. The strength of the evidence that hepatic oxidation of carbohydrate and other substrates is, by contrast, a major fundamental satiety signal was acknowledged in a major review of the role of CCK in satiety [7].

The further complexities of hormonal modulation of basic gastrointestinal and metabolic satiety mechanisms therefore had little place in a chapter that was rendered complex enough by focussing on the key essentials of behavioural physiology.

References

1. Booth DA, Brookover T (1968) Hunger elicited in the rat by a single injection of crystalline bovine insulin. Physiol Behav 3:439–446
2. Booth DA, Pain JF (1970) Effects of a single insulin injection on approaches to food and on the temporal pattern of feeding. Psychonom Sci 21:17–19
3. Booth DA, Pitt ME (1968) Role of glucose in the elicitation of eating by insulin. Physiol Behav 3:447–453
4. Lovett D, Booth DA (1970) Four effects of exogenous insulin on food intake. Q J Exp Psychol 22:409–419
5. Booth DA (1972) Satiety and behavioral caloric compensation following intragastric glucose loads in the rat. J Comp Physiol Psychol 78:412–432
6. Toates FM, Booth DA (1974) Control of food intake by energy supply. Nature 251:710–711
7. Smith GP, Gibbs J (1979) Postprandial satiety. Prog Psychobiol Physiol Psychol 10:179–242

Würsch: I am still hungry after reading this. What I learned is that the mechanism of satiety is very complex and multiple, and there are many ways to measure it. I would like to read examples of studies and results using various methods of assessing satiety. What about the satiating effect of starchy foods of different digestion rates? What, more precisely, is the role of insulin and blood glucose response? (See [1].)

One possible effect of satiety is a change in food intake later in the day. Hill, using lunchtime meals of similar energy content, found that hunger scores 3 hours after lunch correlated closely with subsequent food intake [2].

References

1. Leathwood PD, Pollet P (1988) Effects of slow release carbohydrate in the form of bean flakes on the evolution of hunger and satiety. Appetite 10:1–11
2. Hill AJ, Leathwood PJ, Blundell JC (1987) Short-term caloric compensation in man. Hum Nutr 41A:244–257

Author's reply: The main message of my chapter is *not* that there are many ways of measuring human satiety. There are in fact only two ways, broadly speaking. These are observing eating and asking people whether they want to eat. Investigators have been wrong to claim that highly correlated variables are measuring different phenomena.

Rather, as you say, the complexity is in the mechanisms by which the observed effects on the disposition to eat are produced. I omitted no human study that was adequately designed to measure a particular mechanism of satiety that might be activated by sugars or starches. Hill et al. (reference [2] above) did not identify the effects of carbohydrate. Leathwood and Pollet (reference [1] above) provide evidence that slowly digested starch (from beans) may delay the rise of hunger late after a meal more effectively than does rapidly digested starch (potato). As they point out, however, there are many mechanisms by which the difference might have been generated by the starches and there are other relevant differences between the two menus that they compared.

Macdonald: In view of the current commercial campaign to promote fructose (and, perhaps because of its fructose content, sucrose) in the treatment of obesity, it would be of interest to know more about the rationale behind this, and also Booth's views as to its validity.

Author's reply: The report cited by Macdonald in Chap. 10 is that fructose, unlike glucose, suppresses food intake in women and men an hour or so after absorption. As he mentions, it was suggested that this relates to delayed effects of differences in insulin secretion. It has been proposed that excessive insulin secretion can reduce hunger, but this is unproven in human subjects and Macdonald rightly questions the relevance of the animal paradigm of rapid duodenal infusion. In my chapter I have listed the most plausible ways, in my view, in which sugars and starches could induce satiety. One of the possibilities is that fructose contributes more than glucose to hepatic oxidation (Chap. 13, reference 17) which may be signalled neurally and/or humorally to the brain, with or without sensations. Clearly, though, appetite-suppressing effects of "intestinal hurry" should be considered for substantial doses of free glucose.

Reservations must also be expressed about any treatment for obesity that relies on short-term caloric satiety effects suppressing net food intake in the long term. Such ideas should be contrasted with the mechanism by which low-calorie drink-breaks can aid weight control (Chap. 13, reference 13). This results in an avoidance of the habitual consumption of readily assimilated energy substrates at times when their satiety effect will have passed before it can help to moderate intake at the next meal. Such avoidance seems less likely than self-administration to provoke counteractive effects on repetition for the many months or years needed for weight reduction strategy to be definitely useful.

Subject Index